Environmental Contamination Remediation and Management

There are many global environmental issues that are directly related to varying levels of contamination from both inorganic and organic contaminants. These affect the quality of drinking water, food, soil, aquatic ecosystems, urban systems, agricultural systems and natural habitats. This has led to the development of assessment methods and remediation strategies to identify, reduce, remove or contain contaminant loadings from these systems using various natural or engineered technologies. In most cases, these strategies utilize interdisciplinary approaches that rely on chemistry, ecology, toxicology, hydrology, modeling and engineering.

This book series provides an outlet to summarize environmental contamination related topics that provide a path forward in understanding the current state and mitigation, both regionally and globally.

Topic areas may include, but are not limited to, Environmental Fate and Effects, Environmental Effects Monitoring, Water Re-use, Waste Management, Food Safety, Ecological Restoration, Remediation of Contaminated Sites, Analytical Methodology, and Climate Change.

Nicholas M. Hill · Sarah Hines ·
Nelson J. O'Driscoll
Editors

Wetland Restoration for Endangered Species Recovery

A Multidisciplinary Case Study of Big Meadow Bog, Brier Island, Nova Scotia

Springer

Editors
Nicholas M. Hill
Southwest Nova Biosphere Association
Yarmouth, NS, Canada

Sarah Hines
K.C. Irving Environmental Science Center
Wolfville, NS, Canada

Nelson J. O'Driscoll
Department of Earth and Environmental Science
Acadia University
Wolfville, NS, Canada

ISSN 2522-5847 ISSN 2522-5855 (electronic)
Environmental Contamination Remediation and Management
ISBN 978-3-031-71343-9 ISBN 978-3-031-71344-6 (eBook)
https://doi.org/10.1007/978-3-031-71344-6

This Springer imprint is published by the registered company Springer Nature Switzerland AG
The registered company address is: Gewerbestrasse 11, 6330 Cham, Switzerland

WETLAND RESTORATION FOR ENDANGERED SPECIES RECOVERY

Photo of the hydrologically restored Big Meadow Bog with inset of the Eastern Mountain Avens in flower. (Photo Credit: Mike Dembeck photography)

We would like to dedicate this book to the people and communities of Brier Island both past and present, without whom this work would not have been possible.

Contents

Contributors

Sarah J. Adams Department of Biology, Acadia University, Wolfville, NS, Canada

Alain Belliveau E.C. Smith Herbarium, K.C. Irving Environmental Science Centre, Acadia University, Wolfville, NS, Canada

J. Sherman Boates Department of Biology, Acadia University, Wolfville, NS, Canada

Brianna Bowes Department of Earth and Environmental Science, Acadia University, Wolfville, Nova Scotia, Canada

Molly Bradford Department of Earth and Environmental Science, Acadia University, Wolfville, Nova Scotia, Canada

John C. Brazner Wildlife Division, Nova Scotia Department of Natural Resources and Renewables, Kentville, NS, Canada;
Brazner Ecological Services, Black Rock, NS, Canada;
Fern Hill Institute of Plant Conservation, South Berwick, NS, Canada

Robin D. Browne Department of Biology, Acadia University, Wolfville, NS, Canada;
K.C. Irving Environmental Science Centre, Acadia University, Wolfville, NS, Canada

David L. Burton Department of Plant, Food, and Environmental Sciences, Dalhousie University, Agriculture Campus, Halifax, NS, Canada

Rachel Clarke Department of Earth and Environmental Science, Acadia University, Wolfville, Nova Scotia, Canada

Jeanette C. Denton Brier Island Trails Committee, 197 Water Street, Westport, Brier Island, NS, Canada

Rebecca S. Dodge Department of Biology, Acadia University, Wolfville, NS, Canada

Mia Doncaster Department of Earth and Environmental Science, Acadia University, Wolfville, Nova Scotia, Canada

John Drage Halifax, NS, Canada

Dewey D. Dunnington Voltron, Mountain View, CA, USA

Matt C. Estep Department of Biology, Appalachian State University, Boone, NC, USA

David J. Garbary Department of Biology, St. Francis Xavier University, Antigonish, NS, Canada

Sarah E. Gutowsky Biology, Acadia University, Wolfville, Nova Scotia, Canada

Nikolai M. Hay Department of Biology, Duke University, Durham, NC, USA

Nicholas M. Hill UNESCO Southwest Biosphere Association, Sand Beach, NS, Canada

N. Kirk Hillier Department of Biology, Acadia University, Wolfville, Nova Scotia, Canada

Sarah Hines K.C. Irving Environmental Science Centre, Acadia University, Wolfville, NS, Canada

Audrey Hiscock Dalhousie University, Halifax, Nova Scotia, Canada

Rob Jamieson Dalhousie University, Halifax, Nova Scotia, Canada

Drucie Janes Department of Plant, Food, and Environmental Sciences, Dalhousie University, Agriculture Campus, Halifax, NS, Canada

Lindsay Johnston Dalhousie University, Halifax, Nova Scotia, Canada

Mark O. Johnston Department of Biology, Dalhousie University, Halifax, NS, Canada

Emil H. Jurga Department of Biology, St. Francis Xavier University, Antigonish, NS, Canada

Gavin Kennedy Nova Scotia Environment and Climate Change, Halifax, Nova Scotia, Canada

Jocelyn C. Kickbush Department of Earth and Environmental Science, Acadia University, Wolfville, Nova Scotia, Canada

Katie King Department of Biology, Acadia University, Wolfville, NS, Canada; K.C. Irving Environmental Science Centre, Acadia University, Wolfville, NS, Canada

Sara Klapstein Department of Earth and Environmental Science, Acadia University, Wolfville, Nova Scotia, Canada

Christopher Kozela Department of Biology, Dalhousie University, Halifax, NS, Canada

Craig Lake Dalhousie University, Halifax, Nova Scotia, Canada

Diane LaRue Mersey Tobeatic Research Institute, Mount Merritt, NS, Canada

Juan C. López-Gutiérrez Department of Biology, Acadia University, Wolfville, NS, Canada

Makeala MacIntyre Antipa Minerals Ltd., Perth, WA, Australia

Mark L. Mallory Department of Biology, Acadia University, Wolfville, Nova Scotia, Canada

Emma Manning Department of Biology, St. Francis Xavier University, Antigonish, NS, Canada

Nelson J. O'Driscoll Department of Earth and Environmental Science, Acadia University, Wolfville, NS, Canada

Michael Parker East Coast Aquatics Inc., Bridgetown, NS, Canada

Jennifer Poirier Department of Biology, St. Francis Xavier University, Antigonish, NS, Canada

Jennifer F. Provencher Science and Technology Branch, Environment and Climate Change Canada, Carleton University, Ottawa, Ontario, Canada

Brent M. Robicheau Biology Department, Clark University, Worcester, MA, USA

Josie Rogers Department of Earth and Environmental Science, Acadia University, Wolfville, Nova Scotia, Canada

Andy Sharpe Kejimkujik National Park and National Historic Site, Maitland Bridge, NS, Canada

Dave Shutler Department of Biology, Acadia University, Wolfville, NS, Canada

John P. Smol Paleoecological Environmental Assessment and Research Lab (PEARL), Biology, Queen's University, Kingston, Ontario, Canada

Marlene Snyder Biology Department, Acadia University, Wolfville, NS, Canada

Ian Spooner Department of Earth and Environmental Science, Acadia University, Wolfville, NS, Canada

Kelly Stevens Department of Earth and Environmental Science, Acadia University, Wolfville, Nova Scotia, Canada

Brad E. Toms Mersey Tobeatic Research Institute, Caledonia, NS, Canada

Holly E. Turner Department of Earth and Environmental Science, Acadia University, Wolfville, Nova Scotia, Canada

Deanne van Rooyen Department of Earth and Environmental Science, Acadia University, Wolfville, NS, Canada

Allison K. Walker Department of Biology, Acadia University, Wolfville, NS, Canada

Claire E. Wilson Department of Biology, Acadia University, Wolfville, NS, Canada;
Nova Scotia Department of Natural Resources and Renewables, Kentville, NS, Canada

Part I
Pre-Restoration Biogeochemistry and Ecological Interactions

Chapter 1
Introduction

Nicholas M. Hill, J. Sherman Boates, John C. Brazner, Sarah Hines, and Nelson J. O'Driscoll

Abstract This book is about efforts to restore the habitat of the endangered Eastern Mountain Avens (*Geum peckii*) in Big Meadow Bog, Nova Scotia. *Geum peckii* is one of Canada's most at risk species. In this chapter we overview the rationale for the restoration project, the conservation history for *Geum peckii,* and the concepts of adaptive management which shaped both the restoration project and the structure of this book.

Keywords Adaptive management · Restoration · Eastern Mountain Avens

1.1 Restoration Rationale

This book is about an endangered wildflower, the Eastern Mountain Avens (*Geum peckii*) and the collaborative efforts to prevent its extinction. *Geum peckii* is one of Canada's most at risk species. It is an endemic species; in the world, found only in two areas: Digby Neck and Brier Island in Nova Scotia and in the White Mountains in New Hampshire. The distribution in Canada is centered on Big Meadow Bog, Brier Island, Nova Scotia at the very end of a 50 km long narrow peninsula on the last island projecting into the cooling waters of the Bay of Fundy. The ecological story of the

N. M. Hill
UNESCO Southwest Nova Biosphere Association, 399 Wyman Road, Sand Beach, NS B5A 5G4, Canada
e-mail: nick@swnovabiosphere.ca

J. S. Boates · S. Hines · N. J. O'Driscoll (✉)
Acadia University, 15 University Ave, Wolfville, NS B4P 2R6, Canada
e-mail: Nelson.odriscoll@acadiau.ca

J. S. Boates
e-mail: sherman.boates@acadiau.ca

J. C. Brazner
Wildlife Division, N.S. Department Natural Resources and Renewables, 136 Exhibition St., Kentville, NS B4N 4E5, Canada

N. M. Hill et al. (eds.), *Wetland Restoration for Endangered Species Recovery*, Environmental Contamination Remediation and Management,
https://doi.org/10.1007/978-3-031-71344-6_1

“Avens” (Eastern Mountain Avens) begins tens of millions of years ago when genetic changes eliminated the hooks and barbs that gave its seeds long distance dispersal by attaching themselves to furry mammals. This limited Avens, in common with other specialist and rare plants, to a life of slow-growth and persistence in stress-filled wetland habitats where the risk of being outcompeted by others was low. But the world over, these habitats are being lost due to disturbance and pollution, and a set of generalist species adapted to fast growth, early reproduction and dispersal are replacing these specialists as their habitats are transformed.

While Avens has been listed as Endangered in Canada since the 1980s and has long been protected by both Provincial and Federal legislation, there had been little effort to address the threats to this species. While there was a need to acquire and secure habitat for this species, much of the habitat was severely degraded and large-scale restoration was of Big Meadow Bog was the best option for saving the plant. Avens has faced multiple critical and ongoing threats over decades including wetland ditching, hydrological change, shrubification, nutrient enrichment, disturbance by gulls, and intrusion and competition from weedy plants. Very recently, concern over climate change has increased and with it, an urgent need for the conservation community to consider new extreme and risky interventions such as translocation of some of the remaining plants into new locations.

It was in this spirit that a score of scientists from universities, consultancies, NGOs and government decided not to walk away from Big Meadow Bog. The bog was ditched in 1958, a time before the Environment Act (1994–95) and the Nova Scotia Wetland Conservation Policy (Government of Nova Scotia 2011) when across Canada, wetland “wastelands” were being drained across vast areas. While detailed data on wetland loss across the country is generally lacking, in some of the more rapidly developing areas like southeastern Ontario, 70% or more of the original wetlands were lost primarily to drainage for agriculture (Snell 1987; Ducks Unlimited Canada 2010). In other more urbanized settings such as the Halifax Peninsula, virtually all the original wetlands (18% of the peninsula was estimated to have been covered by wetland prior to European settlement—mostly swamp) had been lost to a combination of drainage and infilling for housing and commercial development projects (Reid 2014). The ditching of Big Meadow was relatively late compared with wetland losses in places like Halifax and the Greater Toronto Area, or the massive losses of salt marshes to Acadian dyking along the Bay of Fundy, which began as early as the 1600s (Hanson and Calkins 1996) or the damming of large (>10^5 ha catchment area) rivers that took place in Nova Scotia in the 1920s and 30s (Hill et al. 1998). It was also a small footprint area of 65 ha but to the traditional way of life of the Brier Island community, and it was discovered, to the security of the Endangered Avens and various other species in the bog, it was substantial.

A decade of study and action, and more than a million dollars, went into the restoration of the Meadow and the conservation of the Avens. Given the sheer scale of wetland alteration and ecosystem disturbances more generally, why restore a bog to recover a single small yellow flower? Why are rare plants important? Rare plants are rare usually because they are specialists that occur in particular habitats with high ecosystem integrity; they are the opposites of the common species and the generalists that signal little other than disturbed conditions. After the ditching, the Meadow had become filled with raspberries and blackberries and the exotic grass, Yorkshire Fog (*Holcus lanatus*), and the specialists and the Endangered Avens were lost or were dwindling (Hill et al. 2018a). The rare plants stood for and required habitats of high ecological integrity, and these were linked to Sphagnum mosses that were a natural water filtration system, and to an intact peatland that absorbed rainfall and runoff to provide a moderated outflow rather than peaks of flooding. The clean water from the intact system did not threaten underground water supplies, the coastal fishery or seawater lobster holding facilities. The intact system also was a community larder for various berries in the bog and for waterfowl from the water lily ponds. More often than not, there are strong societal advantages gained when we maintain the ecosystem integrity needed to maintain rare species (Hill et al. 2018b). In wetlands, these are water quality, flood protection, flood storage capacity, maintenance of base flow for rivers, carbon storage and sequestration, recreation, wild foods and biodiversity (Mitsch and Gosselink 2015). The intact Meadow that supported Avens and a host of other rare species, provided these services, however, in spite of common sense and wetland legislation, it was Species at Risk legislation that made this restoration possible.

The research and the conservation activities reported on in this book provides a sound foundation and a comprehensive knowledge base to direct and guide conservation actions moving forward. The future of *Geum peckii* remains very tenuous, and we must continue to work diligently together to collect new knowledge about this plant and its habitat, track change in the environment and respond with appropriate conservation actions.

1.1.1 Conservation History of Geum peckii

Geum peckii, or Eastern Mountain Avens, is one of Canada's most at risk species. In 1986 COSWIC, assessed Eastern Mountain Avens as Endangered, making it one of the first plant species at risk in the country. COSEWIC confirmed the status of Endangered for this species during it periodic reassessment meetings in 1999, 2000, and 2010 and the plant is again due for another review of its status. Eastern Mountain Avens is also consider to be imperiled in the White Mountains, New Hampshire, the only other location for this plant in the world. At the global level this Avens species was reviewed and again listed as Imperiled (G2) by NatureServe in 2020.

In Canada, there was very little, if any, legal protection for plants and many other wildlife species at risk until legislation was passed in jurisdictions in response to commitments made under the "Accord for the Protection of Species at Risk" in 1996. Nova Scotia passed the first stand-alone Endangered Species Act in Canada in 1998. In 2000, several species were for the first time listed under this legislation. Eastern Mountain Avens was one of these listed species giving it legal protection for the first time, 14 years after it had first been determined to be at risk. This legislation protected the species from harvest and disturbance and established a legal timeline for formal conservation planning to address the threats to the plant. Eastern Mountain Avens received further legal protection when the federal government's Species at Risk Act (SARA) in 2003.

Although formal conservation planning and action did not really begin until species at risk legislation was in place there were some early conservation efforts for Eastern Mountain Avens. In 1986, Cathy Keddy authored the first COSEWIC status report and pulled together all known information on the Canadian population including her field work that resulted in a population estimate of 5450 individuals. Ruth Newell wrote a revised status report in 2000 and despite its estimate of a larger population of about 9000 individuals (the result of exhaustive survey efforts), the subsequent 2010 COSEWIC report by Sean Blaney was a call to arms as it underlined the threats outlined by the previous reports remained unabated and the future outlook for the species' survival in Canada was dire. In the decade or more following Blaney's report, Brad Toms and Diane La Rue at the Mersey Tobeatic Research Institute tracked changes in the Avens population. The entire Canadian population that occurs at the tip of the Digby Peninsula and 95% of which is on Brier Island was estimated to be 6000 individuals in 2022 (see Chap. 4).

Some valuable work was done in the late 1990's and early 2000's, notably June Swift (now June Farnsworth), a Brier Island resident, raised awareness on the island, with visitors, and with calls to government to take action about Eastern Mountain Avens through her photography and informal monitoring of the plant and through guided nature walks. Diane La Rue and Gini Proulx conducted some surveys and monitored parts of the Nova Scotia population.

A long-standing conservation milestone supporting the conservation of Eastern Mountain Avens was the acquisition and designation of the Brier Island Nature Reserve by the Nature Conservancy of Canada in 1993. This Reserve is 400 + hectares, amounts to about one fourth of the island's area and includes some excellent habitat for Avens and critical components of the Brier Island ecosystem. Some conservation activities limiting ATV travel and other vehicles and posting signs and fencing were initiated, but direct actions addressing the key threats to the plants were not yet underway.

An important part of endangered species legislation is that it usually provides regulations that hold the Minister accountable for the timely development of conservation recovery planning for species at risk in the form of recovery strategies, recovery action plans, etc. Although there was some recovery planning work done in the early 2000s, recovery planning documents were not formally completed and approved until 2010 (Strategy/Plan) and 2018 (Action Plan). These plans are well done and layout the key actions that need to be done to begin to address the Eastern Mountain Avens critical situation.

Why was there next to no concerted conservation efforts to address the key threats to Eastern Mountain Avens until 2013? There are now over 800 species at risk in Canada and a small subset of these are very high priority for one reason or another, for example they may be a charismatic bird or mammal like Piping plover or Vancouver Island marmot and there are species of extremely high priority in terms of economic implications and indigenous values like the Caribou and American Eel. Further, there are many degraded high profile ecosystems. There are also very high-priority ecosystems that harbor a suite of at risk species, for example the Gary Oak Ecosystem and the Sydenham River. Unfortunately, Eastern Mountain Avens did not receive the attention and share of resources that could be argued that it merited. It could also be argued that species at risk research and management has been chronically underfunded, albeit there were some limited resources provided nationally and provincially with the implementation of species at risk legislation. In the past few years there has been significant increases in staff and funding for species at risk that has resulted from court cases and auditor general reports that have repeatedly shown slow, or no progress on the implementation of species at risk legislation. Resources continue to be largely consumed by efforts to meet the requirements to reassess the status of species and to complete the legally required recovery planning documents.

The above points to some of the constraints on the lack of progress toward directly addressing threats to Eastern Mountain Avens and other species at risk. What enabled the eventual large-scale collaborative research and conservation action on Avens over the past 10, or so, years? In short, it was essentially a "perfect storm" of largely unrelated factors that set the stage and paved the way for this work. This storm included: the increased availability of funds, the higher priority on restoration as a recovery strategy rather than the purchase of pristine habitats, and the profile of Avens as a globally important Canadian endemic species. Avens only gained real conservation attention when governments realized that eventually pressures would shift from planning to action and implementation. The key factor was the will and interest of researchers, government and non-government conservation practitioners, and the local community on Brier Island; this was the overarching catalyst driving conservation focused on the Eastern Mountain Avens.

In 2011, the first of a series of face to face meetings were held on Brier Island to discuss and plan the work together on the recovery of Eastern Mountain Avens. The first question tackled was: Is it feasible to restore the structure and function of Big Meadow Bog by addressing the ditching and raising the water level of the wetland to the original level. Three aspects of feasibility were considered: (1) Is it environmentally feasible? i.e., Are the overall impacts beneficial from the perspective of the endangered plant and ecosystem? What level of risk was there for Avens and the wetland? (2) Is it feasible from an engineering perspective? i.e., Can we intelligently tinker with the ditches and alter hydrology using small excavators and the resident substrate of the wetland? 3. Is the project socially feasible? i.e., Is the Brier Island community behind the project and willing to work with us on it? After about two years of study and discussion the team agreed that the restoration was feasible from all three perspectives and the team moved from evaluation phase to the implementation phase of the work. However, the project was stalled as we had permission from all the landowners of the Big Meadow Bog but one. A frustrating and difficult year followed with no progress on acquiring this small but critical piece of property. Finally, the Municipality of Digby stepped up and orchestrated and funded a forced purchase of the property for the good of the community, and the formal restoration of the wetland went forward. This book reports on much of the research and the results of the restoration efforts over more than ten years of dedicated fundraising, research and conservation action by our unique team.

Like many small fairly isolated Nova Scotia coastal communities Brier Island is very independent and very much used to do things on their own. The community understood the importance of Eastern Mountain Avens and the bog restoration from the start and to this day individuals and groups provide ongoing support for research and conservation activities. Of particular note the community formed the Brier Island Trail Committee in 2017; this group championed the bog's restoration with information signs, a trailhead garden and an accessible boardwalk where more than 7000 visitors walked over more than a kilometer of trail into Big Meadow Bog (Chap. 2).

Research and conservation action continues in 2024 for Eastern Mountain Avens. Some projects have been completed, but some new work has begun. The risk to Avens from the impacts of relatively rapid climate change on Brier Island is increasingly a concern. Research is currently being conducted on this and in the winter of 2023, a workshop was held to consider and evaluate the benefits and costs of translocating some Avens plants to suitable wetland habitats with more favorable climate futures. This is an extreme but possibly necessary intervention. The Kespukwitk Conservation Collaborative has provided a forum, expertise and links to funding that have helped maintain conservation attention on Eastern Mountain Avens. It is a partnership in Southwest Nova Scotia of Mi'kmaq First Nations, Indigenous organizations, non-government organizations, academic institutions, communities, and federal and provincial government departments. Finally, in Spring of 2024 Brier Island was designated a Global Key Biodiversity Area. While this designation is symbolic in nature, this is important as it will further increase awareness and substantiate the outstanding value of Avens, Big Meadow Bog and the surrounding Bay of Fundy environment.

The future of *Geum peckii* remains very tenuous, and we must continue to work together to collect new knowledge about this plant and its habitat, track change in the environment and respond with appropriate conservation actions.

1.1.2 Adaptive Management and Book Structure

This book summarizes a diverse research and conservation activities that support the recovery of *Geum peckii*. A common problem in conservation is a lack of evidence and other knowledge to direct and guide recovery and restoration actions. Uniquely, scientists, conservation practitioners and local Brier Island community worked closely together on all aspects of this work. As well, an adaptive management approach was taken early on. In an adaptive management approach, collaborators work cyclically through five steps: (1) Conceptualize, (2) Plan and monitor, (3) Implement Actions and monitor, (4) Analyze, use and adapt, and (5) Capture and Share what is learned (Fig. 1.1). Through sequential iterations of this cycle, conservation actions are improved and adapted to changing conditions. This book is an effort to share in

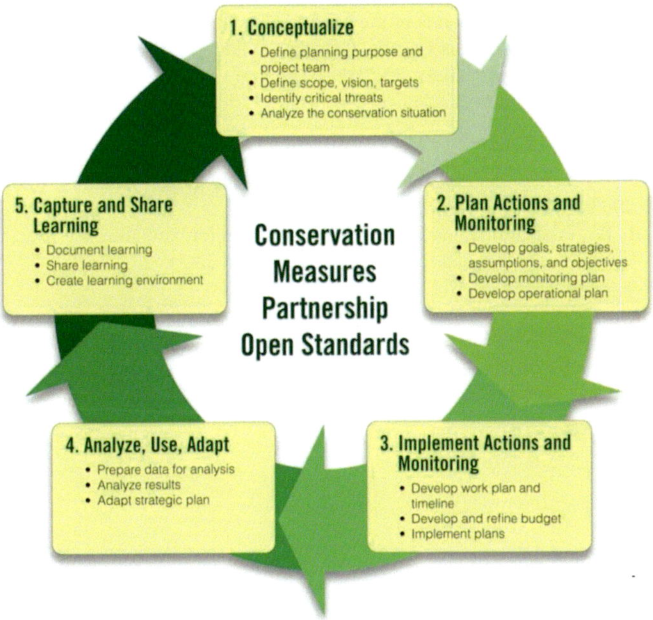

Fig. 1.1 Conceptual diagram of the adaptive management approach (CMP 2007)

one place many of the results of our work so it will be available for future conservation practitioners on Brier Island and for the broader conservation community.

This book is organized into two primary sections; Section I overviews the restoration work to characterize the biogeochemistry of the Bog and the ecological interactions, Section II examines the post restoration temporal and spatial changes in the bog and the need for continued monitoring and intervention.

This book is intended for researchers, policy makers and community members interested in the both the methods and the implications of water table restoration. It will also be of interest to a wide range of scientists interested in the fundamental chemical and ecological interactions within wetland systems.

References

CMP (2007) The conservation measures partnership. Open Standards for the practice of conservation. Version 2.0 October 2007. www.conservationmeasures.org

Ducks Unlimited Canada (2010) Southern Ontario wetland conversion analysis. Duck Unlimited Canada

Final Report, Barrie, Ontario, Canada. https://cbm

Government of Nova Scotia (2011) Nova Scotia Wetland Conservation Policy. Nova Scotia Department of Environment. https://novascotia.ca/nse/wetland/conservation.policy.asp

Hanson A, Calkins L (1996) Wetlands of the Maritime Provinces: revised documentation for the wetlands inventory. Environment Canada, Canadian Wildlife Service, Atlantic Region

Hill NM, Crowell M, LaPaix R, Hicks S (2018a) The rare southern twayblade (*Neottia bifolia*): sentinel of ecosystem integrity for sphagnous swamps. Rhodora 120(982):117–142. https://doi.org/10.3119/17-11

Hill NM, Chinner C, Toms B, Garbary DJ (2018b) Recovery succession of native plant communities at Big Meadow Bog after disturbance and eutrophication from Herring Gulls. Wetlands. https://doi.org/10.1007/s13157-018-1099-1

Hill NM, Keddy PA, Wisheu IC (1998) A hydrological model for predicting the effects of dams on shoreline vegetation of lakes and reservoirs. Environ Man 22(5):723–736

Mitsch WJ, Gosselink JG (2015) Wetlands. John Wiley & Sons

Reid M (2014) The advantages of incorporating historical geographic information systems (H-GIS) into modern coastal management planning. J Map Geogr Libr 10(2):157–172

Snell E (1987) Wetland distribution and conversion in Southern Ontario. Environment Canada, Inland Waters and Land Directorate, Working Paper No. 48, Ottawa, Ontario, Canada

Chapter 2
Big Meadow Bog: A Lost Backyard

Nicholas M. Hill and Jeanette C. Denton

Abstract Big Meadow Bog provided berries and game and medicinal plants for the first people, the Mi'kmaw, and continued to do so for the first settlers. Their livelihoods and island economy were based on the coastal fishery but the Meadow was part of life and tradition. Paths from the village of Westport ran the length of the Meadow leading to berry picking areas, pools with duck, and the beach at the south end of the wetland complex. The Meadow continued to function as a "commons" for people to harvest berries and hunt until it was ditched in 1958. The ditching lowered the water table and altered the ecosystem so that shrubs and brambles grew up making a barrier that separated the village from the Meadow. The bog vegetation was lost and the pools were drained. We argue that when the hydrology of a wetland is fundamentally altered, the traditional uses of that wetland by local people are also lost. The margin, the "lagg", of Big Meadow Bog was a key habitat for the Endangered Eastern Mountain Avens, *Geum peckii.* Conservation concern and the recovery efforts aimed at this plant provided the rationale for the wetland restoration discussed throughout this book. This restoration effort and the boardwalk from the village into the bog have reconnected the village with this lost backyard and have renewed an appreciation of the Meadow and a sense of pride in its globally rare Avens.

Keywords Brier Island · Community · Hydrological disruption · Traditional uses

---

N. M. Hill (✉)
UNESCO Southwest Biosphere Association, 399 Wyman Road, Sand Beach, NS B5A 5G4, Canada
e-mail: nick@swnovabiosphere.ca

J. C. Denton
Brier Island Trails Committee, 197 Water Street, Westport, Brier Island, NS B0V 1H0, Canada
e-mail: jcdenton127@gmail.com

N. M. Hill et al. (eds.), *Wetland Restoration for Endangered Species Recovery*, Environmental Contamination Remediation and Management,
https://doi.org/10.1007/978-3-031-71344-6_2

2.1 Introduction

Brier Island is a special place! For those whose families have lived here for several generations, it is home. For birders, it is known as an important stopover on the spring and fall migratory routes. More recently it has gained fame as one of the few places on earth where one can find the Eastern Mountain Avens, one of the most endangered plants in Canada. Efforts to restore the Avens' natural habitat in the Big Meadow Bog are ongoing and providing a wealth of information on components of successful bog restoration.

Before the written record, Big Meadow Bog was a shallow sea lagoon between two ridges formed by basalt lava flows. The lagoon could have been a fishing area for eels for the first people. The lagoon floor filled in with sands then layers of organic muck before it became a freshwater peatland 2500 years ago (Spooner et al. 2017). The sands, muck and peat so far have revealed no clues about the lives of the first people but during the 1950s and 1960s, a lighthouse keeper discovered numerous projectile points and stone tools along the shores of the Bay of Fundy. In Freeport, on neighbouring Long Island, a resident found evidence of a Mi'kmaq encampment while digging a garden on her property (personal communication to Jeanette Denton). Brier at the mouth of the Bay of Fundy has underwater ridges that create upwellings of cold deep waters that cool the island and sets up good fishing areas. The fish and the clams (Quahogs, *Mercenaria mercenaria;* and soft shell clams, *Mya arenaria*) supported the people, "Lnu", over the millennia as we can see in the clam shell middens near L'sitkuk (Bear River). The middens, the stone tools on Digby Neck, the shoreline on Isle Haute—an island in the middle of Fundy and a stopping place for canoes between New Brunswick and Nova Scotia—which is full of the agate chippings as Lnu made their tools, all call up a deeper human past. The voices of the present reflect the past. Mi'kmaq linguist, Bernie Francis, tells how he never asked the former Chief of L'sitkuk, Richard McEwan, about his accent but noted it had most in common with the Mi'kmaq dialect in Elsipogtog (Big Cove) across the Bay of Fundy in New Brunswick (CBC Mainstreet Interview 2024). The Bay of Fundy connected Lnu. Another former Chief of L'sitkuk, Frank Meuse, recalled his connection with Brier Island when he was a boy:

> "I remember as a young child seeing dried cod and pollock hanging in the basements of people's homes here in L'sɨtkuk. One year my parents said that we were going to go down to the Digby islands to catch fish. I was so excited to go on this trip. We spent the whole day waiting for the tide to come in. We children were allowed to fish off the end of the wharf. We were catching mostly pollock and if they were big enough the adults in the group would take the largest ones to clean them and put them on salt for the trip back home. As it got closer to high tide, the adult fishers would start casting their lines far out to catch the cod, which was what was being sought after for our winter food. The fish stopped biting as the tide receded, which meant it was time to head home to L'sɨtkuk.
>
> From porpoise hunting to catching large ocean fish, our ancestors have been sustaining themselves from the coastline for thousands of years. I was lucky in my generation to participate in the collection of fish to supply for our families and community. I have taken my children down to the islands to tell the stories and let them walk the same steps our ancestors took. And hopefully I can take my grandchildren and great-grandchildren. The

old ones used to say that the land doesn't belong to us, but that we are part of it. We have a responsibility to look after the lands and waters as we will only be as healthy as nature around us." (personal letter to NM Hill, January 2, 2024)

The oldest resident of Brier Island, Raymond Robicheau, remembers a Mi'kmaw couple visiting his father when he was a boy in the early 1930s. There are stories also of a Mi'kmaw couple who canoed from Bear River (L'sitkuk) to summer in Peajack Cove on the western shore of Brier. Maybe they liked the cove because its coastal winds would keep flies at bay, it had a stream and good fishing offshore. The community of L'sitkuk is almost 100 km from Brier and all along the way to Brier are deep sandy shores that keep clam diggers busy today.

The dating for the core taken in the middle of Big Meadow Bog goes back 9000 years into organic muds that lie on top of a bed of dark grey sands that make up the beach at Pond Cove, the southern outfall of the Meadow. It is likely that these marine sands once supported the clams that are harvested in the county today and which supported the Mi'kmaw from earliest times. As the Meadow developed new layers on top of the marine sands, there would have been millennia of saltmarsh followed by freshwater marsh, each the new normal to the people of the area as the herons gave way to duck, muskrat and beaver and as the plant community changed. We can build a misty picture together with western science and traditional knowledge of these lost backyards: the common space of the Meadow enjoyed and valued down through the generations. The peat core extracted and dated from the centre of the Meadow shows that after the marsh phase, Sphagnum mosses became established and a peatland became a treed swamp about 2000 years ago (Spooner et al. 2017). Life was about to change for the people, if we are correct in our interpretation, for the cohort of trees, dated at 1760 years before present, died standing up due to a regional "wetting event" that waterlogged the trees (Spooner et al. 2017). This cool and wet period set up the pattern for the Meadow's development from treed swamp to open fen, then from fen to the raised bog.

The Meadow was a raised bog when the first settlers came to the island in the eighteenth century. At first, fishermen were drawn to Brier in summers by the abundance of fish but throughout the mid eighteenth century, families began to settle, eventually establishing Westport on either side of the northern outflow of Big Meadow Bog (Davis 1997). While fish was the mainstay, there was an agriculture based on the fertility of the marsh at the northern outflow of the Meadow. One of the early settlers, James Morrell had a farm in this northern marshland. There are several ditch lines still visible but this farming provided for families and did not transform the basic ecological processes of Big Meadow. Big Meadow is a quintessential "wetland complex" in time as we have seen through the portal of the core taken in the centre of the bog (Spooner et al. 2017) but also over space. There is a saltmarsh at the northern outflow that visitors can now appreciate from a boardwalk trail. Further along into the Meadow, the saltmarsh becomes a freshwater meadow. In the main body of the Meadow, the wetland was a raised bog that was bordered by fen, a wetter type of bog that supported cranberries and the Endangered Eastern Mountain Avens.

2.2 Settler's Life by the "Meadow"

To the Mi'kmaw, the Meadow may have been known as "pogwin" (lagoon) or "saqpejaq" (a bog), or any of the wetland types it developed through over the 13,000 years since deglaciation (Stea and Mott 1998). By the time colonists settled in the eighteenth century, the dated core tells us that the main bog had been a raised bog for five hundred years (Spooner et al. 2017). At this time, the bog would have supplied a variety of berries in summer that included huckleberries (both the black and the bog huckleberry: *Gaylussacia baccata* and *G. bigelowiana*), blueberries (lowbush and occasional highbush: *Vaccinium angustifolium* and *V. corymbosum*) and chokeberries (black and red: *Photinia melanocarpa* and *P. arbutifolia*) but most memorable was "Jonesberry" (pers. comm. to authors from Eleanor Bailey and David Pugh, two lifelong residents of Brier Island). Jonesberry referred to *Rubus chamaemorus* which is known in other regions as cloudberry or bake-apple. That it is known as Jonesberry only on Brier Island is testament to its local importance. Pugh remembers "ladies would take 50 lb flour bags to pick" and fill with berries (personal email to Nick Hill, December 19, 2023).

From the earliest aerial photos, a line can be seen running down the centre of the Meadow, this was a well-worn path that if "you weren't careful to stay centred, you would stumble over the hummocks on either side" (Pugh, pers comm to NH 2012). The Meadow was part of life for many islanders, in the nineteenth century most notably, Margaret Davis, who was well versed in the medicinal values of the local plants (Stark 2022). Apart from the berries of the bog, there were cranberries (large and small: *Vaccinium macrocarpon* and *V. oxycoccos*) in the fens that are wetter peatlands surrounding the drier central bog. Also, there were the pools in these marginal fens (the "lagg"), the biggest ones in the southern Meadow were known as the "lily ponds". Brier islanders hunted duck in these ponds. Eleanor Bailey told of when she was a child in the 1930s, her father would gather water lily flowers for her mother which she would place on the dining table on Sunday dinners (personal communication to Jeanette Denton, 2020).

2.3 The Ditching

Big Meadow Bog to the Mi'kmaw was a commons with the same bounties of berries and waterfowl and eels available as were used by people of the village of Westport. The first people could also have made use of various plants such as sphagnum mosses, spruce roots, and reed-like plants for textiles and there were many medicinal plants available (see lists in Moerman 1998) in the beach, saltmarsh, freshwater marsh, fen, raised bog and swamp that make up the wetland complex of Big Meadow Bog. The Eastern Mountain Avens was formally documented as a species to science in 1949 (by Roland and Smith, in COSEWIC 2010). The orange rhizomes, likely filled with similar triterpenes and diverse phenolics found in the Water Avens (Orlova et al. 2022), make it possible to suggest it may have been a medicinal plant for the Mi'kmaw. In the early settler period, the Meadow, despite the property divisions, still operated as a commons in practice. Well-worn paths are visible in the aerial photograph from the 1955 (Fig. 2.1); there was a northern entrance to the Meadow, above Jimmy's Pond in the photo, and once in the Meadow, one path follows the western lagg (margin) stream and is a clear light line in the photo, and the other headed down the middle of the bog. This central path appears like a crease in the photo; both paths are clearly in the south of the bog more than a kilometre and half after the entrance. The southern end of the bog ends at the pond above Pond Cove. The Meadow was wetter in the south. A 1928 aerial (Fig. 2.2) shows clearly the beach, the pond and then three lily ponds (labelled E, F and G in pencil on the old photo) where duck were hunted.

At the north end of the Meadow, the bog becomes a marsh. At the right side of the photo (Fig. 2.3) is the sea, Grand Passage between Long Island and Brier. The outflow of Big Meadow Bog appears as a dark line with larger dark areas that are pools. The large pool bounded by a straight line (a berm and a road) is Jimmy's Pond. This pond was seasonally regulated with stop logs to make ice before New Year to fill the ice house in town to refrigerate fish, and then the logs were removed. The pond was a community landmark especially in winter when people skated and

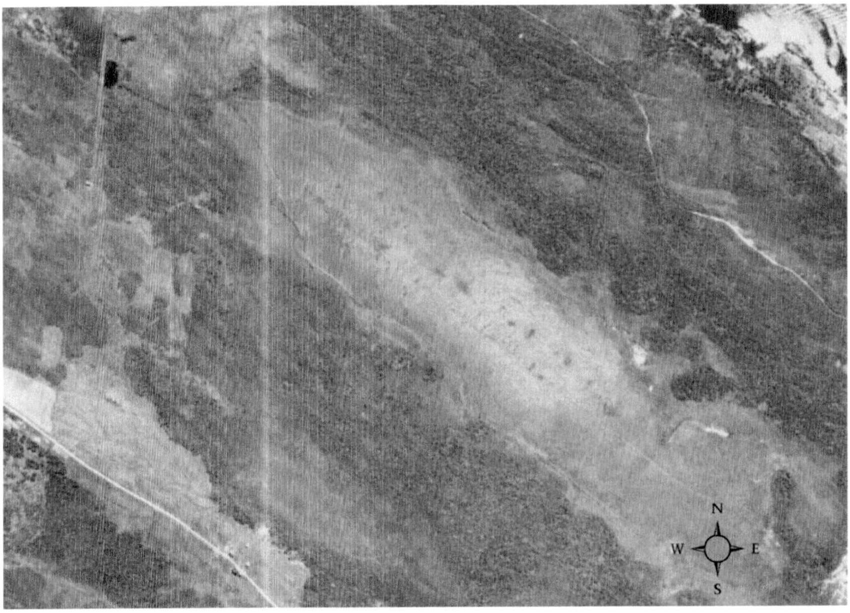

Fig. 2.1 Aerial photograph of Big Meadow Bog taken in 1955. The top left dark shape is "Jimmy's Pond" at the northern end of the bog. The bog, "the Meadow", is an open expanse of low shrub and sedge; note two paths as lines on the figure that run the length of the bog and become apparent in the south. Photo Credit: Nova Scotia Department of Natural Resources and Renewables

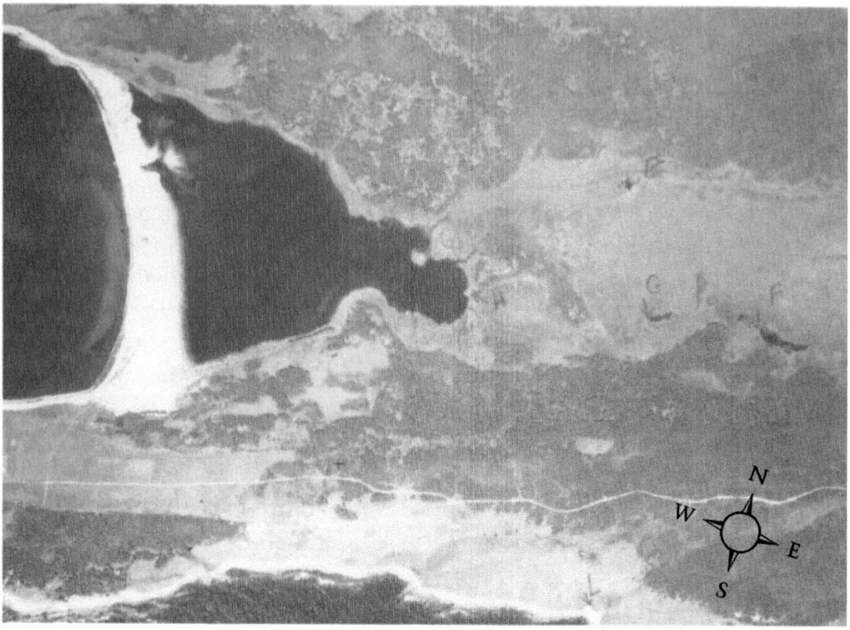

Fig. 2.2 Aerial photograph from 1928 of the southern end of Big Meadow. Note a barrier beach at the left of photo, then a Big Pond followed by the bog (light grey). Letters at the south of the bog, "E,F,G" are the "Lily Ponds" named because they grew the White Waterlily, *Nymphaea odorata*.. Photo Credit: Nova Scotia Department of Natural Resources and Renewables

socialized on the pond. Before the ditching, in a good freeze up, people remember skating the length of the Meadow from the village down to Pond Cove.

While Big Meadow had functioned as a commons, it was not a commons in property terms. In 1958, a drainage project began, and ditches were dug using a Bombardier snow machine frame equipped with an excavator backhoe and two barrels full of gasoline (Fig. 2.4). After two years of excavation, two, 2 km long ditches had been dug the length of the Meadow down to Pond Cove (see Chapter 10). The impetus for the project or its funding is not clear but from many reports on the island, the trial of the agricultural suitability did not last long as carrots were stringy and the celery did not grow well. After a few years, the prospect of agriculture was dropped, and Big Meadow Bog was left to itself in a ditched condition for 55 years.

2.4 Conclusion: The Backyard Lost and Regained

There are a lot of memories of the bog, the "Meadow", as there are in most rural communities. But the dividing lines on the map and in time are sharper at Big Meadow: paths became six-foot-deep ditches which achieved their goal of dropping

Fig. 2.3 1928 Aerial photograph of the north end of Big Meadow Bog. To the right of photo is Grand Passage, the straight between Brier and Long Islands. On either side of the stream outflow of the Meadow are the villages of Westport (top) and "Irishtown" (bottom). The fields (light coloured) closest to the Meadow extend into marsh land. The rectangular pond (dark shape) with a straight pale line to its right is Jimmy's Pond. Photo Credit: Nova Scotia Department of Natural Resources and Renewables

the water table and allowing for plant growth but the village community lost valuable connections. The plants that grew after the drainage were a variety of blackberries and thicket. Hill et al. (2018) examined the ecology and reported that the Meadow had lost its "lagg", the low shrub margin encircling the bog that supported the Avens and a unique set of plants. This lagg has increased density of spruce trees and bushes over the five decades since ditching. This was a biodiversity concern, and these drainage ditches also acted against community tradition, livelihood and recreation. Residents could no longer get into the Meadow at the north end from the village. There were thickets so dense residents recall frustration trying to whack through and blackberry and rose bushes tearing into trousers. The transformation of a bog plant community to chest-high shrubs with patches of white spruce trees was complete. The lower water table from the ditching combined with thousands of nesting gulls that now fertilized the bog catalysed the plant community change. Gulls were seldom found in Big Meadow Bog before the ditching (personal communication from Raymond Robicheau in 2019 to N. Hill). Aerial photos show that Herring Gulls began to nest

Fig. 2.4 The Bombardier-based excavator at work (ca. 1958). Note the low vegetation (sedges and low shrubs) throughout the bog and the far border of shrub and black spruce. Photo Credit: Bernice Powell

in the south of the Meadow in the 1970s and by 2001, they were in the northern half. It is likely that the gulls were drawn to the newly ditched bog by the ditch spoil piles (Hill et al. 2018). These piles gave the gulls an elevated look out for predators as well as a well-drained corridor for nesting.

Better drainage from the ditching and the fertilization brought on by thousands of gulls, transformed the Meadow but the replacement of what ecologists call bog and fen for thorn and thicket meant that the community lost the path from Westport to the bog that had provided fruits and fowl, and the walk to the beach at Pond Cove. How often does human hydrological disruption of natural ecosystems result in a "lost backyard"? Since the loss of glaciers, the landscape of Big Meadow Bog has been governed by the two basalt ridges running along its east and west sides, and by the level of the sea at its south and north ends. With time and changes in sea level, a freshwater peatland was formed approximately 2500 years ago. This began as a treed swamp, but the black spruce trees were drowned during a regional wetting period and the peatland became an open fen over the next thousand years (Spooner et al. 2017). Through the continual growth of Sphagnum mosses which raised the water level of the central peatland above the surrounding water table, a raised "Big Meadow" bog formed 700 years ago. This raised bog and the moat of fen all around it became the Mi'kmaw and settlers depended upon. Ditching efforts that began in 1958 meant that the water flow controlled by the million cubic metres of peat were lost. This led in the loss of the low shrub margins (the lagg), the blocking of the access to the bog by alder thicket, the fertilization of the whole bog by Herring Gulls, the overgrowth of the bog by tall thorns, and the loss of community traditions: berry picking, duck hunting, and walking to Pond Cove. It also resulted in more intense water flows after heavy rains as the outflow was not regulated by migration through peat but rather flowed through ditches. The addition of gull guano also resulted in poor water quality and high nutrients (the bog is now classed as "hyper-eutrophic"). Over time, the restoration is working towards greater flood control as drainage out of the bog is slowed (Chapter 10). A goal of this restoration is improved quality of water in the outflow that empties into Grand Passage, benefiting the coastal fishery. There is still a problem with the numbers of gulls nesting in the Meadow and the community has made clear that this needs attention for water quality and noise reasons.

The impact on communities and their traditions and livelihoods from disruption of the hydrological regime was extensively reviewed for rivers worldwide (Ekka et al. 2020; Richter et al. 2010; Hawley 2011). In these cases, it is most often a damming of rivers that affects communities, their homes and the use the river for recreation or fishing. In some rivers, damming has had global impacts on the carbon cycle (methane production) and in other cases, inundation of shorelines and wetlands has meant that aerobic microbial cycles are replaced by anaerobic pathways that have liberated the toxin, methyl mercury. The risk of methyl mercury accumulating in food chains after flooding terrestrial ecosystems poses a serious risk for humans and wildlife (Eklöf et al. 2021). Chapter 13 makes a holistic biochemical investigation of these pathways and how the fertilization of the bog ecosystem by Herring Gulls may exacerbate this risk.

There are less dramatic impacts that altered hydrological regimes have on how people interact and relate to the natural world. The ditching of Big Meadow Bog resulted in the growing up of a barrier of shrubs and brambles between the village and a natural area which had been part of the collective backyard. Lost access to natural ecosystems can have impact on mental health (Bratman et al. 2019; Mantler and Logan 2015). It can be statistically shown that damming of rivers changes the way people feel about nature (Rivaes et al. 2021). Big Meadow Bog presents an ecological lesson that takes in the strictly ecological but is broader in its social and economic consequences. This book in its most narrow sense is the result of a restoration of a wetland habitat that was deemed critical for the survival and recovery of a nationally Endangered species; it was a result of the Species at Risk Act of Canada. In a broader sense, many researchers and agencies in cooperation with the community were part of a multidisciplinary project to restore a unique and valued wetland ecosystem.

The Big Meadow Bog restoration project was part of a Canadian programme, the National Wetland Conservation Fund (2010–2016). The book showcases the benchmark studies that made the restoration possible, the restoration itself, and our post-restoration understandings. Initiating a boardwalk trail, part of the funding made available by this Environment Canada Climate Change project, might not be considered an essential part of wetland restoration except when one considers that on a community level, the ditching of the bog severed the link to the village. The Brier Island Trails Committee boardwalk has recovered and renewed the link broken by the ditching. The boardwalk allows visitors and community to connect with the Meadow and watch how the Meadow recovers after the restoration of the original hydrological regime (Chapter 10). The year 2023 marked a record year for visitors and locals on the boardwalk; at least 4900 people visited and in 2024, that number surpasses 7000. The boardwalk is a public space where people connect in a shared appreciation of the boardwalk, their community, the wetland and the elusive Endangered Avens not visible from the boardwalk but the ultimate ecological indicator of the success of the restoration. Opening this universal design access to Big Meadow Bog has increased people's ability to interact with nature so building opportunities to improve health and the appreciation of the importance of the natural world (Sandifer et al. 2015).

Acknowledgements This project was principally funded through the National Wetland Conservation Fund administered by Environment Climate Change Canada. It depends upon the close cooperation of ECCC and the Nova Scotia Department of Natural Resources and Renewables, Wildlife Division. The Province has supported this project in many ways as is made evident throughout these chapters. We make a particular call out of appreciation to the Municipality of the District of Digby without whom the restoration project would not have been possible. In like manner, this work was set up by the far-sightedness and intervention of the Nature Conservancy of Canada. Finally, this restoration, the field work, understandings and support could not have happened without the people of Brier Island. There are too many people to thank here but special thanks must go to Danny and Theresa Kenney and to Wally and Joyce Devries.

References

Bratman GN, Anderson CB, Berman MG, Cochran B, de Vries S, Flanders J (2019) Nature and mental health: an ecosystem services perspective. Sci Adv 5(7):eaax0903. https://doi.org/10.1126/sciadv.aax0903

CBC Mainstreet Bernie Francis Interview (2024) https://www.cbc.ca/listen/live-radio/1-37-mainstreet-ns/clip/16033174-linguist-bernie-francis-teaching-mikmaw-future-instructors. Accessed 28 June 2024

COSEWIC (2010) COSEWIC assessment and status report on the Eastern Mountain Avens Geum peckii in Canada. Committee on the Status of Endangered Wildlife in Canada, Ottawa

Davis R (1997) The Davis Family of Connecticut, New York, Brier Island and Yarmouth, Nova Scotia, 1690–1997. A history and genealogy. Self published

Ekka A, Pande S, Jiang Y, van der Zaag P (2020) Anthropogenic modifications and river ecosystem services: a landscape approach. Water 12:2706. https://doi.org/10.3390/w12102706

Eklöf K, Drohan P, Needoba J, Landefeld S, Peterson TD, Hu H, Iavorivska L, Boyer EW (2021) Methyl mercury in lake bed soils following reflooding of an Appalachian reservoir in northeastern USA. Environ Res Comm 3:085004. https://doi.org/10.1088/2515-7620/ac1d83

Hawley H (2011) Recovering a lost river: removing dams, rewilding salmon, revitalizing communities. Beacon Press, Boston

Hill NM, Chinner K, Toms B, Garbary DJ (2018) Recovery succession of native plant communities at Big Meadow Bog after disturbance and eutrophication from Herring Gulls. Wetlands. https://doi.org/10.1007/s13157-018-1099-1

Mantler A, Logan AC (2015) Natural environments and mental health. Adv Integ Med 2(1):5–12. https://doi.org/10.1016/j.aimed.2015.03.002

Moerman DE (1998) Native American ethnobotany. Timber Press, Portland, OR

Orlova A, Kysil E, Tsetkova E, Meshalkina D, Whaley A, Whaley AO et al (2022) Phytochemical characterization of Water Avens (Geum rivale L.) extracts: structure assignment and biological activity of the major phenolic constituents. Plants 11(21):2859. https://doi.org/10.3390/plants11212859

Rivaes R, Couto JS, Schmidt L, Delicado A, Aguiar FC (2021) The influence of river regulation on the affinity for nature and perceptions of local populations. J Environ Mgmt 321:115192. https://doi.org/10.1016/j.jenvman.2022.115992

Richter BD, Postel S, Revenga C, Scudder T, Lehner B, Churchill A, Chow M (2010) Lost in development's shadow: the downstream human consequences of dams. Water Alt 3(2):14–42. https://www.researchgate.net/publication/46093706_Lost_in_Development's_Shadow_The_Downstream_Human_Consequences_of_Dams. Accessed 28 June 2024

Roland AE, Smith EC (1969) The flora of Nova Scotia. The Nova Scotia Museum, Halifax, Nova Scotia

Sandifer PA, Sutton-Grier AE, Ward BA (2015) Exploring connections with nature, biodiversity, ecosystem services, and human health and well-being: opportunities to enhance health and biodiversity conservation. Ecosyst Serv 12:1–15. https://doi.org/10.1016/j.ecoser.2014.12.007

Spooner I, Principato S, Hill N, White H, Dunnington D, Neily T, Stolze S (2017) Late Holocene records of changing moisture regime from wetlands in southwestern Nova Scotia, Canada: implications for wetland conservation and restoration. Northeast Nat 24:331–348

Stark (2022) Naming a ferry for grand passage, Margaret's justice. https://islandshistory.com/2022/02/15/margarets-justice/. Accessed 28 June 2024

Stea R, Mott R (1998) Deglaciation of Nova Scotia: stratigraphy and chronology of lake sediment cores and buried organic sections. Geog Phys Quat 52:1–19

Chapter 3
Genetic Variation, Clonal Structure, and Mating System of the Imperiled Eastern Mountain Avens (*Geum peckii*) and Its Separation from Spreading Avens (*G. radiatum*)

Christopher Kozela, Marlene Snyder, Nicholas M. Hill, Nikolai M. Hay, Matt C. Estep, and Mark O. Johnston

Abstract The Globally Imperiled (G2), yellow-flowered perennial *Geum peckii* occurs on Brier Island and the White Mountains of New Hampshire, and its close relative *G. radiatum* occurs in southern Appalachians along the Tennessee-North Carolina border. We genotyped plants from the three geographically distant locations with the aim of addressing questions of taxonomic status, population structure, propensity to reproduce clonally, and propensity to self-fertilize, using 15 microsatellite loci. We find evidence to support the taxonomic status of *G. peckii* and *G. radiatum* as separate species. The *G. peckii* of New Hampshire and Nova Scotia are genetically similar but exhibit genetic differentiation from each other. The *G. peckii* of

C. Kozela · M. O. Johnston (✉)
Department of Biology, Dalhousie University, Halifax, NS, Canada
e-mail: mark.johnston@dal.ca

C. Kozela
e-mail: ckozela@dal.ca

M. Snyder
Biology Department, Acadia University, Wolfville, NS, Canada
e-mail: marlene.snyder@acadiau.ca

N. M. Hill
UNESCO Southwest Biosphere Association, 399 Wyman Road, Sand Beach, NS B5A 5G4, Canada
e-mail: nick@swnovabiosphere.ca

N. M. Hay
Department of Biology, Duke University, Durham, NC, USA
e-mail: nikolai.hay@duke.edu

M. C. Estep
Department of Biology, Appalachian State University, Boone, NC, USA
e-mail: estepmc@appstate.edu

N. M. Hill et al. (eds.), *Wetland Restoration for Endangered Species Recovery*, Environmental Contamination Remediation and Management,
https://doi.org/10.1007/978-3-031-71344-6_3

New Hampshire show distinct genetic structure at a small scale (10 km), while the *G. peckii* of Nova Scotia are much more uniform genetically. Clonal reproduction is evident, but our results suggest that ramets found at distances greater than 1 m are unlikely to be members of the same clone. We observe a mixed mating system, with partial outcrossing and partial self-fertilization, based on both field and progeny array studies, and lifetime inbreeding depression of 88%. *G. peckii* in Nova Scotia appears to possess ample genetic variation combined with very high inbreeding depression, highlighting the importance of establishing and maintaining a suitable environment for its management and recovery.

Keywords Conservation · Genetic variation · Microsatellite · Allelic richness · Population differentiation · FST · Effective population size · N_e · Inbreeding depression · Mating system · Self-fertilization · Endgangered species · Clonal richness · Clone

3.1 General Background and Chapter Overview

Geum peckii Pursh. (Rosaceae), eastern mountain avens, is a rhizomatous, hermaphroditic, hexaploid herb found in two restricted habitats worldwide, in the White Mountains of New Hampshire USA (NH), and on Brier Island in southwestern Nova Scotia, Canada (NS). In NH, it is found in alpine meadows and along lowland stream banks. In NS, it is found in sea level bogs and fens on Brier Island and in one small mainland population at Harris Lake. This highly restricted global distribution, combined with a number of existential threats in NS, has led to its listing as endangered in Canada (COSEWIC 2010) and imperiled in the United States. The closely related *G. radiatum* Michx., spreading avens or Appalachia avens, inhabits wet microsites at high elevations of the southern Appalachians (Ulrey et al. 2016; Hay et al. 2019), approximately 1200 km from the NH populations.

This chapter is composed of three main sections. Part A: 'Genetic analysis of clonality at sites of varying historical disturbance' is an attempt to characterize the extent of clonal presence and clonal diversity at three sites in NS. It uses focused mapping and genetic characterization of individual plants at sites at the greatest resolution possible without destructive sampling. The sites employed in this study are Brad's Bog (GH6), Coastal Fen (GR3), and Green Head (GH4, also known as Gull Rock Rd.), representing undisturbed, moderately disturbed, and highly disturbed sites, respectively. Part B: 'Mating system analysis and estimation of selfing rate' estimates selfing rates based on two approaches using (1) population survey data and (2) progeny arrays constructed from the same Brier Island sites studied in Part A. Mating system parameters are estimated from the offspring of the array. From analyses of the population survey data incorporating mating system features, Part B also provides estimates of the effective population size (*N*e) of both broad- and fine-scale geographic localities as well as inbreeding depression in Nova Scotia. Part C: 'Genetic diversity of *G. peckii* in Nova Scotia and New Hampshire, and comparison to genetic diversity of *Geum radiatum* in North Carolina' describes genetic diversity between and within the two geographic regions comprising the range of *G. peckii*, and then expands the observations to the relationship between *G. peckii* and *G. radiatum*. It provides several standard measures of genetic diversity and attempts to identify patterns of genetic differentiation at both fine and broad scales.

3.1.1 Principal Findings

3.1.1.1 Part A *Genetic Analysis of Clonality at Sites of Varying Historical Disturbance*

a. Clonal richness (the proportion of plants that are genetically distinct) varies inversely with disturbance level. It is highest at the undisturbed site (Brad's Bog), moderate at the moderately disturbed site (Coastal Fen), and low at the highly disturbed site (Green Head/GH4).
b. At all three sites, clonal genotypes were correlated at the closest distances: ≤ 0.5 m at Brad's Bog and Coastal Fen, and ≤ 5 m at Green Head/GH4.
c. For the sites with the highest spatial resolution (Brad's Bog and Coastal Fen), pooled quadrat data suggest that a 1 m sampling regime is sufficient to avoid sampling clonal genotypes.

3.1.1.2 Part B *Mating System Analysis and the Estimation of Selfing Rate*

a. By both population survey and progeny array, *G. peckii* shows a mixed mating system: partial selfing and partial outcrossing. In NS, the selfing rate is approximately 47% based on progeny array samples and approximately 30% based on population survey data. This difference is consistent with inbreeding depression acting between seedling and later life stages.
b. Significant biparental inbreeding was detected at Brad's Bog.
c. The estimated effective population size, N_e, of *G. peckii* is 57 in NS and 28 in NH.

3.1.1.3 Part C *Genetic Diversity of* G. peckii *in Nova Scotia and New Hampshire, and Comparison to Genetic Diversity of* G. radiatum *in North Carolina*

a. A clear genetic separation of *G. radiatum* and *G. peckii* is evident, supporting the retention of the present classification as separate taxa.
b. Significant deviation from Hardy–Weinberg equilibrium in both Nova Scotia and New Hampshire populations of *G. peckii* (excess homozygosity resulting in a positive inbreeding coefficient) is observed. No significant difference between the two regions is detected.

c. Populations in NH show significantly greater differentiation (by G''_{ST}, D_{EST} indices) than NS populations. These findings are graphically supported by Bayesian clustering analyses in the program InStruct. This supports the previous conclusion that Brier Island is a single population. The NH population differentiation is likely due to topographic landscape features separating populations.
d. Evidence suggests that the NS population is a relict population, not the result of founding by emigrants.
e. There is evidence for isolation by distance. A significant correlation was found between pairwise F_{ST} values and geographic distance.

3.1.2 *Materials and Methods*

3.1.2.1 DNA Genotyping

All three sections of the work were completed using DNA genotyping. Most of the methods are based on standard protocols, and for these only references are provided. Details are presented for methods specific to the work presented here.

3.1.2.2 Field Collection

Leaf samples were collected in Nova Scotia in the summer and fall of 2018 and fall of 2019 from all known populations on Brier Island and the mainland at Harris Lake. In NH, samples were obtained in late September 2018 at five locations: Hancock Branch of the Pemigewasset River, Arethusa Falls, Thompson Falls, Mount Washington, and Cannon Cliff. Samples were collected a minimum of 1 m apart, unless otherwise noted. *G. peckii* population locations are available from the senior author. *G. radiatum* samples were collected from populations 1, 8, and 13 of Hay et al. (2019), which shows approximate locations (populations are labeled with numbers to protect their location).

3.1.2.3 DNA Extraction, DNA Amplification, Library Construction, and Sequencing

Approximately 100 mg of dried tissue was disrupted in 1.1 ml polypropylene tubes containing two 1.5 mm and one 3 mm stainless steel ball bearings using a Retsch MM300 Tissuelyzer. Samples were milled at 30 htz in 1.5-min intervals repeatedly until tissue was fully disrupted. Between each interval, sample tubes were inverted within the apparatus to facilitate even disruption. To each tube was added 650 μl of CTAB extraction buffer (100 mM Tris HCl, pH 8.0; 20 mM EDTA; 1.2 M NaCl; 2% cetyltrimethylammonium bromide) and 1 μl of 20 μg/ml Proteinase K. Lysis tubes were incubated at 55°C for 1–3 h and centrifuged for 15 min at 4000 g to pellet cell debris. We performed a phenol-chloroform-isoamyl (PCI; 25:24:1; pH 8.0) extraction on the lysate. The aqueous phase was then directed into a DNA-silica-binding protocol modified from Holfinger and Till (2003). Loose silica in binding solution was added to a 96-well DNA binding filter plate (Pall Corporation). DNA was eluted in a final volume of 100 μl of TE buffer. DNA was frozen until further use.

Microsatellite loci were amplified using the Qiagen multiplex PCR kit and using the primers listed in Table 3.1. Multiplex products were diluted 100-fold and subjected to another round of PCR to label each sample with a unique barcode. Labeled samples were pooled and cleaned with a magnetic silica bead binding protocol (Speedbeads). Libraries were assessed by quantitative PCR and loaded and run on an Illumina MiSeq (Reagent Kit v.3, 150 cycle, Illumina). Microsatellite genotypes were scored using Megasat code (Zhan et al. 2017) with manual verification.

Table 3.1 Microsatellite primer sequences for *Geum peckii* and *G. radiatum*

Primer ID	Primer sequence
Gpe-02_L	CCCTACACGACGCTCTTCCGATCTTTTCCACCTCCTTGAGCTTC
Gpe-02_R	GTTCAGACGTGTGCTCTTCCGATCTTTGCCATTCCCAGTCTGACC
Gpe-06_L	CCCTACACGACGCTCTTCCGATCTCGGTTGTGGATGGTTAGGAC
Gpe-06_R	GTTCAGACGTGTGCTCTTCCGATCTTCAGTCCTAGTCTCTCAAGCC
Gpe-10_L	CCCTACACGACGCTCTTCCGATCTGCACTCATCAACTCCATCAGG
Gpe-10_R	GTTCAGACGTGTGCTCTTCCGATCTCTCAACACACAAACCATGAACG
Gpe-12_L	CCCTACACGACGCTCTTCCGATCTTGAGGGCTGCGAATGAAGAG
Gpe-12_R	GTTCAGACGTGTGCTCTTCCGATCTTACCCACAATCCCAGCAGAG
Gpe-15_L	CCCTACACGACGCTCTTCCGATCTCTGAACAGAGCCGCTACTTG
Gpe-15_R	GTTCAGACGTGTGCTCTTCCGATCTGTGCTTCCCTCAAAGACTGC
Gpe-16_L	CCCTACACGACGCTCTTCCGATCTAAGTACACGCTTGCTCAGTC
Gpe-16_R	GTTCAGACGTGTGCTCTTCCGATCTTCTCTCATAAGCGAAGAAAGGG
Gpe-19_L	CCCTACACGACGCTCTTCCGATCTTGCACAGAAACATAGATCGGTC
Gpe-19_R	GTTCAGACGTGTGCTCTTCCGATCTGCAAGCGCCTGACTGAAG
Gpe-20_L	CCCTACACGACGCTCTTCCGATCTCTCAAATCAATGACCACGCTG
Gpe-20_R	GTTCAGACGTGTGCTCTTCCGATCTGCGGGTTGTCGATGTGTTC
Gpe-28_L	CCCTACACGACGCTCTTCCGATCTGATAGGAGGAGGCTGTAGGG
Gpe-28_R	GTTCAGACGTGTGCTCTTCCGATCTCACTCACTCTGCTACCTGAAAC
Gpe-29_L	CCCTACACGACGCTCTTCCGATCTTTGAAGGAAGGTGCTGTCCG
Gpe-29_R	GTTCAGACGTGTGCTCTTCCGATCTCAGGTCTAATCACATAAAGCGC
Gpe-31_L	CCCTACACGACGCTCTTCCGATCTACAATGCACAACCTCGCC
Gpe-31_R	GTTCAGACGTGTGCTCTTCCGATCTCTCGCGCACAGTCCATTAC
Gpe-33_L	CCCTACACGACGCTCTTCCGATCTCCCACTGCTCTATCCCGTAC
Gpe-33_R	GTTCAGACGTGTGCTCTTCCGATCTGAGGGAGAGAATACGGGCTG
Gpe-47_L	CCCTACACGACGCTCTTCCGATCTTCCAGCATTGAAAGTTCGTTG
Gpe-47_R	GTTCAGACGTGTGCTCTTCCGATCTACCAACCGTCAAGAAGCATG

3.1.2.4 DNA Fingerprinting

After initially screening 48 microsatellite loci, 15 were found that had the properties required for analysis (disomic inheritance pattern, multiple alleles for each locus, no unusual genotypes). All samples were genotyped using these 15 markers. We used standard population genetic measures to assess the levels of genetic diversity and genetic differentiation across the sampled areas. We employed the software packages Genodive 3.0 (Meirmans and Van Tienderen 2004; Meirmans 2020), InStruct (Gao et al. 2007) and Structure (Pritchard et al. 2000; Hubisz et al. 2009). For detecting genetic structure among k-groups, we used the admixture model, and set allele frequencies to be correlated among populations.

3.2 Part A: Genetic Analysis of Clonality at Sites of Varying Historical Disturbance

3.2.1 *Clonal Diversity is Locally Important in* G. peckii

Prior to this work, the reproductive strategies that *Geum peckii* employs had not been investigated in detail, although it was known that both sexual reproduction and clonal reproduction were possible. Understanding the frequency of clonality and whether it varies based on the history of disturbance in an area is important to the basic understanding of the plant itself. In addition, it is a key aspect for accurately assessing genetic variation in a species overall, as clonal reproduction complicates standard population genetic analyses. The difficulties arise when physiological individuals and genetic individuals cannot be distinguished. The improper inclusion of clones in an analysis inflates particular genotypes in the population relative to others. This violates assumptions in population genetic analyses and leads to biased estimates of parameter values.

Molecular genetic markers allow genets (i.e., unique genotypes representing genetic individuals) to be identified. When populations of genets are then used for population genetic analyses, assumptions built into the analysis methods are more generally satisfied and characterization of the populations is more accurate.

The sampling scheme in any study of clonality is particularly important. Clonal variation may vary at different scales. We have little information on the density of genets at any scale for *Geum peckii*; however, in the congeneric *G. reptans*, 99.5% of genets sampled at least 4m apart were unique (Pleuss and Stöcklin 2004). To determine the frequency of clonality and the scale needed for sampling that would obtain individual genets, we sampled three sites on Brier Island that differ in disturbance history. We ask whether sites differ in clonal diversity at a fine spatial scale. The three sites chosen are Brad's Bog, Coastal Fen, and Green Head (GH4), representing low, medium and high levels of disturbance respectively. Brad's Bog has, to our knowledge, little to no history of disturbance. Coastal Fen has experienced some recent disturbance by burning. Green Head was the previous site of a sheep farm for several decades in the early twentieth century.

3.2.1.1 Sampling Scheme for Clonal Analysis (Fine-Scale Sampling)

To assess clonality, sampling at a scale finer than 1 m is required. We laid out two square 5-m quadrats each at Brad's Bog and Coastal Fen. Choice of quadrat location was not meant to characterize these sites as a whole but rather to identify the fine-scale population genetics where plants are known to occur. Quadrats were laid out by selecting a southeast corner arbitrarily and measuring the quadrat edges along the cardinal directions. GPS coordinates were taken for each southeast corner. Within each quadrat we identified focal clumps of shoots of *G. peckii*. Each clump had one or numerous shoots emergent from the peat mass, but any subsurface connection was unknown. In this way, we defined each sampled shoot as a ramet, despite lack of knowledge regarding physical connection. We sampled a single leaf from every individual ramet within each clump and measured clump size.

At Green Head, the number of plants was limited so we sampled every focal clump we could identify. Quadrat and Green Head sampling maps were produced to assist with further analysis.

3.2.1.2 Clone Determination

Clones were identified by analysis of microsatellite loci using the software program Genodive 3.06 (Meirmans and Van Tienderen 2004). Genodive identifies clones by calculating a genetic distance matrix for all individual samples and pools closely related individuals according to a user-determined threshold.

3.2.1.3 Clonal Richness

Clonal richness (R) is defined as the number of genets (G) minus one divided by the total number of samples (N) minus one (Mandel et al. 2019).

$$R = \frac{G - 1}{N - 1} \tag{3.1}$$

Values were pooled across quadrats for Brad's Bog and Coastal Fen.

3.2.2 *Results and Discussion*

3.2.2.1 Clonal Richness

The assessment of clonality presented here indicates that clonal genotypes are present on the finest scales that can reasonably be made using non-destructive sampling. Examination of the clonal residency at focal plant clusters suggests that clonality is widespread at this scale, generally less than 1 m (Fig. 3.1). From this result, we can conclude that a 1-m sampling regime should be sufficient to avoid sampling clonal genotypes, thus ensuring a relatively reliable collection of genetically distinct individuals for population genetic analysis.

The three sites differ in clonal richness, i.e., the occurrence of multiple clones in the surveyed area. The highest value is at Brad's Bog (0.44), the least disturbed of the sites. At Coastal Fen, the site of medium disturbance, it was 0.20, while at Green Head, the most disturbed, it was 0.25. Thus, the rank order of clonal richness is inversely associated with disturbance level.

3.2.2.2 Spatial Distribution

Clones mapped onto each quadrat show spatially restricted patterns (Fig. 3.1). Brad's Bog quadrat 1 shows sixteen different clonal genotypes arranged diagonally across the quadrat from the southeast corner to the northwest. Most clones were unique to foci but clones 6 and 11 were present in two adjacent focal clumps. Clone 1 was identified across the greatest diagonal extent of the foci in the quadrat in BB01 and BB09. In Brad's Bog quadrat 2, a similar distribution was found. Most clones were unique to foci but clones 26 and 30 were found in adjacent clumps (BB14 and BB15). Clones 17 and 20 were both found in the largest clump, BB11, and also more distant clumps BB13 (clone 20) and BB17 (clone 17).

At Coastal Fen quadrat 1 (Fig. 3.1), some clones were unique to focal clumps but clone 35 was found in five different clumps across the eastern side of the quadrat. At Coastal Fen quadrat 2, smaller focal clumps had generally fewer and more unique clonal types. Clone 38 was found across a wide range within the quadrat from the large clump CF18 to CF21, CF08, CF23, and CF10. This was the largest extent of a clonal genotype found at this site, both in terms of number of foci and greatest expanse (> 5 m).

At Green Head (Fig. 3.1), clonal types were generally restricted to microsites (i.e., waypoint designations) but clones 52, 53, and 54 were present in multiple instances at both 1642 and 1646. Thus, although the generality that clones are spatially contained is supported, some clones appear to become large in rare instances.

The relationship between disturbance and genetic structure in clonal plants is interesting because many clonal plants make use of mixed strategies involving clonality and sexual reproduction. How these strategies are optimized around disturbances is important to characterize to inform recovery strategies in endangered plants such as

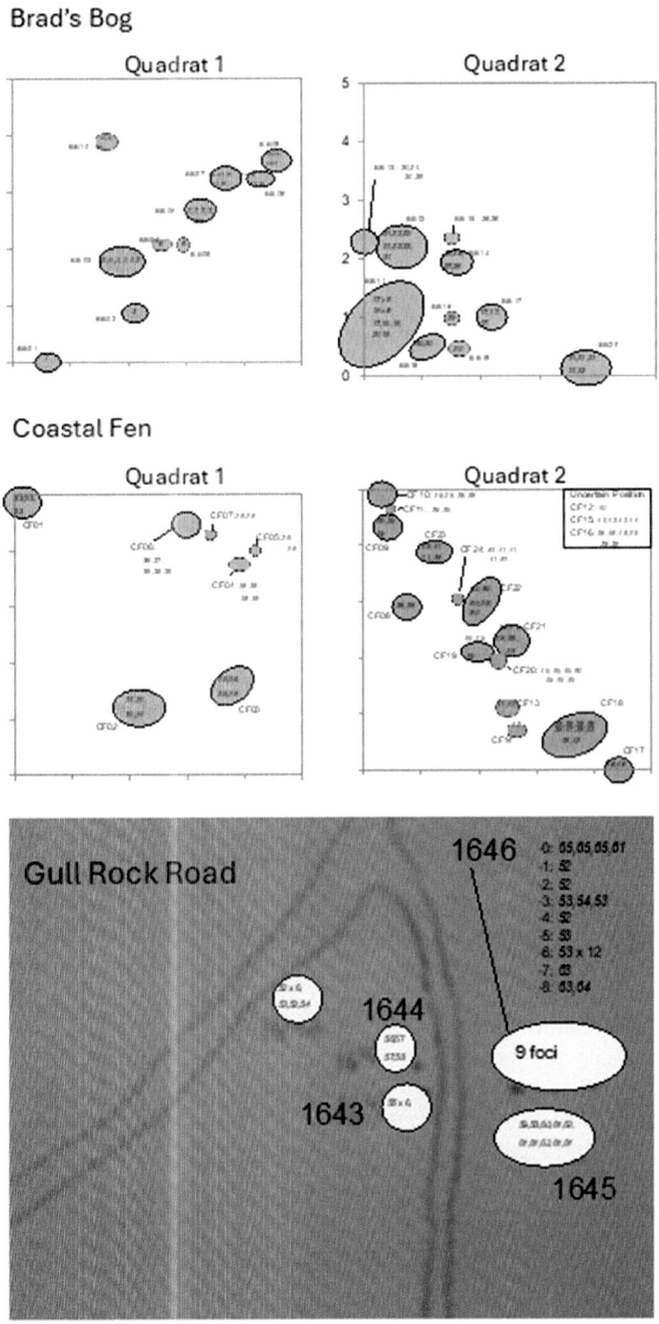

Fig. 3.1 Residency maps of clones at three sites in Nova Scotia. Clone identifications are in italics. Distance scales are in meters

G. peckii. Holt et al. (2020) found that spatial genetic structure increased following disturbance in the aquatic plant *Sagittaria latifolia* following flooding. It is likely that there is a complex interaction between disturbance types, community response, and species' adaptations to particular disturbances. The hypothesis that more disturbed sites on Brier Island will have lower diversity with larger clones is modestly supported here. The differing spatial scales between the most historically disturbed (Green Head) and the other two sites makes some comparisons challenging. The difference in time since disturbance is also a potential confounding factor. Nevertheless, we find that diversity is highest as measured by clonal richness at the undisturbed site (Brad's Bog) when compared to either of the two disturbed sites. Instances of larger individual clones were also found at the disturbed sites. Quantitative analysis of these trends will require further study.

3.3 Part B: Mating System Analysis and Estimation of Selfing Rate

3.3.1 Background

The fate of small populations depends in part on their levels of genetic variation as this variation provides the substrate for adaptive evolution. Small populations are prone to inbreeding, leading to the loss of fitness—inbreeding depression—when harmful recessive alleles come together in the homozygous state (Hedrick and Kalinowski 2000; Amos and Balmford 2001; Charlesworth and Willis 2009). Small populations can also suffer from other costs including limited mating opportunities. The most severe type of inbreeding is self-fertilization, where sexual union occurs between gametes that originate from the same individual. Under self-fertilization (selfing), levels of heterozygosity decline by half each generation, reducing standing levels of genetic variation (Hamrick and Godt 1989; Godt et al. 1996). Various mechanisms have evolved to promote outcrossing (Lloyd 1979; Lloyd and Schoen 1992). Such mechanisms include the spatial (herkogamy) or temporal (dichogamy) separation of male and female organs, dioecy, i.e., the separation of sexes between individuals, as well as other, more elaborate mechanisms of self-incompatibility. However, a substantial number of cosexual plants are self-compatible and thus able to self-fertilize. The genetic consequences notwithstanding, selfing can provide a more immediate benefit through reproductive assurance when conspecifics are rare (Darwin 1862, 1876; Lloyd 1979). Characterization of the mating system in small, at-risk populations can inform decisions to aid their recovery.

The cosexual flowers of *G. peckii* are protogynous, where stigmas become receptive prior to anther dehiscence. The selfing rate is unknown. Another member of the genus, *G. urbanum*, is a predominantly selfing species, while the closely related *G. rivale* is largely outcrossing (Rusham et al. 2010; Jordan et al. 2017). Strategies for recovery management of *G. peckii* could make use of improved understanding of its mating system. Here we compare selfing rates and related measures at both broad and fine geographic scales. In the broad-scale comparison, we measure selfing rates and population parameters between New Hampshire (NH) and Nova Scotia (NS) regions of *G. peckii*. In the fine-scale comparisons, selfing rates and related parameters are estimated at sites within the larger geographic regions. One such parameter of interest is the level of biparental inbreeding, arising from matings among relatives. In addition to important mating system information, the analytic methods employed here allow for the estimation of effective population size, N_e.

3.3.2 Selfing-Rate Estimation: Two General Approaches

The estimation of selfing rates can be performed in a number of ways. Two general approaches will be used in this study that estimate slightly different aspects of selfing within the *G. peckii* population genetic system.

The first approach uses population-level genetic survey information. These data provide a multigenerational window on the populations of interest and, with certain assumptions, can be used to infer selfing rates. Several analytical methods have been developed to estimate selfing rates from population survey data (Bürkli et al. 2017). The simplest makes use of the deviation from Hardy–Weinberg equilibrium (DHW). These deviations arise from decreased heterozygosity in selfed individuals, often expressed through the inbreeding coefficient (GIS). This method makes use of within-locus information on heterozygosity, potentially across multiple loci (Hartl and Clark 1997). Another method, Identity Disequilibrium (ID), makes use of correlations of heterozygosity among loci to estimate selfing (Weir and Cockerham 1973). Newer methods make use of both kinds of information (Wang et al. 2012).

The second approach is to construct a progeny array. This consists of offspring from known family groups collected from the wild. The genotypes of offspring, with maternal types if available, can be used to infer the number of selfing events within the array. Two analyses will be employed to estimate the selfing rate from a *G. peckii* progeny array. The method of Ritland (2002) evaluates a mixed selfing-random mating model by maximum likelihood. The method of Wang et al. (2012) also uses maximum likelihood and infers selfing rates from a reconstructed pedigree.

The two general approaches—population survey and progeny array—provide different but complementary estimates of the selfing rate. In particular, population surveys integrate information from several generations in a perennial plant and provide an estimate of selfing that has contributed to the extant adult population. Progeny array-based estimates, on the other hand, provide information on only the most recent generation at early life stages where less-fit inbred offspring may not contribute to the adult population. A lower selfing rate estimated from the population survey than from the progeny array indicates inbreeding depression acting between the seedling and mature life stages (Ritland 1990).

3.3.3 *Selfing-Rate Analyses*

3.3.3.1 Population Surveys

The population survey and progeny array genotypic data were each subjected to multiple analyses. Population genotypic data were analyzed by three methods:

(i) Deviation from Hardy–Weinberg equilibrium (DHW):

$$s = \frac{2G_{\mathrm{IS}}}{1 + G_{\mathrm{IS}}}, \tag{3.2}$$

where s is the selfing rate and G_{IS} is the inbreeding coefficient (Wright 1969). Inbreeding coefficients for each population are presented in Part C.

(ii) Identity Disequilibrium as implemented by the Robust Estimate of Multilocus Selfing (ID-RMES; David et al. 2007) with s estimated using the maximum likelihood method. Precision of LnL was set at 0.001, and the maximum number of selfing generations for a heterozygote (kmax) was set at 10. Starting values of s were computed from g_2.

(iii) Sibship Assignment by maximum likelihood as implemented in the software program Colony (version 2.0.6.6; SA-Colony; Wang et al. 2012). Genotypic data were input as offspring data. Mating systems were set as both female and male polygamy with inbreeding and without clones. Identical multilocus genotypes were removed from the data set. Species setting was set to monoecy. The full likelihood method of analysis was used. Variable numbers, lengths, and precision levels of runs were used depending on the computational demands of the analysis. In all cases, a minimum of three runs was performed to ensure convergence at the maximum likelihood. Runs were performed with updates to allele frequencies and sibship scaling. The sibship prior was set to 'weak prior' and the paternal and maternal sibship size settings were set to 1.0. Initial locus error rates were input.

3.3.4 *Progeny Arrays*

For progeny arrays, leaf tissue and fruits were collected from 43 maternal families on Brier Island at Brad's Bog ($n = 21$), Coastal Fen (17), and GH4 (5) in October 2019. Leaf tissue was dried overnight in coin envelopes with silica gel. Seeds were sown on wet soil in self-watering seed starter trays (Lee Valley Tools, Canada), wrapped in opaque plastic and placed at 4 °C for six weeks' stratification. Following cold treatment, trays were moved to a greenhouse room at 22 °C under an 18-h daylight regime with natural light supplemented by artificial illumination. Seedlings were harvested at various points in time once they reached sufficient size for DNA extraction. Trays were relocated to an alternative growth facility in March 2020 and seedlings were periodically harvested over the following six months. Harvesting consisted of pulling whole seedlings from their tray, taking care to minimize disturbance to remaining seedlings. Soil was gently removed from roots, and whole seedlings were placed into tissue disruption tubes for DNA extraction.

Across the three sites, we obtained 343 offspring representing 42 families (Brad's Bog 20 families and 192 offspring, Coastal Fen 17 families and 135 offspring, GH4 5 families and 16 offspring). Progeny array data were analyzed by two methods:

(i) Multilocus t and r (MLTR; Ritland 2002). Data were considered grouped by site and as a single group for all NS. The selfing rate $(1 - t)$, correlations of selfing among families (r_t), correlations of paternity among siblings (r_p) and the paternal inbreeding coefficient (f) were jointly estimated. The Newton-Raphson method was used to evaluate the likelihoods. Standard errors of estimates generated with 1000 bootstrap replicates resampling families were used to construct 95% confidence intervals. Biparental inbreeding was estimated by comparing the single locus to multilocus outcrossing rates. Pollen and ovule gene frequencies were assumed to be equal, and the most-likely parent chosen.
(ii) Pedigree Reconstruction-Colony (PR-Colony). Settings were identical to the SA-Colony analyses described above. Offspring data were the progeny array offspring. The parental generation included the progeny array parents in addition to the population survey data, for a total of 245 potential paternal types in addition to the known maternal types comprising the array.

3.3.5 Results

3.3.5.1 Broad-Scale Analyses: Population Surveys

In New Hampshire, the selfing rate estimated by DHW in NH was 0.222 (Table 3.2). By ID-RMES, the estimate was 0.115 (95% CI: 0, 0.273) and by SA-Colony s was 0.454 (0.358, 0.550) (Table 3.2). Visual inspection of the sibship assignments (Fig. 3.2) showed grouping within each location with the exception of the three sites within the Mount Washington population. The estimate of the effective population size (N_e) for all NH was 28 (17, 48).

In Nova Scotia, the selfing rate estimate based on DHW for all populations was 0.301 (Table 3.2). The ID-RMES estimate for NS was 0.326 (0.206, 0.426), and the SA-Colony estimate was 0.280 (0.222, 0.338). In NS, the sibship assignment (Fig. 3.3) showed little structure by site, with the exception of CRS and Harris Lake. The N_e estimate for all NS was 57 (40, 83).

3.3.5.2 Broad-Scale Analyses: Progeny Array

Selfing rates were estimated from 343 progeny from 42 families across the three NS sites (Table 3.2). The selfing rate estimated by MLTR was 0.541. Biparental inbreeding was significantly greater than zero, estimated to be 0.147 (0.053, 0.241).

Table 3.2 Broad-scale selfing rate (s) and effective population size (N_e) estimates for *Geum peckii* in New Hampshire and Nova Scotia

Sampling method	Analysis method	Parameter	New Hampshire	Nova Scotia
Population survey				
	DHW	s	0.22	0.30
	ID-RMES	s	0.12 (0, 0.273)	0.33 (0.206, 0.426)
	SA-Colony	s	0.45 (0.36, 0.55)	0.28 (0.22, 0.34)
	SA-Colony	N_e	28 (17, 48)	57 (40, 83)
Progeny array				
	MLTR	s	–	0.54 (0.38, 0.70)
	MLTR	Biparental inbreeding	–	0.15 (0.053, 0.24)
	PR-Colony	s	–	0.40 (0.35, 0.46)

Values in parentheses are lower and upper 95% confidence limits. Analysis methods and abbreviations are explained in the text

The selfing rate as estimated by PR-Colony was 0.402 (0.349, 0.456) and N_e was estimated at 34 (21, 55).

3.3.5.3 Fine-Scale Analyses: Population Surveys

Selfing rates were estimated for each of the seven NH and eight NS sites (Table 3.3). For the NH sites we pooled Huntington Ravine, Tuckerman Ravine, and Alpine Garden Trail to calculate estimates for Mount Washington. Similarly, we pooled individuals for Big Meadow Bog on Brier Island to determine a single estimate for this site.

New Hampshire: Selfing estimates based on DHW ranged from 0.198 at Tuckerman Ravine to 0.347 at Hancock Branch, though only three estimates were based on significant values of G_{IS}. Estimates of s based on ID-RMES ranged from zero at several sites to 0.267, although none was statistically significant. Selfing rate estimates based on SA-Colony ranged from 0.048 at Tuckerman Ravine to 0.368 at Arethusa Falls. All estimates based on SA-Colony in NH were significantly different than zero except for the estimate at Tuckerman Ravine.

Nova Scotia: In NS, selfing rates estimated by DHW ranged from -0.090 at Coastal Fen to 0.547 at GH4. Four DHW selfing rates were based on significant G_{IS} values. NS estimates from ID-RMES ranged from zero at several sites to 0.693 at GH4. Significant non-zero selfing rates were found by ID- RMES at Camp Rd. South and

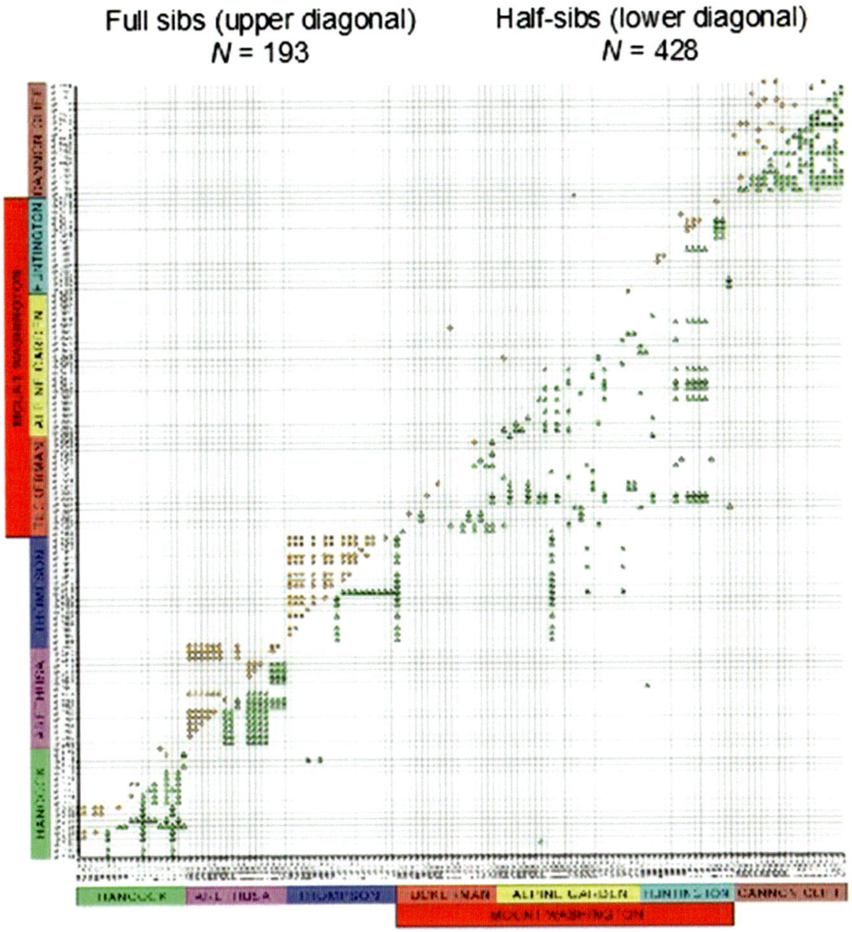

Fig. 3.2 Maximum-likelihood sibship assignment for 130 unique genets in five New Hampshire populations

GH4. Estimates of selfing at fine-scale sites in NS based on SA-Colony ranged from 0.121 at Big Meadow Bog to 0.546 at GH4. Significant selfing was detected at all NS sites except at BM5.

3.3.5.4 Fine-Scale Analyses: Progeny Array

Estimates of selfing rates and biparental inbreeding at the three sites comprising the progeny array (Table 3.4) were made by MLTR. At Brad's Bog and Coastal Fen the selfing rates are significantly greater than zero, at 0.534 (0.293, 0.775) and 0.623 (0.419, 0.827) respectively. For GH4, all individuals of the array were identified as

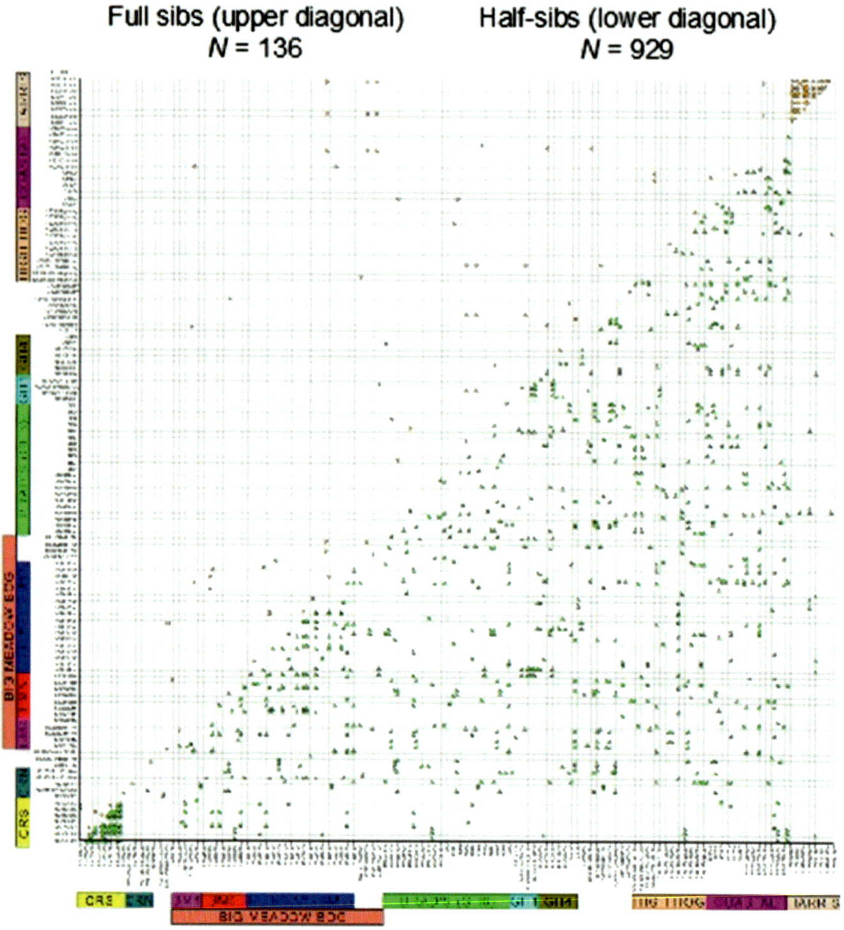

Fig. 3.3 Maximum-likelihood sibship assignment for 245 unique genets across the species range in Nova Scotia, including Brier Island and Harris Lake

products of selfing as the selfing rate was effectively 100%. No confidence interval was calculated as the standard error estimated by the program was zero. Estimates based on PR-Colony analysis are 0.453 (0.381, 0.525) at Brad's Bog, 0.348 (0.266, 0.430) at Coastal Fen, and 0.313 (0.062, 0.563) at GH4.

Paternity assignment at Brad's Bog indicated that most identified fathers were assigned to the cohort comprising the progeny array (data not shown). This assignment identified 128 individuals across Brier Island as the highest probability paternal genotype. The remaining 64 highest probability paternal types were inferred. In addition to assignments made at Brad's Bog, additional paternity assignments to the Brad's Bog progeny array were made from BM5, High Hog, Coastal Fen, Runway Bog, and Harris Lake sites. At Coastal Fen, similarly, most assignments of highest

Table 3.3 Fine-scale selfing rate (*s*) and population parameters for *Geum peckii* in New Hampshire and Nova Scotia estimated from population surveys

		DHW	ID-RMES	SA-Colony	
	N	*s*	*s*	*s*	N_e
New Hampshire					
Hancock branch	19	**0.35**	0.238 [0, 0.483]	0.263 [0.051, 0.475]	11 [6, 28]
Arethusa Falls	19	0.26	0.266 [0, 0.531]	0.368 [0.134, 0.603]	10 [5, 28]
Thompson Falls	20	0.26	0 [0, 0.375]	0.200 [0.009, 0.391]	6 [2, 21]
Huntington Ravine	18	0.23	0.060 [0, 0.358]	0.333 [0.059, 0.607]	11 [5, 31]
Tuckerman Ravine	21	0.2	0 [0, 0.170]	0.048 [0, 0.188]	14 [7, 32]
Alpine Garden Trail	24	**0.29**	0 [0, 0.338]	0.208 [0.027, 0.389]	16 [9, 34]
Mount Washington*	63	**0.24**	0 [0, 0.126]	0.191 [0.084, 0.297]	25 [15, 43]
Cannon Cliff	20	0.2	0.267 [0, 0.536]	0.200 [0.011, 0.389]	11 [6, 28]
Nova Scotia					
Brad's Bog	23	0.150	0.258 [0, 0.461]	0.191 [0.012, 0.369]	18 [10, 38]
Coastal Fen	14	– 0.090	0.483 [0, 0.719]	0.429 [0.143, 0.714]	14 [7, 40]
GH4	11	**0.547**	0.693 [0.547, 0.810]	0.546 [0.209, 0.882]	8 [4, 24]
Harris Lake	20	– 0.038	0 [0, 0.456]	0.300 [0.088, 0.512]	8 [4, 24]
Runway Bog (BM3)	37	**0.229**	0.153 [0, 0.432]	0.222 [0.074, 0.370]	23 [13, 43]
BM5	15	0.115	0 [0, 0.380]	0.200 [0, 0.426]	13 [6, 30]
Big Meadow Bog*	69	**0.232**	0.156 [0, 0.375]	0.121 [0.037, 0.205]	24 [15, 42]
Camp Rd South	25	0.350	0.471 [0.379, 0.554]	0.304 [0.107, 0.502]	4 [2, 15]
High Hog	25	**0.321**	0 [0, 0.433]	0.360 [0.164, 0.556]	12 [6, 28]

Numbers in parentheses are lower and upper 95% confidence limits

Numbers in bold are estimates calculated from significant values of G_{IS}

*Mount Washington includes individuals from Huntington Ravine, Tuckerman Ravine, and the Alpine Garden Trail. Big Meadow Bog contains individuals from BM1, Runway Bog (BM3), BM4, BM5, BM6

Table 3.4 Fine-scale estimates of selfing rates and biparental inbreeding from progeny arrays at three sites in Nova Scotia. Numbers in parentheses are lower and upper 95% confidence limits

		MLTR		PR-colony
Site	*N* (families, offspring)	Selfing rate, *s*	Biparental inbreeding, $(t_m - t_s)$	Selfing rate, *s*
Brad's Bog	20, 192	0.534 (0.293, 0.775)	0.171 (0.016, 0.326)	0.453 (0.381, 0.525)
Coastal Fen	17, 135	0.623 (0.419, 0.827)	0.130 (-0.015, 0.275)	0.348 (0.266, 0.430)
GH4	5, 16	1	-0.061 (-0.186, 0.064)	0.313 (0.062, 0.563)

probability paternity were from Coastal Fen among the progeny array families. Other sites assigned highest probability paternities are BM4 and Runway Bog. At GH4, the PR-Colony analysis identified six highest probability fathers across the 16 offspring in the array. All six assigned types belonged to the progeny array parental cohort as no assignments were made from other sites.

3.3.5.5 Effective Population Size

Sibship assignment in the Colony program (Wang 2009) produced estimates of N_e for all populations analyzed by this method (Tables 3.2 and 3.3). At the broad scale, the N_e value for NH was 28 (17, 48) and for NS was 57 (40, 83). At the fine scale, among the NH sites, N_e values ranged from 6 (2, 21) at Thompson Falls to 25 (15, 43) for Mount Washington. Among NS sites, N_e ranged from 4 (2, 15) at Camp Rd. South to 24 (15, 42) at Big Meadow Bog.

3.3.6 Discussion

3.3.6.1 Broad-Scale Selfing Estimates

Selfing rate estimates in the eastern mountain avens, *G. peckii*, were made with several methods in the two geographic regions where this species is found. In general, the results presented here indicated that *G. peckii* is a mixed-mating species. In NH, where only population survey-based approaches were used, the broad-scale point estimates from the various methods ranged from ~ 12 to ~ 45% selfing. The same methods used for the broad-scale NS estimates range from ~ 28 to 33% selfing.

These differences could be attributed to several factors. The methods used to infer selfing rate are known, to varying degrees, to be sensitive to factors such as sample size, population size and population structure (Wang et al. 2012). Our sample size in NS is roughly twice that in NH. Also, results from Part C of this study show strong population structure among NH sites but weak structure within NS, aside from the mainland site at Harris Lake. Those results are mirrored in the sibship assignment analyses presented here (Figs. 3.1 and 3.2) and probably arise from some of the same population genetic substructure. Non-random mating, such as self-fertilization, produces correlations of alleles both within and among loci (Wright 1965) and the resultant information is what the various analytic methods used here attempt to parse according to specific population genetic models. Inconsistencies can arise when factors outside of the models affect the information contained in the population or sample thereof. For example, it has been recognized that the DHW method often overestimates selfing rates because any genotyping and scoring errors usually lead to an excess of homozygotes, mimicking selfing (David et al. 2007). While the ID-RMES method is robust to this bias, both DHW and ID-RMES will underestimate s if the populations are not in inbreeding equilibrium (Wang et al. 2012).

The more sophisticated methods, while generally performing better under most cases, will suffer when the information content of the system is low due to their high parameterization (Wang et al. 2012). For these reasons, it is important to consider several analytical methods.

3.3.6.2 Fine-Scale Selfing Estimates from Population Surveys

At the finer spatial scale, selfing rate estimates appear more variable between the methods used and only in a few cases between sites. DHW estimates are generally in agreement with those produced by SA-Colony. ID-RMES estimates are more variable, sometimes indicating no selfing where significant selfing was detected by other methods. In some cases, such as Arethusa Falls, the three methods are in good agreement. In most cases, the 95% confidence limits are broad, suggesting insufficient power to assess selfing rates at these scales with high precision. In the two fine-scale sites with larger sample sizes, namely Mount Washington, NH ($N = 63$), and Big Meadow Bog, NS ($N = 69$) we see some evidence of improved precision by their narrower confidence limits.

One intriguing observation among the NS results is the apparent inverse relationship between the SA-Colony selfing rate estimates and census size data from Toms and LaRue (2017) (Fig. 3.4a). Such a relationship could arise if pollinators are limiting and act on relatively small spatial scales in NS. Pollinator limitation in *G. peckii* on Brier Island was also suggested by Zinck (1996) to explain the observed simultaneous flowering of *Dasiphora fruticosa* through floral mimicry. Given several considerations mentioned, this relationship should be considered a preliminary finding. Imprecise estimates of s at this scale and some disagreement between the analytic methods suggest more study is needed. In addition, while we do not currently have census data for the NH sites, field observations of population sizes suggest that such a relationship is weak or non-existent among the NH populations (Fig. 3.4b). This may reflect methodological differences but also may be explained by biotic or abiotic differences between NH and NS.

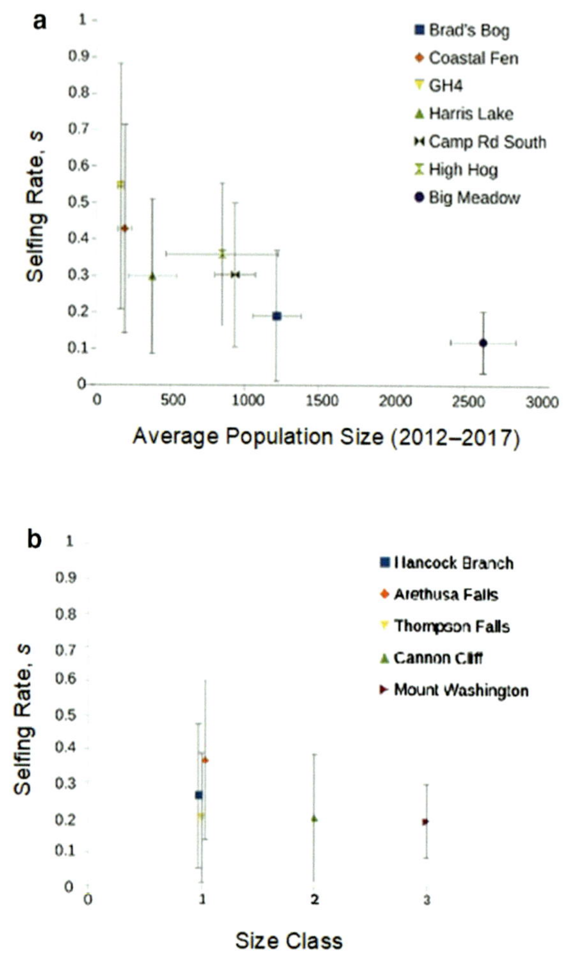

Fig. 3.4 Relationship between census population size and selfing rate. **a** Nova Scotia populations. Population sizes are averages from 2012 to 2017, taken from Toms and LaRue (2017), and selfing rates are estimated by sibship assignment in Colony. Y-error bars are 95% confidence intervals, X-error bars are standard deviations of population sizes. **b** New Hampshire populations. Size classes are assigned based on field observations, and selfing rates are estimated by sibship analysis in Colony. Y-error bars are 95% confidence intervals. Size class 1 sites are separated for visual clarity

3.3.6.3 Selfing Estimates from Progeny Arrays

The results of this study are generally in alignment with the expectation that progeny arrays will estimate higher selfing rates than estimates based on population surveys. For the broad-scale analysis of NS, we pooled the three progeny array sites into a single analysis each for the MLTR and PR-Colony. Between these two methods, the progeny array-based estimate of selfing rate in NS is ~ 47%. This compares to the population survey-based estimate of ~ 30% in NS. In other words, nearly half of seeds produced in the 2019 seed crop were the product of selfing. The adult population they arose from however, was constituted differently, with a lower fraction of individuals produced by selfing. One explanation is that selfed offspring are less fit, possibly due to greater effects of inbreeding, so that early life mortality reduces their proportion of the population as they mature. Indeed, in addition to uniparental inbreeding, i.e., selfing, significant biparental inbreeding was found at Brad's Bog. Another possibility, however, is that the populations are not at selfing equilibrium, a key assumption of the population survey methods. These differences then may reflect differences in selfing rates that are changing over time. These two possibilities are also not mutually exclusive, and additional work would be needed to disentangle the causal factors. An additional caveat to this discussion is to ask whether the three sites chosen to construct the progeny arrays are representative of the whole of NS *G. peckii*. Additional analyses (not shown) of Brad's Bog, Coastal Fen, and GH4 pooled population data do give similar values to the broad-scale estimates of NS.

The fine-scale estimates from the progeny array display some disagreement. Estimates from MLTR follow the same general trend as the population survey estimates for the three sites. The estimates by PR-Colony show the opposite trend across the sites. The reasons for this discrepancy are uncertain, however a few considerations may be relevant. The estimates from GH4 are based on only a small number of families and individuals, compared to the other sites. In general, the confidence intervals are wide, particularly for the MLTR estimates. One important difference between the two methods, is that allele frequencies are estimated from within the array for the MLTR method. For the PR-Colony analysis, allele frequencies are estimated from the parental generation data, comprised of the entire NS population data. Thus, instances of outcrossing may be detected in the PR-Colony analysis that may be interpreted differently by MLTR, with more limited information. Despite these inconsistencies, estimates from Brad's Bog for both methods are in general agreement.

3.3.6.4 Effective Population Size of *G. peckii*

The estimates of N_e presented here are generally comparable to those of other species of conservation concern (Palstra and Ruzzante 2008). Interestingly, estimates are lower in NH despite field observations and anecdotal evidence that the total *G. peckii* population size is larger in NH than NS. Several factors could contribute to this result. First, despite attempting to cover the geographical range over which *G. peckii* is found in NH, there are many identified sites of occurrence where it was not sampled. Second, the level of sampling between the two sites was not equal, NS having a larger sample size. In NH, several sites were small and had limited numbers of individuals that were sufficiently spaced to maintain our sampling regime. This was true to some extent at NS sites as well, however several NS sites had more extensive opportunities for sampling, especially given logistical constraints experienced in the field. Thirdly, the observed population substructure in NH could bias estimates of N_e (Nunney 1999). Importantly, these possibilities probably affect our estimates in the same ways, such that the N_e estimate for NH is probably underestimated.

3.4 Part C: Genetic Diversity of *G. peckii* in Nova Scotia and New Hampshire, and Comparison to *G. radiatum* in North Carolina

3.4.1 Background

The eastern mountain avens, *Geum peckii* (Rosaceae) is confined to two highly restricted regions in northeastern North America. In Nova Scotia, *G. peckii* is found in boggy habitat at sea level. In contrast, *G. peckii* in New Hampshire grows in the White Mountains in a wide variety of habitats, including streamside, alpine meadow, and rock faces. This range of habitats suggests that this species is capable of extraordinary flexibility. The closely related spreading avens, *G. radiatum*, inhabits wet microsites at high elevations of the southern Appalachians (Ulrey et al. 2016; Hay et al. 2019), approximately 1200 km from the NH populations. The two species are hexaploid ($2n = 6x = 42$; Gajewski 1959) and morphologically similar, generating disagreement over whether *G. peckii* should be considered as a species separate from *G. radiatum* (Zinck 1996). Both species are of special conservation concern. Eastern mountain avens is listed as *endangered* in Canada (COSEWIC 2010) and *imperiled* in New Hampshire. Spreading avens is federally listed as *endangered* in the United States (Murdock 1993).

Paterson and Snyder (1999) reported on the population genetics of mountain avens, *G. peckii*, and spreading avens, *G. radiatum*, using RAPD markers. Dominant markers such as RAPDs have limited use in population genetics because they are usually unable to resolve heterozygote genotypes. The application of microsatellite, or simple sequence repeat (SSR), markers to study the genetic system of these two species enables the use of a wide variety of tools designed for population genetic analysis (Hodel et al. 2016). Hay et al. (2019) used microsatellite markers to study genetic variation in *G. radiatum*.

In this study we use 15 microsatellite loci to characterize population genetic parameters of both *G. peckii* and *G. radiatum*. We focus on levels of genetic diversity within *G. peckii* and the apportionment of diversity within and among populations of the two species.

3.4.2 Genetic Variation Within Populations of G. peckii *and* G. radiatum

The two species were broadly similar across several measures of genetic diversity (Table 3.5). Across all loci examined, the average number of alleles was 4.4 and 4.6 for *G. peckii* and *G. radiatum*, respectively. The effective number of alleles was 1.3 and 1.6, reflecting the presence of uncommon alleles at most loci. The observed heterozygosity, H_O, in the two species was 0.18 and 0.24, the gene diversity (expected

Table 3.5 Genetic diversity in *Geum peckii* and *G. radiatum* across 15 microsatellite loci

Statistic	*G. peckii* Nova Scotia $n = 147$	*G. peckii* New Hampshire $n = 108$	*G. peckii* All populations $n = 255$	*G. radiatum* All populations $n = 111$
Average number of alleles	3.5 (2.7, 4.3)	3.4 (2.8, 4.0)	4.4 (3.5, 5.4)	4.6 (1.7, 5.3)
Effective number of alleles	1.4 (1.2, 1.6)	1.3 (1.2, 1.4)	1.3 (1.2, 1.4)	1.6 (0.73, 1.6)
Observed heterozygosity (H_O)	0.17 (0.09, 0.26)	0.18 (0.13, 0.24)	0.18 (0.14, 0.22)	0.24 (0.08, 0.28)
Expected heterozygosity (H_S)	0.21 (0.12, 0.31)	0.22 (0.15, 0.29)	0.22 (0.17, 0.26)	0.30 (0.10, 0.35)
Corrected total heterozygosity (H'_T)	0.28 (0.15, 0.41)	0.43 (0.30, 0.56)	0.48 (0.36, 0.58)	0.48 (0.16, 0.56)
Inbreeding coefficient (G_{IS})	0.19 (0.11, 0.33)	0.16 (0.06, 0.24)	0.17 (0.10, 0.28)	0.21 (0.14, 0.28)
Corrected standardized fixation index (G''_{ST})	0.31 (0.06, 0.52)	0.63 (0.49, 0.74)	0.69 (0.59, 0.77)	0.54 (0.30, 0.65)
Jost differentiation (D_{est})	0.09 (0.01, 0.19)	0.27 (0.16, 0.40)	0.33 (0.22, 0.45)	0.0 (0.0, 0.33)

Number of populations is 2 in Nova Scotia, 5 in New Hampshire and 3 in *G. radiatum*. 95% confidence intervals are in parentheses

heterozygosity), H_S, was 0.22 and 0.30, and the corrected total heterozygosity, H'_t, was 0.48 (Table 3.5). The two regions of *G. peckii* did not differ in H_O ($P = 0.28$) or H_S ($P = 0.70$).

Statistically significant deviations from Hardy–Weinberg equilibrium, quantified by the inbreeding coefficient, $G_{IS} = 1 - H_O/H_S$, were found in all three regions and in most populations (Tables 3.5 and 3.6). G_{IS} was 0.19 in *G. peckii* in NS, 0.16 in NH and 0.21 in *G. radiatum* ($P < 0.001$ for all; Tables 3.5 and 3.6). Inbreeding coefficients did not differ among the three regions.

3.4.3 Population Differentiation

We quantified population genetic differentiation in several ways, including (1) variance components from analysis of molecular variance (AMOVA; Michalakis and Excoffier 1996) and the associated metrics F_{ST} and R_{ST}, (2) the corrected (unbiased) standardized fixation index, G''_{ST} (Meirmans and Hedrick 2011), and (3) D_{est} (Jost 2008). When all 366 individuals from the three regions were included in the analysis, the largest fraction (62%) of genetic variance was accounted for by differences between *G. peckii* and *G. radiatum*, while population differences accounted

Table 3.6 Multilocus inbreeding coefficients, G_{IS}, for *G. peckii* and *G. radiatum* populations across 15 microsatellite loci

Species and region	Population	*N*	G_{IS}	Multilocus *P*
G. peckii				
Nova Scotia	Harris Lake	20	– 0.003	0.52
	Brier Island	127	0.29	0.001
	NS Overall	147	0.19	0.001
G. peckii				
New Hampshire	Hancock Branch	19	0.19	0.005
	Arethusa Falls	19	0.20	0.02
	Thompson Falls	20	0.07	0.20
	Mt. Washington	30	0.17	0.003
	Cannon Cliff	20	0.13	0.064
	NH Overall	108	0.16	0.001
G. radiatum				
North Carolina	1 {Genodive 9}	23	0.07	0.29
	8 {Genodive 10}	66	0.32	0.001
	13 {Genodive 8}	22	0.10	0.17
	Overall	111	0.21	0.001

for 22% and individual differences for 6% (AMOVA, Table 3.7). Within *G. peckii*, geographic region (NS vs. NH) accounted for 21% of genetic variation, population for 25%, among-individual for 11% and within-individual for 43% (Table 3.8). These results show much greater genetic differentiation between *G. peckii* and *G. radiatum* than between the NS and NH regions of *G. peckii*.

Multiple metrics showed greater population differentiation within NH than within NS. First, NH had a higher corrected standardized fixation index, G''_{ST}, than NS (0.63 versus 0.31, respectively, $P = 0.031$) as well as a higher level of population differentiation ($D_{est} = 0.27$ versus 0.09; $P = 0.050$; Table 3.5). Second, the proportion of total genetic variance occurring among populations was nearly twice as great in NH than in NS (47% versus 25%, Table 3.8). Third, the average pairwise genetic distance as measured by standardized Nei's G''_{ST} was twice as high within NH (0.48, $n = 10$) than within NS (0.24, $n = 1$; Table 3.9 above diagonal).

Average pairwise genetic distances corroborated the conclusions above (Table 3.9). Mean pairwise G'_{ST} was 0.70 between *G. peckii* and *G. radiatum*, 0.71 between NS *G. peckii* and *G. radiatum*, 0.69 between NH *G. peckii* and *G. radiatum*, and 0.60 between *G. peckii* in NS and NH. Average values of the unbiased standardized measure G''_{ST} showed a similar pattern (Table 3.9, below diagonal).

In the analyses above, the NS plants were considered as two populations—Brier Island ($n = 127$) and Harris Lake ($n = 20$). Results from more fine-scale analyses where Brier Island plants were divided into several populations did not alter the

Table 3.7 Apportionment of genetic variation in *G. peckii* and *G. radiatum* from analysis of molecular variance (AMOVA)

Source of variation	*N*	Variance proportion	*F*-statistic	*F*	SD	95% CI	*P*
Within individual	–	0.105	R_{IT}	0.895	0.060	0.751, 0.955	–
Among individual (population)	366	0.057	R_{IS}	0.352	0.114	0.145, 0.555	0.001
Among population (species)	10	0.218	R_{SC}	0.572	0.113	0.359, 0.674	0.001
Among species	2	0.620	R_{CT}	0.620	0.174	0.210, 0.801	0.001

Populations from all three regions representing the entire extent of both species are included. Sample sizes: 2 species, 10 populations, 366 individuals. For *F*-statistics, standard deviations (*SD*) of were obtained by jackknifing over loci, 95% confidence intervals (CI) of *F*-statistics by bootstrapping over loci and *P*-values by 999 permutations. Parentheses indicate nesting

conclusions. Significant population differentiation was found in *G. radiatum* when measured as G''_{ST} (0.54) but not as D_{est} (0.0; Table 3.5).

Within *G. peckii,* population differentiation (F_{ST}) increased with geographic distance over large spatial scales (Fig. 3.5a). The correlation between the genetic and geographic distance matrices was statistically significant (Mantel test, $P = 0.001$). Across both regions, F_{ST} fell sharply at scales of less than 10 km (Fig. 3.5b).

3.4.4 Population Structure

Analyses of population structure using software program InStruct (Gao et al. 2007) revealed a clear separation of *G. peckii* and *G. radiatum* assuming both $K = 2$ and $K = 3$ groups (Fig. 3.6). Similarly, the NS and NH regions were distinct within *G. peckii* at $K = 3$ (Fig. 3.6). Within NS, little population structure was apparent except for possibly between Brier Island and Harris Lake (Fig. 3.7). In contrast, structure across the NH populations was apparent with various *K*-values including $K = 5$, which differentiated the five populations (Fig. 3.8). Population structure was also found within *G. radiatum*, where $K = 2$ separated Population 1 from the others and $K = 3$ differentiated all three populations (Fig. 3.9).

Table 3.8 Apportionment of genetic variation in *G. peckii* from analysis of molecular variance (AMOVA)

Source of variation	N	Variance proportion	F-statistic	F	SD	95% CI	P
Nova Scotia and New Hampshire							
Within individual	–	0.426	R_{IT}	0.574	0.078	0.440, 0.703	–
Among individual (population)	255	0.108	R_{IS}	0.202	0.051	0.114, 0.308	0.001
Among population (*peckii* region)	7	0.251	R_{SC}	0.320	0.075	0.187, 0.467	0.001
Among *peckii_* region	2	0.214	R_{CT}	0.214	0.124	-0.025, 0.405	0.017
Nova Scotia							
Within individual	–	0.602	R_{IT}	0.398	0.140	0.155, 0.609	–
Among individual (population)	147	0.149	R_{IS}	0.198	0.061	0.092, 0.334	0.001
Among population	2	0.249	R_{ST}	0.249	0.132	0.044, 0.439	0.001
New Hampshire							
Within individual	–	0.427	R_{IT}	0.573	0.086	0.416, 0.709	–
Among individual (population)	108	0.104	R_{IS}	0.195	0.069	0.090, 0.328	0.001
Among population	5	0.470	R_{ST}	0.470	0.087	0.322, 0.616	0.001

Parentheses indicate nesting. Populations are nested within region in the analysis including all populations, and individuals are nested within populations. For F-statistics, standard deviations (SD) of were obtained by jackknifing over loci, 95% confidence intervals (CI) of F-statistics by bootstrapping over loci and P-values by 999 permutations

3.4.5 Discussion and Management Implications

This portion of the study provides an overview of the population genetics of eastern mountain avens, *G. peckii*, and its separation from spreading avens, *G. radiatum*. Working with a suite of novel genetic markers, we have estimated a series of genetic parameters that may inform future management decisions for *G. peckii* in Nova Scotia. Of particular note is that the Nova Scotia populations were not genetically depauperate despite the small total area occupied and distance from the much larger area occupied by the New Hampshire populations. All measures of genetic diversity are similar in Nova Scotia and New Hampshire. These measures included number of alleles, effective number of alleles, gene diversity (expected heterozygosity), and

Table 3.9 Genetic distances measured by Nei's standardized G'_{ST} (above diagonal) and unbiased standardized G''_{ST} (below diagonal) for all 10 populations of both species

Species	Population	Harris Lake	Brier Island	Hancock Branch	Arethusa Falls	Thompson Falls	Mt. Washington	Cannon Cliff	Pop. 13	Pop. 1	Pop. 8
G. peckii	Harris Lake	0	0.244	0.669	0.725	0.670	0.511	0.714	0.807	0.763	0.676
G. peckii	Brier Island	0.308	0	0.544	0.561	0.560	0.437	0.563	0.732	0.690	0.586
G. peckii	Hancock Branch	0.846	0.744	0	0.495	0.531	0.392	0.468	0.693	0.703	0.570
G. peckii	Arethusa Falls	0.872	0.741	0.643	0	0.591	0.394	0.516	0.765	0.769	0.594
G. peckii	Thompson Falls	0.795	0.717	0.677	0.724	0	0.260	0.666	0.757	0.729	0.612
G. peckii	Mt. Washington	0.663	0.614	0.551	0.524	0.340	0	0.526	0.662	0.689	0.553
G. peckii	Cannon Cliff	0.847	0.719	0.598	0.626	0.797	0.689	0	0.777	0.787	0.648
G. radiatum	Pop. 13	0.977	0.980	0.914	0.968	0.910	0.898	0.944	0	0.574	0.264
G. radiatum	Pop. 1	0.926	0.916	0.943	0.990	0.877	0.954	0.979	0.749	0	0.347
G. radiatum	Pop. 8	0.921	0.873	0.851	0.860	0.818	0.855	0.888	0.415	0.516	0

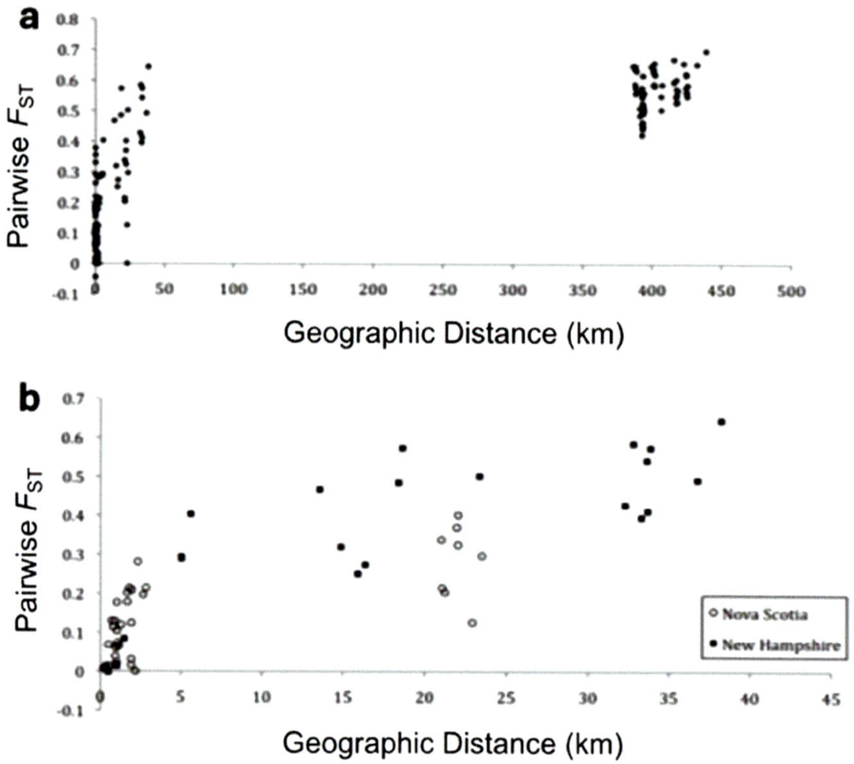

Fig. 3.5 Genetic differentiation (F_{ST}) as a function of geographic distance in *G. peckii*. **a** All pairwise comparisons. **b** Comparisons at within-region scale. Open circles are within NS, closed squares are within NH

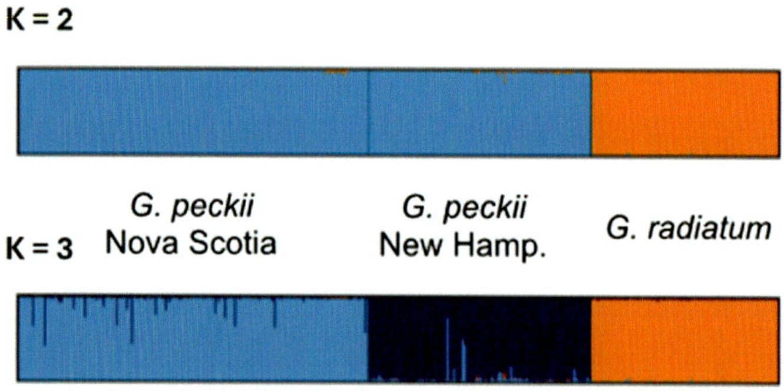

Fig. 3.6 Genetic structure analyses for all samples from *G. peckii* and *G. radiatum*

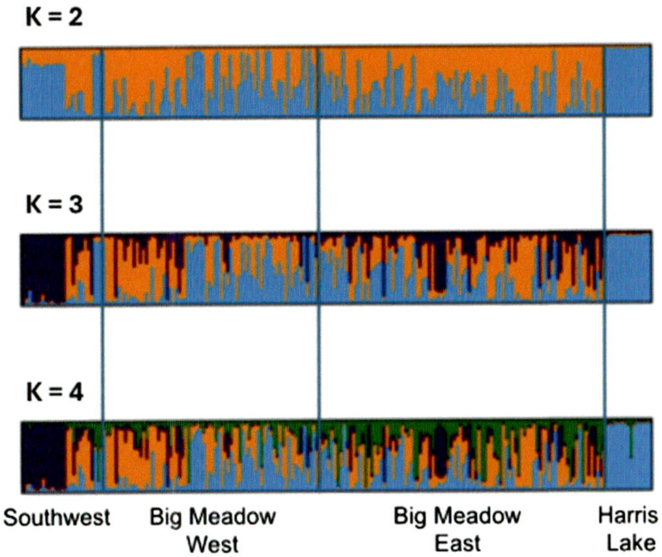

Fig. 3.7 Genetic structure analyses for *G. peckii* in Nova Scotia

observed heterozygosity. This pattern suggests that the Nova Scotia plants comprise one or more relict populations and are not the result of founding by emigrants.

In contrast to genetic diversity, measures of differentiation were higher among the New Hampshire population than between the two Nova Scotia populations. This difference cannot be explained by differences in geographic distance within each region. First, the New Hampshire populations were separated on average by 23.7 km, similar to the 21.2 km distance between the Brier Island and Harris Lake populations in Nova Scotia. Furthermore, five of the 10 pairwise geographic distances in New Hampshire were smaller than the distance separating the two Nova Scotia populations, and all showed greater genetic differentiation as measured by G'_{ST} and G''_{ST}. The greater genetic divergence would appear instead to result from landscape features—including elevational changes and intervening forest—that inhibit pollinator movement and seed dispersal among populations.

Although we found no evidence that *G. peckii* in Nova Scotia is genetically depauperate, the populations exhibit substantial inbreeding depression. Inbreeding depression, δ, is the proportional reduction in fitness of selfed compared to outcrossed offspring (Charlesworth and Charlesworth 1987) and can be quantified using fitness data from hand pollinations in Nova Scotia (Chapter 4). Seed number per flower was 53 for outcrossed and 39 from self-pollinated flowers, resulting in $\delta = 1 - 39/53 = 0.26$ at the seed stage. Germination rates were 0.214 for selfed seeds, 0.342 for outcrosses within bogs, 0.50 between bogs and 0.540 between Brier Island and Harris Lake. Therefore, inbreeding depression acting between seed maturation and germination was 0.37 within bogs, 0.57 between bogs and 0.60 between the Brier

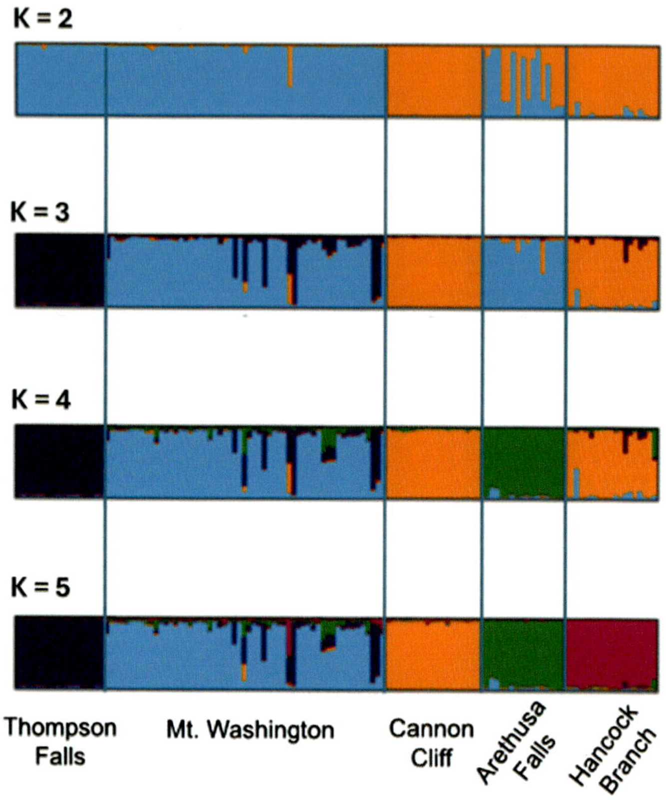

Fig. 3.8 Genetic structure analyses for *G. peckii* in New Hampshire

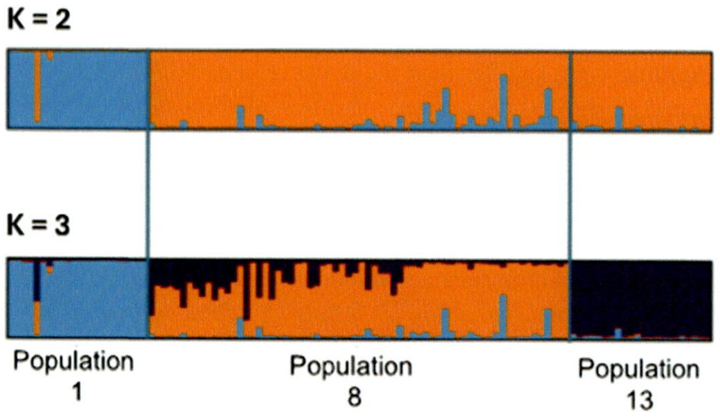

Fig. 3.9 Genetic structure analyses for three populations of *G. radiatum* in North Carolina

Island and Harris Lake populations. A direct measure of lifetime inbreeding depression would require many years of observing fitnesses of selfed and outcrossed plants from hand pollinations in the field. We can indirectly estimate lifetime inbreeding depression by comparing the selfing rate measured at the seedling stage, s_1, to that measured in mature plants in the field, s_2.

$$\delta = \frac{s_2 - s_1}{s_1(s_2 - 1)} \tag{3.3}$$

This equation is equivalent to Ritland's (1990) formulation using s_1 and the inbreeding coefficient of mature plants. Selfing rates were approximately 0.54 from the progeny array at the seedling stage and 0.30 from the population survey of mature plants. This difference suggests a $\delta = 0.63$ lower survival of selfed individuals. Lifetime inbreeding depression is therefore $0.88 = 1 - (1 - 0.26)(1 - 0.57)(1 - 0.63)$. *G. peckii* in Nova Scotia thus appears to possess ample genetic variation combined with very high inbreeding depression, indicating the presence of deleterious alleles. This combination highlights the importance of establishing and maintaining a suitable environment for the management and recovery of the eastern mountain avens.

Acknowledgements We thank Daniel Sperduto and Thomas Rawinski for help in locating, accessing and collecting from the New Hampshire populations. We thank Caleb Beck, John Collins, Elizabeth Mack and Sophia Mekarski for help with rearing progeny array offspring and entering data; Megan Fraser and Jackie Bradbury for assistance in the field in Nova Scotia; and Ian Paterson and Greg McCracken for help with genotyping. This study was supported by grants or contracts from the Natural Sciences and Engineering Research Council Discovery Grants Program, Environment and Climate Change Canada, Southwest Nova Scotia Priority Place and the Nova Scotia Museum of Natural History.

References

Amos W, Balmford A (2001) When does conservation genetics matter? Heredity 87(3):257–265. https://doi.org/10.1046/j.1365-2540.2001.00940.x
Bürkli A, Sieber N, Seppälä K et al (2017) Comparing direct and indirect selfing rate estimates: when are population-structure estimates reliable? Heredity 118:525–533
Charlesworth D, Charlesworth B (1987) Inbreeding depression and its evolutionary consequences. Annu Rev Eco Evol Syst 18:237–268. https://doi.org/10.1146/annurev.es.18.110187.001321
Charlesworth D, Willis JH (2009) The genetics of inbreeding depression. Nat Rev Genet 10(11):783–796. https://doi.org/10.1038/nrg2664
COSEWIC (2010) Assessment and status report on the eastern mountain avens *Geum peckii* in Canada. Government of Canada Report
Darwin C (1862) The various contrivances by which orchids are fertilised. Murray (ed). London
Darwin C (1876) The effects of cross and self-fertilisation in the vegetable kingdom. Murray (ed). London
David P, Pujol B, Viard F et al (2007) Reliable selfing rate estimates from imperfect population genetic data. Mol Ecol 16:2474–2487
Gajewski, W (1959) Evolution in the genus *Geum*. Evol 13:378–388

Gao H, Williamson S, Bustamante CD (2007) A Markov chain Monte Carlo approach for joint inference of population structure and inbreeding rates from multilocus genotype data. Genetics 176:1635–1651

Godt MW, Johnson BR, Hamrick J (1996) Genetic diversity and population size in four rare southern Appalachian plant species. Cons Biol 10(3):796–805

Hamrick JL, Godt MJ (1989) Allozyme diversity in plant species. In Brown AHD, Clegg MT, Kahler KL, Weir BS (eds) Plant population genetics, breeding and genetic resources. Sinauer, Sunderland, MA, pp 43–63

Hay NM, Ulrey C, Kauffman G et al (2019) The genetic diversity and population structure of *Geum radiatum*: effects of a past augmentation of an endangered hexaploid. Castanea 84(2):273–288. https://doi.org/10.2179/0008-7475.84.2.273

Hartl DL, Clark AG (1997) Principles of population genetics. Ver. 3. Sinauer, Sunderland, MA

Hedrick PW, Kalinowski ST (2000) Inbreeding depression in conservation biology. Annu Rev Ecol Evol Sys 31:139–162

Holfinger B, Till B (2003) Low cost DNA extraction for use in TILLING and ecotilling assays, version 1.3. Plant Breeding and Genetics Laboratory. IAEA.org

Holt R, Kwok A, Dorken ME (2020) Increased spatial-genetic structure in a population of the clonal aquatic plant *Sagittaria latifolia* (Alismataceae) following disturbance. Heredity 124:514–523

Hodel R, Gitzendanner M, Germain-Aubrey C et al (2016) A new resource for the development of SSR markers. APPS 4:1600024

Hubisz MJ, Falush D, Stephens M et al (2009) Inferring weak population structure with the assistance of sample group information. Mol Ecol Res 9:1322–1332. https://doi.org/10.1111/j.1755-0998.2009.02591.x

Jordan CY, Lohse K, Turner F et al (2017) Maintaining their genetic distance: little evidence for introgression between widely hybridizing species of *Geum* with contrasting mating systems. Mol Ecol 27:1214–1228

Jost L (2008) G_{ST} and its relatives do not measure differentiation. Mol Ecol 17:4015–4026

Lloyd DG (1979) Some reproductive factors affecting the selection of self-fertilization in plants. Am Nat 113:67–79. https://doi.org/10.1086/283365

Lloyd DG, Schoen DJ (1992) Self- and cross-fertilization in plants. I. Functional dimensions. Intl J Plant Sci 153:358–369. https://doi.org/10.1086/297040

Mandel JR, Major CK, Bayer RJ, Moore JE (2019) Clonal diversity and spatial genetic structure in the long-lived herb, Prairie trillium. PLoS ONE 14(10):e0224123

Meirmans PG (2020) Genodive v.3.0: easy-to-use software for the analysis for genetic data of diploids and polyploids. Mol Ecol Resour. https://doi.org/10.1111/1755-0998.13145

Meirmans PG, Van Tienderen PH (2004) Genotype and Genodive: two programs for the analysis of genetic diversity of asexual organisms. Mol EcoNotes 4:4. https://doi.org/10.1111/j.1471-8286.2004.00770.x

Meirmans P, Hedrick P (2011) Assessing population structure: F_{ST} and related measures. Mol Ecol Resources 11:5–18

Michalakis Y, Excoffier L (1996) A generic estimation of population subdivision using distances between alleles. Genetics 142:1061–1064

Murdock N (1993) Recovery plan for spreading avens (*Geum radiatum*). https://ecos.fws.gov/docs/recovery_plan/930428.pdf

Nunney L (1999) The effective size of a hierarchically structured population. Evolution 53:5–10

Palstra FP, Ruzzante DE (2008) Genetic estimates of contemporary effective population size: what can they tell us about the importance of genetic stochasticity for wild population persistence? Mol Ecol 17:3428–3447

Paterson IG, Snyder M (1999) Genetic evidence supporting the taxonomy of *Geum peckii* (Rosaceae) and *G. radiatum* as separate species. Rhodora 101:325–340

Pleuss AR, Stöcklin J (2004) Population genetic diversity of the clonal plant *Geum reptans* (Rosaceae) in the Swiss Alps. Am J Bot 9:2013–2021

Pritchard JK, Stephens M, Donnelly P (2000) Inference of population structure using multilocus genotype data. Genetics 155:945–959

Ritland K (1990) Inferences about inbreeding depression base on changes of the inbreeding coefficient. Evol 44:1230–1241

Ritland K (2002) Extensions of models for the estimation of mating systems using *n* independent loci. Heredity 88:221–228

Rusham M, Hollingsworth PM, Squirrell J et al (2010) Significant differences in outcrossing rate, self-incompatibility, and inbreeding depression between two widely hybridizing species of *Geum*. Biol J Linn Soc 101:977–990

Toms B, LaRue D (2017) A census of Eastern Mountain Avens (Geum peckii) in Nova Scotia 2017. Mersey Tobeatic Research Institute Report

Ulrey C, Quintana-Ascencio P, Kauffman G et al (2016) Life at the top: Long-term demography, microclimatic refugia, and responses to climate change for a high-elevation southern Appalachian endemic plant. Biol Cons 200:80–92

Wang J (2009) A new method for estimating effective population sizes from a single sample of multilocus genotypes. Mol Ecol 18:2148–2164

Wang J, El-Kassaby YA, Ritland K (2012) Estimating selfing rates from reconstructed pedigrees using multilocus genotype data. Mol Ecol 21:100–116

Weir BS, Cockerham CC (1973) Mixed self and random mating at two loci. Genet Res 21:247–262

Wright S (1965) The interpretation of population structure by f-statistics with special regard to systems of mating. Evolution 19:395–420

Wright, S (1969) Evolution and the genetics of populations. Vol 2. The theory of gene frequencies. University of Chicago Press, Chicago

Zhan L, Paterson IG, Fraser B et al (2017) Megasat: automated inference of microsatellite genotypes from sequence data. Mol Ecol Res 17:247–256

Zinck M (1996) Numerical evaluation of the taxonomy of *Geum radiatum* and preliminary studies of the pollination biology of its Nova Scotian population. BSc Thesis Acadia University, Wolfville, NS

Chapter 4
Ecology and Conservation of the Eastern Mountain Avens in Nova Scotia

Nicholas M. Hill, J. Sherman Boates, Mark O. Johnston, Claire E. Wilson, Brad E. Toms, Emma Manning, Jennifer Poirier, and David J. Garbary

Abstract The ecology of the Eastern Mountain Avens (*Geum peckii*), Endangered in Canada and Imperiled globally, was examined on Brier Island, Nova Scotia, to inform its conservation management and prospect for long-term survival in situ in Canada. Over ten years of study, a picture emerged of a plant adapted to growth in naturally stressed wetland habitats where the interspecific competition (e.g., from shrubs) is low. In-field measurements and experimentation at the adult (transplantations) and regenerative (pollination and germination) stages have provided an understanding of the survival strategy of the plant, and how this may be affected by alterations in drainage, nutrient availability, disturbance and climate (temperature). In the Avens' principal Canadian habitat, the evolution of the landscape of Big Meadow Bog on Brier Island created a lagg (peatland margin) where the Avens thrived in low biomass, waterlogged fens. In 1958, ditching lowered the water table which brought on a suite of ecological changes, including the colonization of the bog by thousands of Herring Gulls, increasing nutrient inputs. Our studies have informed the restoration of the

N. M. Hill
UNESCO Southwest Biosphere Association, 399 Wyman Road, Sand Beach, NS B5A 5G4, Canada
e-mail: nick@swnovabiosphere.ca

J. S. Boates · C. E. Wilson
Department of Biology, Acadia University, 15 University Ave, Wolfville, NS B4P 2R6, Canada
e-mail: sherman.boates@acadiau.ca

M. O. Johnston
Department of Biology, Dalhousie University, Halifax, NS, Canada

B. E. Toms
Mersey Tobeatic Research Institute, Kempt, 9 Mt Merritt Rd, Caledonia, NS B0T 1B0, Canada
e-mail: brad.toms@merseytobeatic.ca

E. Manning · J. Poirier · D. J. Garbary (✉)
Department of Biology, St. Francis Xavier University, Antigonish, NS B2G2W5, Canada
e-mail: dgarbary@gmail.com

C. E. Wilson
Nova Scotia Department of Natural Resources and Renewables, 136 Exhibition Street, Kentville, NS B4N 4E5, Canada

N. M. Hill et al. (eds.), *Wetland Restoration for Endangered Species Recovery*, Environmental Contamination Remediation and Management,
https://doi.org/10.1007/978-3-031-71344-6_4

bog's hydrology as well as an adaptive management approach to augmenting the population of the Avens. They have also, in combination with the monitoring of climate change on the island, raised the specter that without interventions and ex situ techniques, the future climate and resultant shrubification of its remaining habitat could eradicate Avens from Canada.

Keywords Eastern Mountain Avens · Rare plants · Realized niche · Wetland restoration

4.1 Introduction

Rare plants are notoriously fussy. They are usually specialists (Espelland and Emam 2011) with narrower niches than their common associates (Griggs 1940; Hill and Keddy 1992; Edwards and Weakley 2001). When we want to test their growth rates against those of common plants, they prove more difficult to grow (Gaudet and Keddy 1988). When we look at their distributions, we see many of them restricted to particular habitats, and in the case of the Eastern Mountain Avens, we are faced with one population growing at 5000 feet along streams, beside waterfalls, and in tundra and scree of the White Mountains of New Hampshire, and the other, our subject here, growing in fens or successional sheep pasture at sea level in Nova Scotia. A closer look reveals that in Canada this species, the Eastern Mountain Avens (*Geum peckii*, hereafter "Avens") is restricted to open communities on one rock type, basalt, and that the main wetland habitats occupied are laggs, the margins of raised bogs. While sheep pastures come and go, the fens at Big Meadow Bog started forming 1700 years ago after a "regional wetting event" (Chap. 6). This was the primary habitat for the species in Canada, supporting an estimated 10,000 plants, until it was ditched for agriculture in 1958. Ditching led to cascading ecological and social impacts: trees colonized the open fen margins and set up in the bog, thousands of Herring Gulls (*Larus argentatus*) came to the bog for nesting each summer, bog berry plants (e.g., cloudberries, *Rubus chamaemorus*, and cranberries and blueberries, *Vaccinium* spp.) were displaced, bog pools dried up and no longer supported waterfowl, and people from the nearby village stopped going into Big Meadow for foraging or recreation (Hill et al. 2019). Here, we dig into the ecology of the Avens to understand the factors that control its growth, reproduction, and survival to guide the restoration and recovery of Big Meadow Bog and help bring back the native peatland biodiversity.

The Eastern Mountain Avens, *Geum peckii*, is a perennial herb of the Rose family. Geum and the other genera of the tribe, Colurieae, are mostly herbaceous plants whose seeds (achenes) are dispersed by hooked styles that stick into the fur of mammals (Gajewski 1959). Our focal Avens (Eastern Mountain Avens) and other members of the tribe have abandoned hooked styles that afford animal dispersal of seeds (ectozoochory) and while some of this group have achenes that are wind dispersed because their styles bear plumose hairs, *Geum peckii* and a few other montane species lack structures that enable either type of long distance dispersal of seeds. We show here how the Avens focuses on survival, with occasional recruitment from seed, at the local level. This strategy has worked since deglaciation and Avens found two markedly different habitat solutions, both at 44.1°N, in the White Mountains of New Hampshire and on the Digby Peninsula of Nova Scotia. Due to major anthropogenic disruptions of the hydrological regime of the fens along the entire margin of Big Meadow Bog, Brier Island (20 ha; ditched) and of fens at Harris Lake (19 ha; bermed and flooded) on the Digby Peninsula, in combination with the anthropogenically-related eutrophication at Big Meadow (Chap. 12 and 13), the Canadian populations of the Avens run a serious risk of being extirpated. The hydrological restoration of Big Meadow Bog was a positive step to reverse this trend (Chap. 7 and 10). These efforts were made in full recognition that climate warming of this tundra-like landscape that supports the Avens may ultimately thwart this recovery (Chap. 5 and 15). But lastly, we make the case that one must do more than one thing in conservation, and both in situ and ex situ strategies are necessary.

4.2 Methods and Materials

4.2.1 Setting the Stage: Range and Habitat of the Eastern Mountain Avens

The Avens occurs in only two places in the world, at high elevation in the White Mountains of New Hampshire, and in Nova Scotia where it grows near sea level mostly in fens on Brier Island. Its range in New Hampshire is in the order of 400,000 ha which is 200 times the size of Brier Island. At a global level, the species is Imperiled (G2), a ranking mirrored at the national (N2) and state (S2) levels in the United States and New Hampshire, respectively. In Canada and Nova Scotia it is ranked Critically Imperiled, both nationally (N1) and provincially (S1) (Nature Serve 2024). The majority (96%) of the Avens (Eastern Mountain Avens) plants in Canada occurs on Brier Island but there is a small population at Harris Lake, 20 km away at the end of the Digby Peninsula. The Brier Island map shows three linear, north–south arrays of plants: those labeled GR and GH (Green Head and Gull Rock Road) to the east, and then a linear array on either side of Big Meadow Bog in the east and west lagg (margin) of this wetland (Fig. 4.1).

In the White Mountains, Avens grows at high elevation amid dwarf shrubs at stream sides, and in bogs and heaths (Bliss 1963), as well as at the edges of waterfalls, in scree and in tundra. Each of these habitats has its own set of local environmental stresses but all of them are exposed to extreme winter low temperatures, high winds and a short growing season exacerbated by snowpacks (Bliss 1963). The majority of woody plants in these high elevation communities are dwarf shrubs (10–30 cm high, Fig. 4.2). In contrast, on Brier Island, the majority of shrub associates range from 30 to 50 cm in height and waterlogging stress reduces the vigor of shrubs and the potential for competitive exclusion.

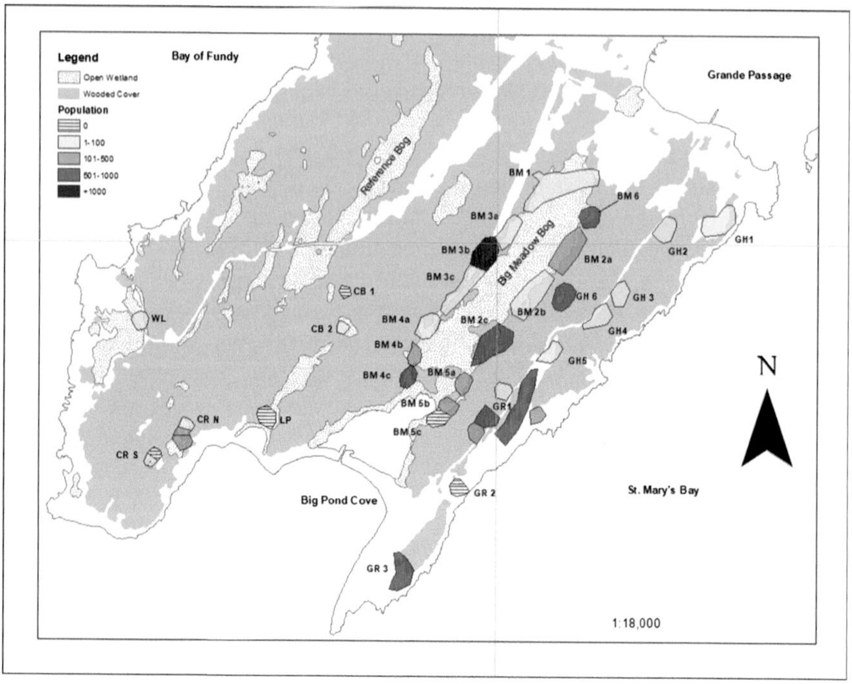

Fig. 4.1 Map of Brier Island showing Big Meadow Bog with its satellite population areas (BM) of Avens, and the Reference Bog. Other areas with Avens populations extant and extirpated are shown for Gull Rock Road (GR), Green Head (GH), Central Brier (CB), Camp Road (CR), Western Light (WL), and Little Pond (LP)

Fig. 4.2 Percent of three size classes in the woody taxa of Eastern Mountain Avens habitats of Brier Island and Mount Washington. Dwarf shrubs are usually prostrate and less than 30 cm tall, shrub heights were 30–60 cm on average, and tall woody species had maximum heights greater than 2 m. Shrub information for Mount Washington was taken from Bliss (1963) and for Brier Island from Hill (2015)

Avens was grouped into the Arctic-Alpine Plant Element by Roland and Smith (1969) and the survival of this low-stature, slow-growing rare plant of infertile wetlands matches the conservation concerns for the persistence of the Atlantic Coastal Plain Flora on lakeshores in Nova Scotia (Keddy and Wisheu 1989). The vulnerability of the rare coastal plain herbs on lakeshores or in the present case, the Endangered Avens in Brier Island fens, is rooted in evolutionary trade-offs that accompany the stress-tolerant strategy typified by an obligate slow growth rate and a lack of investment in reproduction (Grime 1979). The threat from eutrophication in remote, rural southwest Nova Scotia seemed academic thirty years ago when we published a model that predicted that most (67%) of the variation in rare coastal plain species richness on lakeshores was linked to watershed area (Hill and Keddy 1992). But in the last decade, nutrient flows from mink ranches have threatened the coastal plain flora on lakeshores via leaching into river water (Brylinsky and Sollows 2014). Similarly, eutrophication via avian dispersal of food wastes by Herring Gulls into Big Meadow Bog (Chap. 12; Shlepr et al. 2021) has severely altered the principal habitat of the Eastern Mountain Avens in Canada (Hill et al. 2019). Understanding the life histories and strategies of rare stress-tolerant plants is becoming more important as eutrophication of wetlands increases (Keddy 2010; Chislock et al. 2013; Hwang 2020), particularly when wetlands are promoted as ecosystems capable of purifying nutrient-rich runoff (Verhoeven et al. 2006).

The studies on the ecology of Avens reported herein, all take place in open peatlands—fen or bog—however Avens is not an Obligate Wetland plant (Reed 1988) because 8% of its occurrences throughout its sole location in Canada, the Digby Peninsula, are in upland sites that were formerly (until the 1980s) grazed by sheep (Hill and Belliveau 2014). The balance (92%) of Avens occurrences on Brier Island (yellow symbols) lie in wetland areas (Fig. 4.3) where competition from the growth of shrubs will be reduced by waterlogging (Kozlowski 1997). It is not clear how Avens avoids competitive exclusion in these upland areas; it is possible that grazing plays a role but this requires further study. Grazing trials in some other studies have shown benefits on overall biodiversity (Rysiak et al. 2021; Anonymous 2022).

Avens occurs in various wetlands on the eastern side of Brier Island but this book focuses on Big Meadow Bog because of its size and the intermittent marginal fens around a raised bog (Chap. 6). Big Meadow Bog is unique because it is the only large, raised bog on basalt in mainland Nova Scotia, and it is the only basalt-based bog to have a lagg that could support Avens. The lagg of a raised bog forms along its margins. There is a centrifugal drainage down the slopes of the raised bog and this meets a centripetal drainage from the surrounding upland or slope swamp toward the raised bog. These two flows are very different. That flowing from the bog down the "rand" has a low pH and low nutrient concentration since the raised bog is ombrotrophic getting its only nutrients from precipitation and deposition. That flowing toward the bog from the surrounding landscape has dissolved minerals from the rocks and ground and has a higher nutrient concentration. The lagg is an ecotone between the bog with its low pH and low calcium waters and the surrounding landscape. Thus, the lagg is frequently a fen and swamp margin to the raised bog, with distinct vegetation denoting a greater availability of calcium and a low N:P ratio with

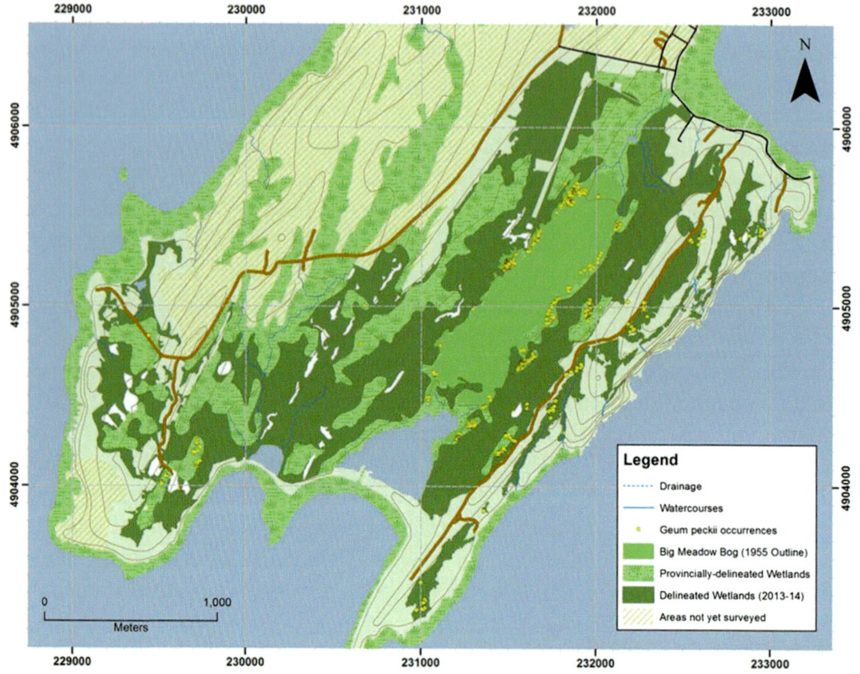

Fig. 4.3 Wetlands (all green areas) on Brier Island and the occurrences of Eastern Mountain Avens (yellow dots). Upland occurrences lie to the south of Gull Rock Road in the lower left quadrant of map (adapted from Hill and Belliveau 2014)

a relatively high water table marked by water level fluctuations (Howie and Tromp van-Meerveld 2011; Paradis et al. 2015). There are no records of plant community composition for the Meadow before the ditching, and after ditching, Avens has only been found in the lagg area. This lagg is distinct from both the remnant heath on Big Meadow (Hill et al. 2019) and the narrow, raised bog to the west of Big Meadow (the "Reference Bog") which support distinctive ombrotrophic species (*Eriophorum vaginatum* and *Rubus chamaemorus*) or most fen vegetation in other parts of Brier Island. Distinctive lagg specialists at Big Meadow include the shrubs (*Dasiphora fruticosa* and *Lonicera villosa*), sedges (*Carex livida*, and *Carex buxsbaumii*), and forbs (*Menyanthes trifoliata* and *Geum peckii*).

4.2.2 Shrub and Water Table Relationships with Avens

(a) The Realized Niche: Pattern in the Field

In November of 2012, a 250 m long transect was established across the north end of Big Meadow Bog from BM1–BM6. At each position, measurements were made of the change in surface elevation, of the depth to water table, and maximum height of shrubs. The change in elevation was measured using a string and line level over the transect interval that ranged from 10 to 20 m. Depth to water table was measured 15 min after a hole was made in the peat surface with a soil auger. The 2012 transect across Big Meadow Bog revealed that Avens occurred where the water table was near the surface and shrub growth was least. In June of 2016, within-fen transects were established in two marginal fens (laggs) of Big Meadow Bog (BM3, Runway Bog; and BM6, Beauty Bog) to assess the relationships between the presence of Avens, the depth to water table, and the cover or height of shrubs. Multiple line transects were set up in the area of each fen where Avens occurred and measurements ($n = 132$ at BM3 and 124 at BM6) were made of depth water table, shrub height and cover, and observations made of the presence of Avens and of associated rare plants, at one meter intervals along each transect.

Water samples were collected from two wells at the Reference Bog and two wells from each of two Avens fens, BM1 (northwest lagg) and BM3 (Runway Bog) on August 21, 2013, to differentiate bog and fen chemistry (AGAT labs, order # 13X750634, Standard Water Analysis). Peat cores (10 cm diameter, 0–25 cm depth) were removed from the Northwest Lagg (BM1) and from the Reference Bog on the same day (August 21, 2013). Cores were placed in plastic bags and kept cool on ice and taken to the Applied Geosciences and Technology laboratories in Dartmouth, Nova Scotia for analysis of total nitrogen and phosphorus (AGAT labs, order # 13X750806, Inorganic Analysis).

(b) Transplantation of Avens Over a Range of Moisture and Nutrient Availabilities

The Action Plan for Eastern Mountain Avens in Canada (Environment and Climate Change Canada 2018) ranked as "high priority" an experiment to determine the growth and survival response of Avens when transplanted over a wide variation of soil moisture regimes. This would inform the hydrological restoration of Big Meadow Bog, with regard to the water level target and its impact on the Endangered species. In 2015, 36 plants were removed from the northwest lagg (BM1) of Big Meadow Bog, where disturbance by Herring Gulls threatened to eliminate the Avens population. Plants were dug out of the peat and associated loose peat was carefully removed as was biomass of other plants in the community. The plants were arranged in groups of three that best minimized overall differences in the composite size of all three plants. Groups of three plants each were transplanted to high, medium, and low water table areas in: (1) healthy lagg areas of Big Meadow Bog, (2) the isolated fens away from Big Meadow Bog, and (3) along a water table gradient from the lagg to the center of a nearby raised bog (the Reference Bog). Three lots of three plants were replanted back into the source fen as a control, i.e., the eutrophic northwest lagg (BM1). In each destination habitat, a hole slightly larger than the soil and root mass of the plant was dug, the plant inserted so its shoots were at moss level, and excess peat was used to snug the plant into place. After transplantation, the above-ground vegetation of the habitat was clipped to form a 15 cm competition-free buffer around the transplanted Avens. After all transplants were in place, two small plants were left over, having detached from larger transplants; these were laid into the shallow margin of a pool of Runway Bog (BM3).

The transplants were monitored over three years (2016–2018) and were removed from the transplant areas as required by permit, in August 2018. In each year, survivorship was tracked and monthly measurements were made in summer of leaf widths (of 3 widest leaves per plant) and small (< 2 cm) and large (> 2 cm) healthy leaves, dead or discolored leaves, and red-petioled leaves were counted. In addition, once a month, fluorometry measurements using a portable pulse amplitude modulated fluorimeter (Diving-PAM; Heinz Walz, Effeltrich, Germany) were made on three representative leaves of each plant. Fluorimetry offers a non-destructive means to assess photosynthetic efficiency; its quantum yield measure relates to the efficiency of electron passage from Photosystem II to Photosystem I and so it has been used as a measure of health in plants (Flynn et al. 2018). Biomass was measured by removing the community above-ground and below-ground biomass from a 0.625 m^2 quadrat placed around the Avens transplant area at each site. Peat and dead rhizomes were removed from the below-ground material, but no attempt was made to separate dead from live roots. Similarly, leaf litter, dead stems and twigs were removed from the above-ground material but attached dead litter was not removed. The biomass samples were dried for three days in paper bags and then weighed. Observations at the mesic fen site of Big Meadow (BM5c) were discontinued after it was flooded throughout 2016 after the outlet of Big Pond to the sea had been blocked after an overwinter storm. Year three data presented for the mesic fen, therefore, includes only that from the isolated fen site.

4.2.3 *Regeneration Biology*

(A) Vegetative Growth

Avens reproduces vegetatively according to a monopodial pattern of growth. The apex of the stem is the growth axis and leaves are produced from the stem produced in the current year. Larue (2016a) discussed the extensive rhizomes and how Avens could be propagated from rhizome material. In the present study, we made non-destructive measurements of Avens rhizomes and their spatial arrangement in two habitats (GR1 and GH6). The largest Avens plants were selected at 10 hummocks and 10 hollows at each of these sites in July 2017. Moss and peat was carefully pulled away from the base of the Avens leaves and from rhizomes, taking care not to damage the plant. Measurement of the vertical depth of each rhizome and its length led to the trigonometric calculation (cosine) of the angle of rhizome growth in each area.

(B) Controlled Pollination and Subsequent Effect on Seed Germination

A controlled pollination experiment was set up at two fens (Runway Bog, BM3b, and Brad's Bog, GH6) in June 2018. At each fen, five bagged, hand-pollination treatments were carried out at ten locations, each with control untreated flowers that were bagged and not bagged for a total of seven treatments per location. At each location, seven plants bearing inflorescences with a minimum of three flowers were chosen within a radius of 5 m and each inflorescence was randomly assigned to one of the treatments or controls: self-pollinated flower, self-pollinated plant, within-fen and between-fen crosses, Harris Lake cross, and the unbagged and bagged controls. Bags made of bridal veil were supported by a wire frame that consisted of a ring (10 cm diameter) at the top to ensure flowers were kept free of the fabric, and a single wire that ran the length of the veil (25 cm). The bag was fitted over the inflorescence and secured from wind movement by attachment to two bamboo garden poles inserted on either side of the plant into the peat. Pollinations were made by synchronizing the time of anthesis of the donor flowers to the female stage of the recipient flower. Donor anthers for the within-fen crosses and for self-pollinations were removed by tweezers to effect pollination. Whole flowers were placed in vials for transport to make the between-fen and the Harris Lake far (2 km) crosses. Dehiscing anthers were rubbed against the stigmatic surfaces of the gynoecium and flowers inspected with a hand lens to ensure adequate transfer of pollen. The pollinated flower heads were allowed to ripen in the field until October 25, 2018, when they were collected, and the contents of hypanthia separated by size and plumpness into fertilized and unfertilized achenes.

Seeds from the pollination crosses were cold stratified and tested for germinability at Dalhousie University (Johnston lab) after a preliminary trial at St. Francis Xavier failed due to fungal contamination. Due to the loss of seed numbers involved in the first test, the second germination trials pooled the seeds for each pollination treatment so that the results were analyzed using the *G*-test likelihood-ratio.

(C) Seed Stratification and Germination

A study of factors affecting seed (achenes) germination of Avens compared seed collected in fall versus winter and the influence of pre-treatment stratification. Seeds were collected from the field at two sites (Brad's Fen and High Hog Fen) 10 November 10, 2017, and January 04, 2018. The November seeds were air-dried and stored in vials at room temperature until the January collections were in hand. Both seed lots were then compared for the percent of seeds that germinated and the speed of germination. Seeds from seven individual seed heads (hypanthia) from High Hog and from Brad's Fen were separated into plump versus shrunken (spoon shaped) achenes for each collection date material. Two lots of the plump, good seeds from each seed head were sown in separate cells (15 by 10 cm surface area) of plastic trays that were filled with 8 cm of moist potting medium (Pro-mix). One lot was taken to the greenhouse for a trial of direct germination, the trays of the other lot were wrapped loosely in a plastic bag and incubated for two months at 4 °C in a fridge to test cold stratification. Germination rates following cold stratification between replicate trays were quite variable (0–87%). The statistical distribution of numbers of germinants per tray was measured and a chi square test for difference was performed. The same set up was followed for the shrunken seeds but no viable seedlings emerged with or without stratification.

(D) Seed Bank Potential

This study used peat cores from three peatlands with large population densities of Avens to determine the likelihood of reestablishment of Avens from a buried reserve of germinable seed—the seed bank—if former Avens habitats were restored. Could degraded habitats of Big Meadow Bog have a reserve of dormant seed that could start new vegetative populations? Two lagg fens, Runway Bog (BM3) and Northwest Lagg (BM1), and the Camp Road Bog (CR S) were sampled in 2013. Sites were chosen to represent a gradient of habitat conditions; Runway Bog appeared relatively pristine whereas Northwest Lagg and Camp Road Bog were both disturbed by ATVs (all-terrain vehicles) and Northwest Lagg was additionally disturbed by water table changes from the ditching of Big Meadow, a fire in the 1980s, and by Herring Gulls. All habitats contained populations of Avens (206, 1134, and 998 individuals in Runway Bog, Northwest Lagg and Camp Road, respectively; 2012 counts in Toms 2023). The broad area of open fen was sampled in the two lagg fens whereas sampling was confined to two linear open areas [at the Camp Road Bog] that matched the pattern of the former vehicle traffic and supported all of the Avens population. The two lagg fens each had populations of the rare boreal sedge, *Carex livida* (S1) and rare shrub, *Betula michauxii* (S3) (S-ranks from NatureServe 2024).

Peats cores were collected on 15–16 November 2013 from the above sites. At each site, peats were collected from 21 randomly selected points and then from 15 points within 10 cm of Avens individuals. Fifteen large, flowering Avens individuals were chosen so that points were well-separated and dispersed throughout the colonized area of the peatland. The current Avens seed (achenes) crop was ripe but had not yet been released from the receptacles at the time of collection. Peats were collected with a bulb planter (7.5 cm aperture diam.) that was inserted into the peat with the aid of a sharp knife cutting the passage along the outside of the planter to sever shrub and sedge rhizomes. Each peat core was cut into shallow (0–5 cm depth) and deep (5–10 cm) sub-cores that were bagged and set up for germination as separate samples. The shallow and deep cores were processed individually within three days of collection. Intact leaf litter was removed and adhering fragmented material was washed into the top of the sample peat. Large rhizomes were removed and discarded. Each core sample was placed on a 2.5 cm base of sand in a 50 cm^2 cell of a 10-compartment plastic tray liner placed in a plastic seed tray. One control cell per tray consisted of sand and potting soil that was monitored to identify seed contaminants. Trays were watered and then put inside folded over plastic bags and incubated on soil in a slat house exposed to ambient winter temperatures. Bagged trays were covered with a layer of tree litter and conifer boughs, overwintered, and then set up in an unheated, small greenhouse in Berwick, Nova Scotia on March 15, 2013.

Seedlings were noted as they emerged. Representatives of each seedling type were transplanted into potting soil in pots and identified as they matured. Trays were maintained by regular watering. Each peat sample was stirred and mixed on 30 June 2013, to stimulate germination and the emergence of seedlings was followed until 17 September 2013. Contaminants found in control cells included *Cerastium fontanum*, an unidentified *Salix* sp., *Betula populifolia* and *Viola* spp. Several species emerged from rhizome fragments as well as from seed (e.g., *Vaccinium oxycoccos*, *Muhlenbergia glomerata*, *Trichophorum cespitosum*, *Chamaedaphne calyculata*) but these were distinguished based on evidence of seed or seed leaves or whether there were fine rhizome attachments.

4.2.4 Population Dynamic Trends

Monitoring of Avens populations at different sites was carried out over various years between 2012 and 2022 (Toms 2023). These are field count data based on identifying separate plant crowns regardless of possible rhizome connections belowground (Toms and LaRue 2012). The data record the fluctuations in local populations and associated variables such as numbers of flower stems, the number of plants overgrown by shrubs and those affected by Herring Gulls. We have used these population surveys to show both areas of population stability and rapid change and have shown associated levels of shrub overgrowth and impact by gulls. Shrub overgrowth is a concern for the survival of the coastal plain rare flora on lake shores (Keddy 2010) and observations based on monitoring data (Toms and LaRue 2012) suggested that a similar relationship existed between shrub growth and growth of Avens.

4.2.5 Ecological Interventions:

(a) Shrub Removal Experiment: Unintended Impact on Efficiency of Photosynthesis

Shrub removal experiments were carried out in November 2012 in nine 15 × 15 cm quadrats around Avens plants in four fens (BM1, BM3b, BM5c, CRS) leaving an equal number of quadrat areas with untouched Avens as controls. In the following summer (July 2013), there was indication of an increase in Avens cover, but there was an increased frequency of reddened leaves in the shrub removal quadrats. A follow-up experiment a week later at two of the sites (BM3b and BM1) tested whether there were photosynthetic differences between the Avens leaves in shrub removal areas and controls. Dark adaptation clips (Flynn et al. 2018) were applied to the two greenest and two most discolored leaves in each quadrat. A two-factor analysis of variance tested for overall significance of differences. After three years, the percent cover of Avens was measured in all four sites.

(b) Facilitating Seedling Establishment: Scarification and Sowing

A preliminary trial of seedling establishment was made at three Avens habitats (CR S, BM1 and BM3b) in 2015. Small quadrat areas (0.0625 m^2) at a hummock and a hollow at six locations in each fen were scarified and adjacent hummock and hollow areas were left untouched as controls. Scarified areas were cleared of above ground shrub biomass and of *Sphagnum*, and loose peat was removed as needed to provide a consolidated seed bed. At each cleared or control quadrat, the seeds of two Avens inflorescences collected from the same respective sites, were scattered over the area in November 2015, and a count of the established seedlings was made the following August. A follow-up experiment was done at Brad's Bog (GH6) in November of 2016 using the same experimental design (site = hummock or hollow, treatment = scarified, control = not scarified) but wood (poplar) ash from a stove was applied to one half of each quadrat area in an attempt to mimic post-fire conditions in peatland. Scarification and sowing was carried out when Avens seeds were fully mature and sowing in November ensured seeds sowed on a moist peat surface become stratified over winter as has been documented to optimize speed and percentage of germination (Sect. 4.2.3 c). In addition, in early June 2017, two lengths of 20-gauge wire with a crimped middle section were inserted into each half of each quadrat for measurements of *Sphagnum* growth (Clymo 1970). The bottom Sect. (20 cm) of a 33 cm piece of wire was inserted into the peat until the following 3-cm horizontal section nestled on top of the capitula of the *Sphagnum* mosses after which a 10 cm vertical length of wire extended into the air so it could be relocated. The extension of *Sphagnum* capitula above the horizontal wire sections inserted in June 2017 was measured on August 20, 2017.

In 2021 through 2023, a series of large scarifications of 1 m^2 quadrats in the lagg area of Big Meadow Bog were established and each quadrat was sown with the contents of 15 seed heads. In 2022, seedlings were counted in the 1-m^2 scarification plots seeded in November 2021 and the elevations of the seedlings above standing water at the low point of the quadrat were measured.

4.3 Results and Discussion

4.3.1 The Realized Niche of the Eastern Mountain Avens

On Brier Island which supports 95% of all Avens plants in Canada, 92% of Avens plants occur in wetlands (Fig. 4.3) and most of this habitat can be classed as fen. Avens occurred at each end of a 400 m long transect across Big Meadow Bog (yellow squares on Fig. 4.4). These areas are fens at each margin of the raised bog, the "lagg" of the bog (Howie and Tromp van Meerveld 2011). The figure shows that the shrub heights in the areas of Avens occurrences are relatively low and that the water table is close to the surface (Fig. 4.4). The difference in depth to water table between Avens and non-Avens stations for values over the transect length was not significant; however, the difference in shrub height was (Table 4.1).

The relationship between water table and shrub growth was examined at the within-fen level for multiple points at BM6, the lagg fen at the east margin of Big Meadow Bog. All the low shrub cover (5–30%) points ($n = 18$) occur where water tables are within 20 cm (in June 2016) of the wetland surface (Fig. 4.5). The scatter of points reflecting the range in shrub cover, on Fig. 4.5, decreases as depth to water

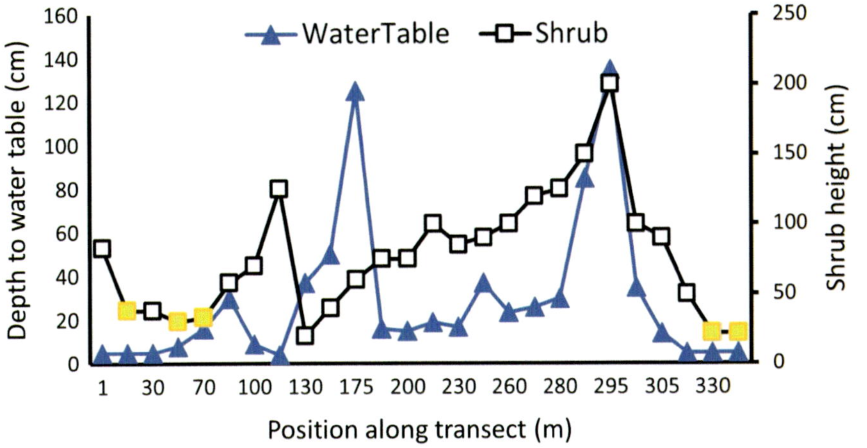

Fig. 4.4 Depth (cm) from surface to water table and height (cm) of shrubs along a transect across the north of Big Meadow Bog. Shrub height markers filled in yellow are areas that supported Avens

Table 4.1 Depth to water table and shrub height for positions along a transect (Fig. 4.2) across Big Meadow Bog with ($n = 5$) and without ($n = 20$) Avens. Values are means ± s; significance based on Student's t-test

	No Avens	Avens	*P* value
Water table (cm)	29.7 ± 29.4	7.80 ± 4.8	NS
Shrub height (cm)	82.7 ± 33.0	29.6 ± 6.4	< 0.005

table increases. The variation (coefficient of variation) in shrub cover is greatest at the wettest portions of the fen and declines three-fold as depth to water table increases (Fig. 4.6). The same assessment of water table and shrub growth was carried out at BM3b (Runway Bog), one of the wet fens used in the transplant experiment. The regression between water table and shrub cover, as was found for Beauty Bog (Fig. 4.5) showed a statistically significant positive relationship between depth to water table and shrub cover, however the equation accounted for little of the total variation ($R^2 = 0.127$, $n = 132$ points, $p < 0.001$).

At both fens, Avens microsites had lower average shrub cover and shrub height although the difference was not significant at Runway Bog for shrub cover (Table 4.2). The water table relationship to the presence of Avens was significant (albeit at $p < 0.01$ at Beauty Bog) but it is noteworthy and instructive that while Avens occurred on average at wetter sites at Beauty Bog, the drier of the two fens (see water table differences in Table 4.2), it was associated with drier microsites in the wetter fen, Runway Bog. Taken together, Tables 4.1 and 4.2 suggest that avoidance of shrubs is the primary determinant of Avens occurrences in the lagg fens of Big Meadow Bog. In addition, the shrub cover associated with Avens and with each of two rare species studied at Beauty Bog, was significantly lower than areas of that fen where no rare plants were found (Fig. 4.7).

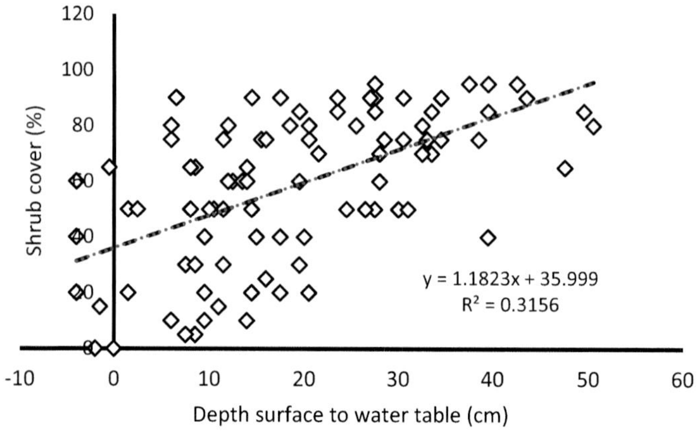

Fig. 4.5 The relationship between percent shrub cover and depth to water table (June 2016) at the eastern lagg fen (BM6, Beauty Bog) of Big Meadow Bog for multiple transect positions. Linear regression significant at $p < 0.001$, $n = 126$ points

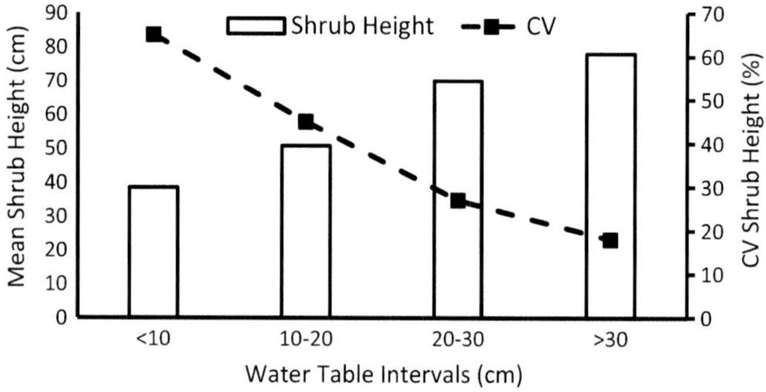

Fig. 4.6 Mean shrub height over four water table depth intervals from the eastern lagg fen (Beauty Fen) of Big Meadow Bog and the corresponding variation (CV, coefficient of variation: % s of Mean). Data rearranged from Fig. 4.5

Table 4.2 Depth to water table, shrub height, and percent cover for multiple positions assessed in Junew 2016 along transects across two lagg fens of Big Meadow Bog: A. BM6, Beauty Bog and B. BM3b, Runway Bog. Values are means ± s; significance based on Student's *t*-test

BM6 Beauty bog	No Avens	Avens	*P* value
Water table (cm)	20.8 ± 12.8	14.2 ± 10.2	< 0.1
Shrub height (cm)	50.0 ± 21.2	36.2 ± 6.4	< 0.05
Shrub cover (%)	62.7 ± 25.8	35.0 ± 22.4	< 0.001
BM3b Runway bog			
Water table (cm)	11.8 ± 11.1	17.4 ± 10.8	< 0.05
Shrub height (cm)	31.7 ± 19.6	20.4 ± 16.0	< 0.05
Shrub cover (%)	44.3 ± 25.7	29.9 ± 20.9	ns

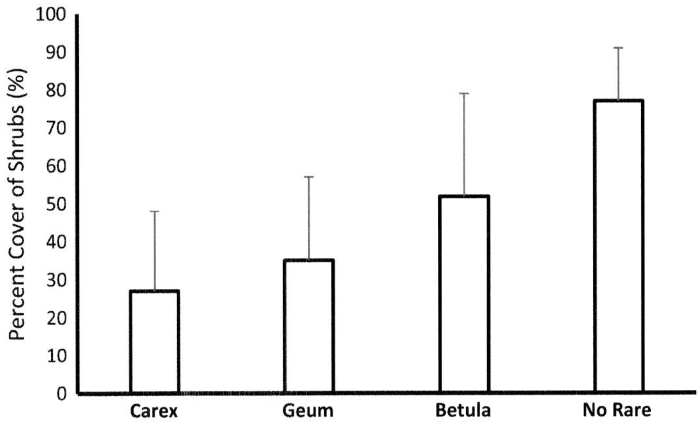

Fig. 4.7 Shrub cover (mean ± s) in Beauty Bog associated with points ($n = 124$) on transects with three rare plants (Avens, *Carex livida,* and *Betula michauxii*; $n = 13, 14, 65$ points, respectively) or no rare plants ($n = 26$). Each Student's t-test between rare species data and the no rare species data significant at $p < 0.001$

4.3.2 Transplantation Experiment: Insight into Hydrology and Nutrient Availability

The correlation in the field between the occurrence of Avens and low biomass shrub communities has been noted for rare Atlantic Coastal Plain plants on lakeshores in Nova Scotia (Keddy and Wisheu 1989) and this has become a conceptual model: rare plants occupy stressed and disturbed, post-competition niches (see Fig. 4.4 in Wisheu and Keddy 1992). Given the potential risks to the population of Avens in Big Meadow Bog from the proposed blocking of the drainage ditches (Chap. 10), a transplantation experiment was conducted in 2015 to test the response of Avens to water table differences and habitat nutrient regimes.

Many variables were measured over the course of the three-year experiment including Avens metrics, water levels, community associate data, and a test of photosynthetic health using a non-destructive technique, the Pulse Amplified Modulated fluorescence technique (Poirier 2016). For Avens health, however, leaf width proved to be the most reliable indication of the growth of the plant (Fig. 4.8) and ultimately, survivorship (Table 4.3, Fig. 4.9).

Avens survival and growth (leaf width) was greatest in healthy fens and was in each case, about 60% lower in gull-enriched, eutrophic fen and in the ombrotrophic bog positions of the Reference Bog (Table 4.3). We have categorized all the lagg positions of Big Meadow Bog as fen as well as flat peatlands which receive inflow from adjacent upland. Fens are peatlands that receive groundwater flow and are geogenous while bogs are ombrogenous since precipitation and atmospheric deposition are their sole sources of water and nutrient (Vitt 2006). On Brier Island, drainage from uplands typically moves through swamps and transports dissolved minerals from basalt into

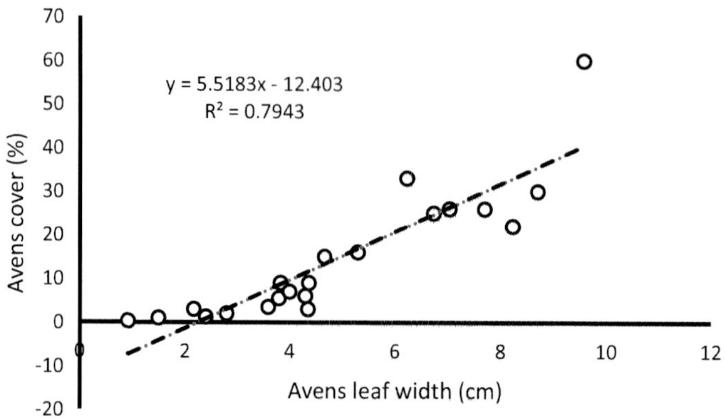

Fig. 4.8 The relationship between maximum leaf width and cover of Avens in 25 × 25 cm quadrats after two years of transplant growth in 2017 for 21 extant stations in Big Meadow Bog, Brier Island (p value for regression < 0.001)

Table 4.3 Performance after three years of transplants into peatlands varying in nutrient content

	Leaf width* (cm ± s)	Survival (%)
Ombrotrophic bog	1.91 ± 1.1	50
Healthy fen	5.01 ± 1.2	87
Eutrophic fen	1.9 ± 1.3	33

*Leaf widths of Avens at healthy fens (2 Big Meadow sites and 3 isolated fens = 15 plants) were significantly larger (Student's t-test < 0.05) than those at the ombrotrophic bog (2 sites = 3 plants) and the eutrophic fen (one site, 3 plants). Note that these values are calculated only for the plants that survived three years ($n = 21$ versus $n = 36$ of the original transplants)

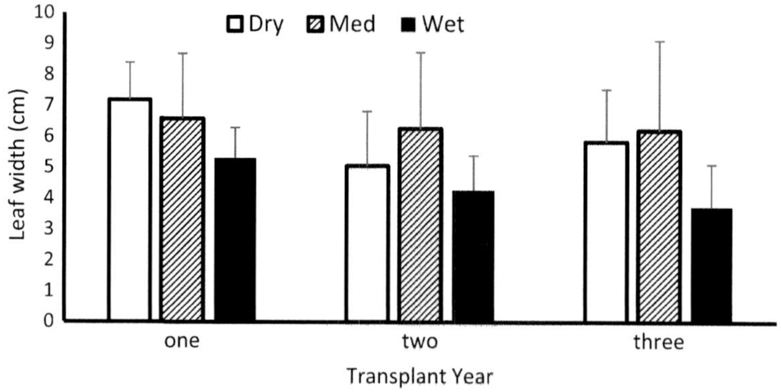

Fig. 4.9 Average leaf width (± s) of Avens transplants (3 transplants at 2 sites per moisture treatment ($n = 6$), except in year 3 where $n = 3$ and 4 for the mesic and dry sites

the fens that support Avens. In a few cases, such as at Western Light Bog (WL), the water table is relatively deep and the peat surface is more ombrotrophic than the majority of the Avens peatlands that can be described as geogenous in water chemistry (Chap. 8). Another Avens habitat, Brad's Bog (GH6), has supported a healthy and stable Avens population for over a decade since its discovery by Brad Toms who has monitored the number of Avens there since 2012 (Toms 2023); this flat peatland appears to be bog-like in summer but in winter, there is a flow through and over the peatland from south to north. Finally, Avens does not occur in the only extant ombrotrophic, raised bog on Brier Island, the Reference Bog, and its poor performance in this ecosystem relative to the healthy fens (Table 4.3) informs the conservation planning for the Eastern Mountain Avens.

Big Meadow Bog was a raised bog throughout most of its central north through central-mid zone; there was abundant "Jonesberry" (cloudberry, *Rubus chamaemorus*) as now occurs only in the Reference Bog, and in the remnant patch of heath in the north of Big Meadow Bog (Chap. 2; Hill et al. 2019). Avens is a fen species in Canada and its associates in the lagg of the Meadow (e.g., *Dasiphora fruticosa* and *Carex livida*) suggest that these peatland communities are enriched in cations as has been discussed for fens in Newfoundland (Wells and Pollett 1983). Chemical analysis of the surface waters of the Reference Bog and two Avens fens (BM1 and Runway Bog) in wells highlight the greater acidity of the former and that the greater concentrations of multivalent cations in the latter (e.g., Ca, Mg, Mn, Fe; Table 4.4).

In the healthy fens, Avens leaf width remained constant for the plants moved into the two mesic stations (Fig. 4.9) but declined linearly over the 3 years for the wet sites ($r^2 = 0.97$, $p < 0.001$, leaf width versus years after transplantation). The apparent health of plants in the drier sites in Fig. 4.9 is in part an artifact of data analysis since the two plants with smallest leaves in year two died and could not be included in the year three average. For the dry site Avens, the leaf width average of the three-year surviving plants measured in year two was 40% greater than that

Table 4.4 Average values for chemical variables in water of wells at the reference bog ($n = 2$) and at two Avens fens (BM1 and BM3; 2 wells each, $n = 4$)

Variable	Units	Reference bog	Avens fens
pH	Units	3.98	5.62
Conductivity	μmho/cm	82.5	114.3
Nitrate and nitrite as *N*	mg/L	0.28	0.19
Ammonia as *N*	mg/L	0.04	0.17
Ortho-phosphate as *P*	mg/L	0.06	0.09
Total calcium	mg/L	2.70	5.93
Total magnesium	mg/L	1.80	3.58
Total phosphorus	mg/L	0.35	0.39
Total iron	mg/L	851	10,335
Total manganese	mg/L	34.0	104.3

of the two that died in the winter between years two and three (5.9 ± 1.6 versus 3.4 cm, respectively). Other key deaths included all plants of one mesic site of Big Meadow Bog that were flooded when the drainage to the ocean of a barrachois pond at Pond Cove was blocked after a storm of winter 2016–2017 and the two extra, unused transplants that were placed in a fen pool of BM3b and died after a month in the summer of 2015. This sensitivity to rapid changes in water levels would later be observed in the lowest lying lagg fen (BM6) after the hydrological restoration.

Transplanting Avens into the two wet fens also showed that plant growth (leaf width) was reduced in comparison with that in the less waterlogged stressed fens with lower water tables (Fig. 4.9). In this case, all plants transplanted into the healthy wet fens survived unlike the 33% mortality found with transplants into the dry fens. While the transplants had a year with limited community competition, shrubs were allowed to encroach in the second and third years. The percent of the transplant area that had shrub encroachment was much smaller for the wet fen transplants than for the mesic and dry fen transplants (7.07 ± 4.83 versus 25.8 ± 14.7 percent area encroached respectively, $n = 6$ and 12; $p < 0.01$, Student's t-test). The death of the dry fen transplants (one of three per site) also suggests greater competitive intensity in better-drained areas.

Avens transplanted back into their area of origin, where the Herring Gull nesting population had increased from 2015 onwards, began to die after the first year (2 of 9 transplants) and in the third year only 3 of 9 remained. The second-year average leaf width of the three-year survivors was almost 50% greater than that of the group that died (7.3 versus 4.9 cm, Fig. 4.10). While this difference is not significant, when we calculated this leaf width difference for plants of the two driest fens and the two Reference Bog stations that had second winter deaths, the average leaf width of plants that would survive was 79% greater (5.7 ± 1.6 versus 3.2 ± 0.6 cm; n = 7 and 4, respectively) than those that would later die overwinter ($p < 0.05$, Student's t-test).

From the above data, we suggest that plants that make very small leaves are either juveniles or have run out of resources and are about to die. In the eutrophic site at

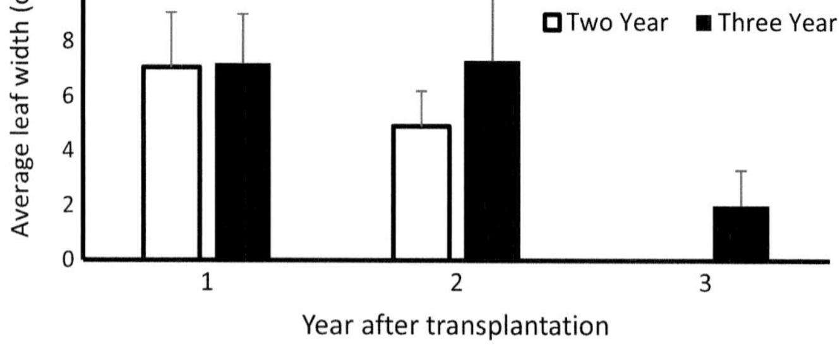

Fig. 4.10 Means ± s of leaf sizes of three-year survivor transplants versus two-year survivors in a fen made eutrophic by Herring Gulls (BM1 source area of all transplants)

BM1, the structure of the vegetation was very different from that of the healthy fens or the Reference Bog. In addition, most of the plant cover in the enriched site was in the high canopy fraction (Fig. 4.11). Herring Gull wastes have a major impact on nutrient cycling in these peatlands (Chap. 13). Peat samples from BM1, the northwest lagg of Big Meadow, taken two years before the transplantation experiment, revealed that phosphorus concentrations in the surface peat at an early stage of gull occupancy (heath vegetation supporting Avens intact with occasional nests) were already 50% greater than those in the lagg of the Reference Bog (979 ± 480 s versus 651, $n = 2$; units are ppm total P, see Fig. 4.12). The greatest phosphorus concentration (1400 ppm, Fig. 4.12) occurred where the impact of gulls was greatest and where the vegetation was dominated by the exotic grass, *Holcus lanatus*, and annuals (*Epilobium ciliatum, Bidens frondosa,* and *Impatiens capensis*). Despite the 44% increase in P from the low impact to high gull impact zone, this was not statistically different although a 17% decline in total nitrogen (25,467 ± 1078 s versus 21,067 ± 2255.6 s ppm total nitrogen) between the low gull impact and high impact zone was significant ($p < 0.05$, t-test). While there may be an interplay between P and N as limiting nutrients as the concentration of one (i.e., P) increases, the elaboration of such an interplay will require further study.

The transplantation experiment gave insight into the mechanism behind the loss of rare plants to eutrophication. As phosphorus levels increases, the original matrix vegetation of small ericaceous shrubs and fine tussock sedges is replaced by faster-growing herbaceous plants that allocate most of their biomass aboveground. Avens is a plant of low stature with a rhizome base that is impregnated with tannins to protect the store of carbohydrates it slowly accumulates. Our observations above from field measurements show a consistent association with low shrub zones in wet or mesic fens. It appears that it forages for light openings in patchy shrub habitats. At high nutrient levels, however, there is a wholesale shift in the surrounding plant community as resources are allocated to shoots that support a competitive canopy (Fig. 4.11, fen versus eutrophic fen). The low-growing Avens then becomes competitively excluded and shaded out as the plant community changes in response to nutrient enrichment. There may be an additional mechanism at work that may account for the speed of loss of the Avens transplants (only 3 of 9 Avens survived after three years after

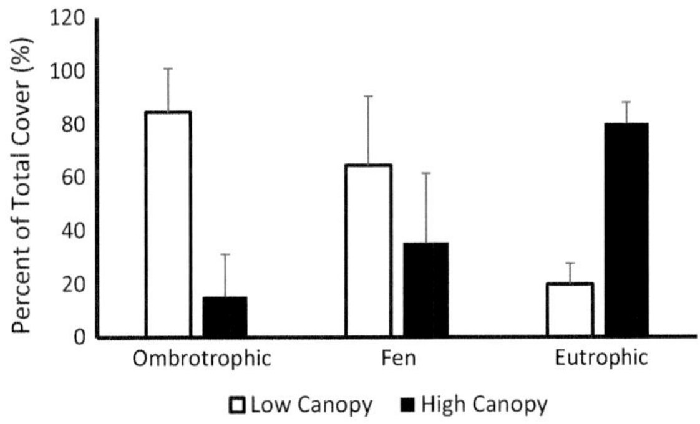

Fig. 4.11 Differences in the structure of the above-ground plant community between the ombrotrophic raised bog (Reference Bog, $n = 2$), the fens ($n = 5$ sites) and the nutrient-enriched eutrophic fen ($n = 3$ sites). Averages ± s are percentages of total plant cover in the low (< 25 cm) and high (> 25 cm) canopy fractions (3 replicates per site)

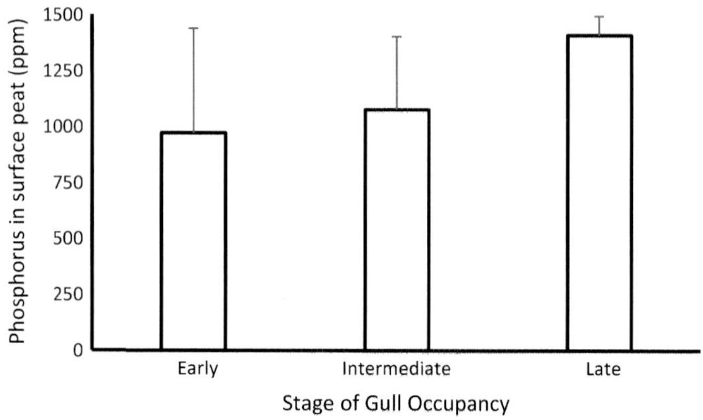

Fig. 4.12 Phosphorus concentrations at three areas of the Western Lagg (BM1) that differed in the impact of gull nesting on the vegetation. The early occupancy zone was dominated by low heath and sedges and the late stage zone by the exotic grass, *Holcus lanatus*, and annual herbs (e.g., *Epilobium ciliatum, Bidens frondosa,* and *Impatiens capensis*). Phosphorus levels (mean ± s of 3 replicates) are ppm per kilogram in the surface (0–25 cm) peat

transplantation into BM1 nesting area). Mycorrhizae and the microflora of Avens in healthy peatlands (Chap. 15) would likely be lost with eutrophication of the habitat, a generalization supported by a recent meta-analysis (Ma et al 2021). In addition, the Ericaceae, the dominant shrubs in Avens fens, depend upon ericoid mycorrhizae that allow them to obtain nitrogen in low nutrient peatlands (Read 1996). We do not yet understand the microbial interaction between Avens and its mycorrhizal associates, but changes in the microflora may have contributed to the loss of the Avens transplants.

4.3.3 Avens Regeneration Biology

4.3.3.1 Vegetative Growth

Avens plants grow in fens with *Sphagnum* moss that can bury their leaves, and with shrubs that can overtop them and shade the whole plant. Avens shoots stay at the surface of the *Sphagnum* moss and as they grow, they leave behind a storage rhizome. Avens leaves flush out after bud break of the shoot in spring and before bud break of the associated deciduous shrubs (e.g., *Myrica gale, Gaylussacia* spp., *Betula michauxii, Lonicera villosa* and *Dasiphora fruticosa*). Leaves are produced from the current year's shoot growth. The rhizome/shoot system is monopodial, a main stem is formed by the shoot apical meristem, and it bears leaves and continues the growth of the stem. In these growth systems, a reserve of axillary buds remains dormant to form branch shoots from the main axis or inflorescences. It is important for arctic-alpine plants in climates where the growing season is short to develop buds as soon in spring as possible and these plants are noted for the initiation of organs a year or more prior to the current season (Mauracher and Wagner 2021).

Avens survives through conservative growth via the shoot/rhizome system. Branching of rhizomes is infrequent but needs study, and flowering ranges from once a decade to every 4 years depending on resources (see Sect. 4.4.4 below). The shoot must be responsive to spatial changes in light availability as is suggested by three separate field studies showing that Avens occurs in areas of reduced shrub cover or height (Tables 4.1 and 4.2). The Avens shoot also may keep vertical track of the growth of *Sphagnum* moss and the water level. A non-destructive investigation of the rhizomes in hummocks and hollows in June of 2017 revealed that in a mesic fen, there was 29 and 15 cm of peat depth potentially available for the Avens shoot and root system in hummocks and hollows, respectively (maximum depths of deepest rhizome), whereas in a wet fen (Coastal Fen), the thickness of these respective layers was 15 and 7 cm thick. The difference between these maximum values for rhizome depths in hummocks and hollows may reflect need for the rhizome tissue of *Geum peckii* to be maintained in an aerobic peat zone in summer. Seasonal vertical growth of *Sphagnum sp.* moss capitula, determined using Clymo's bent wire technique (1970), in the wet and the mesic fen was 1.18 and 1.94 cm ($\pm$s of 0.77 and 1.06 cm; $n = 28$ and 22 at Coastal Fen, GR3 and Brad's Bog, GH6, respectively) 1.2 and 2.0 cm ($\pm$ SD, respectively). At the within-fen scale, *Sphagnum* extension rates were greater for hollows, the wetter areas, than for hummocks. *Sphagnum* and shrubs are the competitive arena in which Avens survives. Over time, the survival of an Avens shoot/rhizome system will depend on its ability to spatially find higher light patches as well as, perhaps, to keep up with the vertical growth of *Sphagnum sp.* The rhizome investigation showed that on average rhizome lengths were 17.9 ± 5.2 and 11.4 ± 4.27 cm in hummock and hollows, respectively ($n = 6$ in each case), in the mesic fen and correspondingly, 16.5 ± 0.92 and 12.2 ± 0.57 cm in the wet fen ($n = 10$ in each case); lengths for hummocks versus hollows were significantly different at each fen ($p < 0.01$ and 0.05, t-tests). In each fen, the rhizome angle, the angle (upward growth relative to fen surface) at which the shoot tip grows to maintain position, was greater for the hummock rhizomes (56° and 43° for mesic and wet fen, respectively) than the hollow rhizomes (38° and 29° for the same fens, respectively).

The rhizome tissue is orange and its bitter taste suggests that secondary compounds protect the rhizome tissues from herbivory. Ellagic acid tannins have been documented as being organ specific in *Geum rivale,* occurring only in the rhizomes (Owczarek et al. 2017). These secondary compounds may be elaborated to protect this source of carbohydrate in a fleshy tissue from herbivory. Deer browsing on inflorescences is a frequent observation but we (authors NH and BT) have not noted herbivory (apart from insect damage) on the leaves or the rhizome/shoot system.

4.3.3.2 Flowering: Pollination Implications

Inflorescences are formed from axillary buds on shoots. Most flowering occurs from late-June to mid-July (Zinck 1996). The inflorescence is usually a branched, leafy scape, from 10 to 40 cm tall with one to five flowers (COSEWIC 2010). Flowers typically have 5 petals, are bisexual and multiple stamens surround many (40–60 usually) single-seeded carpels in the center of the flower. Flowers are protogynous, the stigmas of the carpels in the center of the flower are receptive before the anthers that surround them have shed pollen (Fig. 4.13). A pollination study examined the impact of various treatments on seed set and germination. These results are discussed here and are integrated into an analysis of lifetime inbreeding depression in Chapter 3.

There was no significant difference in the proportion of achenes that were fertilized between the natural state—unbagged and open to pollinating insects—and the pollination treatments. There was a significant difference between both self-pollination treatments and the bagged flowers that were not manipulated. The significant result indicates that while self-pollination can result in fertilized achenes, without mechanical disturbance of the anthers, the likelihood of pollination and fertilization is low (8% fertilization for Bagged only treatment versus 22% and 39% for the Selfed Flower and Selfed Plant treatments, respectively).

There was no significant difference for seed set between the manipulated pollination treatments when we omitted the Selfed Flower treatment since its lower seed set may have resulted from dichogamy, the temporal separation between the development of the stigma and the anthers in the protogynous Avens. The seed set value for flowers that are self-pollinated with flowers from the same plant (geitonogamy) but in a different developmental stage is similar to that for natural, unbagged

Fig. 4.13 Flower stem of Avens showing open flower in female phase (central gynoecia) with the outer whorl stamens with anthers not dehisced. Photo credit: Nicholas Hill

plants (39% versus 42% respectively, Fig. 4.14). The similarity in seed set between Selfed and Natural is expected if a large proportion of the pollination of achenes is due to selfing (Chap. 3). When the seeds resulting from different pollen sources, (i.e., self, within-bog, between-bogs, between islands, no bag unmanipulated) were tested for germinability, the benefit of cross-pollination on performance was striking (Table 4.5): there is an almost linear increase in germination with distance in Avens.

The substantial inbreeding depression (Chap. 3) revealed for this Canadian population of the Eastern Mountain Avens at Big Meadow Bog underlines the importance of maintaining gene flow. The ditching caused trees to grow up throughout the bog's margin—the lagg—which cordoned off local populations. Prior to ditching, Big Meadow was the sole site where a long linear expanse of open fen with patches of Avens was available to pollinating insects. As this linear lagg became patches of fen separated by tall woody growth (Toms 2015), there could have been a reduction in the insect pollination of flowers and so a decrease in seedling fitness. While we have

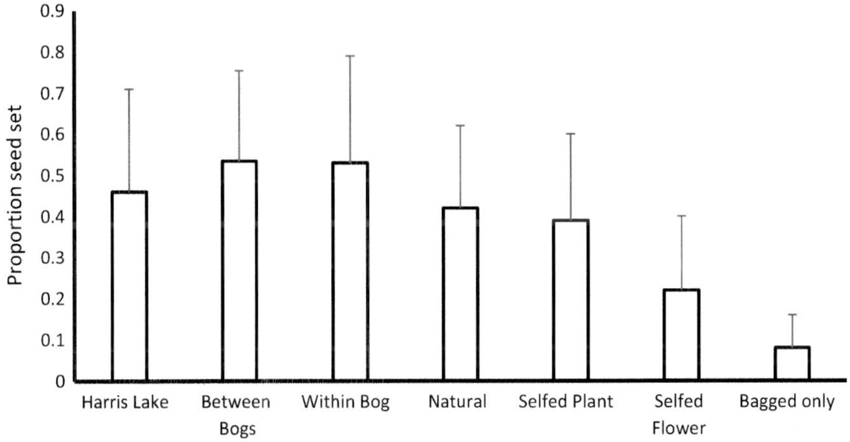

Fig. 4.14 Seed set per total achenes for seven pollination treatments (mean ± s, $n = 8$–15). Natural (no bag) and bagged only were not hand pollinated

Table 4.5 Analysis of proportional germination for seeds set from five pollination sources

Treatment (distance of pollen donor)	Proportion not germinated or dead	Proportion germinated and alive	SE of proportion
Selfed (0 m)	0.79	0.21	0.029
Natural (unknown)	0.78	0.22	0.060
Within bog (10 m)	0.66	0.34	0.039
Between bogs (200 m)	0.50	0.50	0.035
Harris Lake (2 km)	0.46	0.54	0.044

Note that the seeds from within flower and within plant selfing treatments have been combined and tested as one group. *G*-test heterogeneity = 56, degrees freedom = 4, $p = 1.37 \times 10^{-11v}$

shown the ability to grow plants using field-collected seeds, the "natural" seed source of Table 4.5, ex situ conservation programs must keep in mind that there is a genetic load of deleterious genes in Avens and the fittest progeny will be those produced by outcrossing. The Village of Westport may also help the fitness of Avens individuals and the population by programs aimed at encouraging native pollinators. There has been no systematic study of the pollinators of Avens in Nova Scotia (COSEWIC 2010). We observe mostly syrphids and small dipterans but have seen a variety of wasps, *Bombus* species and ants. Crab spiders in Avens flowers are common but their impact on outcrossing is not known.

4.3.3.3 Seed Dispersal

Seeds develop after ovules inside carpels are fertilized in summer. The carpels remain green until the end of September and ripen in October. Dispersal of individual seeds as well as clusters of seed held together by the mat of hairs over the carpel and style, occurs in November and over winter, while some of the seed-bearing inflorescences collapse under snow. A minority of inflorescence stems remain standing into the following summer and the papery hypanthium may contain a few seeds; however, study has shown most of the seeds that remained on standing stalks over a year were not viable. Deer eat some of the inflorescences in summer and into fall (observations by authors, NH and BT). Deer could theoretically disperse seeds of Avens over long distances after they ingest seed heads. Deer have long gut retention times but the overwhelming majority of the seeds shown to be viable in white-tailed deer feces that was collected over a year, were small seeded weedy species (Myers et al. 2004). Deer may also transport seeds in their hooves and in their fur (Petersen and Bruun 2019). The 1980s, according to residents of Brier Island, was the peak time for all-terrain vehicle (ATV) use all over the island. ATVs were implicated in moving seeds of the exotic *Rosa rugosa* along the coast on Brier Island based on the pattern of colonies, colony age and trails, (Garbary et al. 2013). In similar fashion, we noted in 2013 the coincidence of several linear arrays of Avens with former ATV trails through fens at sites at Camp Road, Green Head, and particularly, along Gull Rock Road (Toms 2023).

Many weedy and common species of *Geum* have achenes with hooked styles that project from the receptacle. These are adapted to ectozoochory—the hook of the burr gets entangled in animal fur—and a comparative study showed that *Geum allepicum,* a common woodland herb, was best adapted to sticking to mouse fur (Kulbaba et al. 2009). Other *Geum* species, like the grassland Prairie Smoke (*Geum triflorum*) have plumose hairs along the length of the style so adapting them to wind dispersal (Voss and Reznicek 2012). In contrast, *Geum peckii* in common with other congeners adapted to montane habitats (e.g., *G. rossi*) has no evident structural features that adapt the seed units (achenes) to either wind or animal dispersal. Limited wind dispersal would occur when flower stems with hypanthia and seeds remaining project above a snow or ice layer overwinter. Greene and Johnson (1997) documented that birch seeds were blown 40 m across a snow surface and suggested that this mechanism could be more effective for herbs. This dispersal syndrome for low-growing plants was dubbed "chamaechory" (Vittoz and Engler 2007). The advantage that chamaechory would provide would partly depend upon an intact, open fen landscape and be reduced as the open lagg around Big Meadow Bog was colonized by tall shrubs and trees (Toms 2015).

4.3.3.4 Seed Germination

Laboratory experiments have shown that percent germination is greater for seeds that have been cold stratified then incubated at warm temperatures in comparison with those immediately incubated at warm temperatures to test for direct germination (mean, respective, percent germinatio $n = 31.2 \pm 13.9$ versus 7.5 ± 3.7; $p < 0.01$, Student's t-test of 4 lots of seeds; Manning 2018). This test, based on the mean of four average germination rates masks the large variation among seven replicates of seed lots of 15 seeds each, collected at two sites in late fall (November 10, 2017) and mid-winter (January 4, 2018), a variation that takes in absence of germination due to experimental conditions. The percent germination for stratified seed lots collected in fall and winter based on the trays from each of the two sites with the two highest germination rates, is 66.6 ± 12.2 and 66.5 ± 17.2 for fall and winter, respectively. While direct germination in absence of a cold stratification period was low, the period of germination was almost ten-fold longer than the compressed period of germination of the stratified seeds (61 versus 7 days respectively). LaRue (2016a) observed lower germination (24%) and in some collections, no germination. Some of the variation between studies could result from differences in collection times, and while Manning (2018) found no significant difference between the November and January collection, the seeds in the September collections in LaRue (2016a) may not have fully ripened.

Seedling occurrence in the field is infrequent and unless observed in the dicot stage, difficult to document as some small individuals are ramets with connection to established plants. Seedlings are observed in hollows particularly along deer or human paths where mechanical disturbance reduces *Sphagnum* and shrub growth. LaRue (2016b) noted that 30 seeds had germinated in late spring from a single hypanthium of a fallen inflorescence. Seedlings were tiny and this gives insight into the intraspecific competition at the establishing phase and the potential for winnowing out of genotypes with slower seed germination or seedling growth rates. The genetic fingerprinting results (Chap. 3) confirmed that at the patch level in three different habitats, what might have been interpreted as an Avens pattern derived from clonal growth and fragmentation was in each case a combination of that and recruitment via sexual reproduction. These strategies, asexual and sexual reproduction, can be thought of as working together to maintain the evolutionary unit. The potential for genetic recombination and population mixing is maintained during clonal growth until the conditions—climatic, competitive dynamics, natural disturbance regime frequencies (e.g., fire, disease, herbivory)—permit an allocation of plant energy to sexual reproduction and the appropriate regeneration niche, the "window of opportunity" (Eriksson and Froberg 1996; VanderKloet and Hill 2000).

4.3.3.5 Seed Bank

Buried reserves of dormant seeds—soil seed banks—are often reservoirs of diversity within a plant community and an insurance in case of the extirpation of the vegetation. When the ecosystem is disturbed, plants dormant in this seed bank can emerge and diversity is maintained and the disturbance is repaired as the germinating seeds form the new community (Keddy and Reznicek 1986; Vandvik et al. 2016). Our random soil sampling of three Avens fens—BM3b, BM1, and CRS—revealed a seed bank largely composed of herbaceous plants with little representation of woody species (11 species of graminoids and tiny forbs, 3 species of shrubs, Table 4.6). There was no detectable seed reserve of Avens from soils collected by random sampling (Table 4.6) although there was a small Avens seed bank at one fen in the shallow and not the deep layer (BM1 = 15 seeds. m^{-2}) when soils were sampled immediately adjacent to Avens plants (data not shown). The data show that the seed bank contains a set of graminoid seeds that can rapidly stabilize the peatland when its vegetation is disturbed. This was most evident along all the ditches of Big Meadow after the restoration by ditch blocking. The former ditches became dark green with the biomass of rushes (*Juncus* spp.), species also found to persist long after gull disturbance in the central bog area of Big Meadow (Hill et al. 2019). In addition, in the Avens fens, the seed bank suggests there would be a reestablishment of "deergrass" (*Trichophorum cespitosum*), the dominant graminoid co-occuring with shrubs in healthy Avens fens. The poor representation of shrub diversity in the Avens fen seed bank was also observed in seed bank study of the bog portion of Big Meadow and of the Reference Bog (Hill et al. 2019).

The absence of an Avens seed bank in the random point soil collection, and its extremely low density in the Avens focused collection, underlines that the seeds in nature are short-lived. Avens germinates readily after cold stratification of seeds. Seed bank assessment without Avens representation means that its seeds germinated in the spring and there was no dormant fraction (Baskin and Baskin 2014). If Avens is extirpated from the vegetation, the probability that its achenes, equipped neither for wind nor for animal dispersal, would be able to colonize unoccupied sites on Brier through seed dispersal is low, and very low for dispersal from the island to the rest of the Digby Peninsula. Taken together with its near absence in the seed bank, these results support the reintroduction of Avens via stratification plots and sowings.

4.3.4 Population Dynamics

In 1958, Big Meadow Bog was ditched which lowered the water tables adjacent to the ditches although some lagg features such as the "Lily Ponds" took years to drain (Chap. 10). Orthorectified aerial images from before and after the ditching revealed no change in the width of the bog in the decade before ditching (1945–1955 widths = 507 versus 508 m, respectively, $n = 12$ transects; Toms 2015). Between 1955 and 2015, however, the width of open peatland declined by more than 90 m (Fig. 4.15,

Table 4.6 Range in mean seed bank density (germinants/m^{-2}) for shallow and deep soils for three fens supporting *Geum peckii* populations. Ranking based on maximum average densities from random sampling at 15 stations per fen

	Shallow (0–7.5 cm)		Deep (7.5–15 cm)	
	rank	Density (seeds m^{-2})	rank	Density (seeds m^{-2})
Graminoids				
Juncus brevicaudatus		0–32	9	90–108
Juncus canadensis	4	86–323	2	86–376
Juncus effusus		0	4	0–290
Juncus tenuis		0	1	0–538
Muhlenbergia uniflora	5	52–247	3	11–323
Trichophorum cespitosum	2	182–559	1	64–538
Forbs				
Drosera intermedia	7	0–161	6	0–172
Drosera rotundifolia	1	32–764	7	11–161
Epilobium ciliatum	10	0–65		0
Hypericum canadense		0–43	8	0–129
Viola sp.	6	75–226	8	11–129
Shrubs				
Rhododendron groenlandicum	8	0–119		0–11
Rubus hispidus	3	108–355	5	11–280
Vaccinium oxycoccos	9	11–108		0–21
Rare species				
Betula michauxii		0–15		0–15
Geum peckii		0		0

data from Toms 2015). The widths were measured from the start of open lagg on one side of the bog to the end of open lagg on the other side. An average loss of 90 m in width, therefore, suggests that 45 m of Avens fen habitat was lost on either side of the Meadow over almost 60 years (1958–2015). The lowered water table enabled trees to colonize the lagg as well as the center of the bog. The increased area of tall woody cover in the lagg zone displaced Avens and other small stature plants. It also fragmented Avens lagg areas because they became surrounded by woody vegetation. The wet fen of Runway Bog (BM3b) had been part of the lagg, open fen landscape surrounding Big Meadow Bog, but after the ditching it became cordoned off by a thick barrier of coniferous trees. Avens seeds (achenes) start to become loosened from the hypanthium (the flower cup that holds in the achenes because of the persistent sepals) in early winter. Throughout winter, seeds are blown out of the cup in small groups that are entangled by the hairs on the achenes. These hypanthia are elevated 25–60 cm above the fen surface and may be wind dispersal of the seeds over snow in open fens but not through a forest barrier.

Accurate loss estimates in Avens populations in the lagg of Big Meadow Bog stemming from the loss of 45 m of lagg habitat is not possible. A calculation can be made that if a bog 1.5 km long and 500 m wide lost a margin habitat 45 m in width, this would represent a loss of 14 ha of fen that supported Avens. COSEWIC reports provided important estimates of population yet these were not intended to be used as benchmarks for population monitoring. The first detailed monitoring of the numbers of Avens ramets for Big Meadow Bog was performed by Toms and Larue in 2012 and these surveys are brought together in Toms (2023). From the above estimate of loss of open habitat, likely was caused by a lowering of water table because of the ditching in 1958, there was added the threat of Herring Gulls (COSEWIC 1986).

Avens count data for populations that have declined and those that have been stable or have increased since the beginning of systematic monitoring (2012) are shown in Table 4.7 (data from Toms, 2023). The first two are: BM1, the northwest lagg of Big Meadow Bog, and BM6, Beauty Bog, in the northeast of the Bog. BM1 was the source of the transplant material and a permit was provided the study in 2015 because of the imminent threat of Avens extirpation due to Herring Gull nesting. In the late 1980s, a large population of Avens became established in this sloping lagg after a fire (COSEWIC 1986). In 2012, gulls were nesting, and Table 4.7 shows the early (2012–2015) and late (2017–2021) Herring Gull impact (data from Toms, 2023). Notice that shrub overgrowth is at a maximum (94% overgrowth of Avens) in the late phase as also are the losses of Avens individuals (– 452 and – 733 for early and late phases). The second loss event occurred at Beauty Bog (Fig. 4.14). This was a healthy fen with low shrub overgrowth (Fig. 4.16), but the rapid restoration of

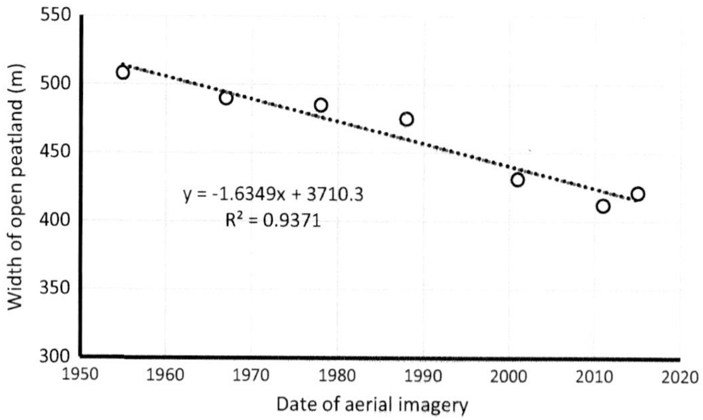

Fig. 4.15 Width of open peatland over 60 years measured on orthorectified aerial imagery. Each data point is the average for 12 transect measurements across Big Meadow Bog (data from Toms 2015)

the whole bog hydrology flooded this low-lying lagg area and the northeast of Big Meadow in particular (Chap. 7). Flooding kills most woody plants but opportunistic obligate wetland plants such as *Myrica gale* are fully adapted to flooding (Skene et al. 2001). *Myrica gale* was a minor part of the vegetation at Beauty Bog in 2015. In 2023, it dominated this low-lying lagg and increased in cover from 5.6% to 85% and in height from 43 to 109 cm. Table 4.7 shows a shrub overgrowth of 94% in the last phase. The community dominance of *M. gale* in flooded situations was reported by Schwintzer (1983) in Massachusetts who found it accounted for 66% of the community biomass and had the highest tissue nitrogen values with N_2-fixation contributing 43% of its annual budget. For contrast, we present a Big Meadow lagg, Runway Bog (BM3b) and an isolated fen Brad's Bog (GH6) where populations have been stable and maximum overgrowth by shrubs ranges from 15 to 24% (Table 4.7).

Over the decade of study at Big Meadow Bog, 2012 to present, we have observed the near extirpation of Avens from the east and the west lagg positions surveyed in the 2012 transect (Fig. 4.4) due to very different processes. The Avens population at the northwest lagg, BM1, was lost because of eutrophication, a major concern for wetland biodiversity (Keddy 2010). On the eastern end of the same transect, Beauty Bog which was a low biomass fen (Figs. 4.4, 4.5 and 4.6) with Avens visible beside pools (Fig. 4.16), was in a low-lying position in Big Meadow and was flooded in the hydrological restoration of the whole wetland complex (Chap. 10). The obligate wetland shrub, sweet gale (*Myrica gale*) has taken advantage of the rapid increase in water table resulting from the ditch blocking and increased from 5 to 85% cover,

Table 4.7 Avens population dynamics: extremes of instability (A and B) and stability (C and D)

A	BM1, $N_{2012} = 1205$, $N_{2021} = 14$	2012–2015	2017–2021
No. gull impacted	275	8	
Percent overgrown	17	94	
Change in number	– 452	– 733	
B	BM6, $N_{2012} = 57$, $N_{2021} = 109$	2012–2017	2018–2021
No. gull impacted	37	0	
Percent overgrown	27	93.5	
Change in number	+ 358	– 306	
C	BM3b, $N_{2012} = 207$, $N_{2021} = 1029$	2012–2017	2018–2021
No. gull impacted	22	0	
Percent overgrown	8	15	
Change in number	+ 822	– 40	
D	GH6, $N_{2012} = 1336$, $N_{2017} = 1104$	2012 and 2017	
Number gull impacted	34		
Percent overgrown	24		
Change in number	– 232		

Fig. 4.16 Beauty Bog, BM6, in 2014 (left) and a permanent plot in same area, revisited in 2023 (right), showing conditions before and after the restoration. Note pool and low sedgy vegetation with Avens (yellow flowers) at left and dominance of chest high shrubs (1.25 m, *Myrica gale*) and head high larch (> 2 m, *Larix laricina*) in right photo

reaching heights of 1–2 m. The habitat degradation and loss of Avens in both of these cases underlines how rare species are sentinels of ecological integrity and when anthropogenic disturbance disrupts hydrology, the rare species are the first to go (Hill et al. 2018).

4.3.5 Ecological Interventions

(a)Testing the Removal of Shrubs to Release Avens from Competition

Research presented the rare Atlantic Coastal Plain Flora as competitive subordinates dependent on the removal of shrubs by the flooding regime of the lakeshore wetlands for survival (Wisheu and Keddy 1992). A four-year shrub removal experiment gave evidence of the competitive release of four small lakeshore in the clipped areas (Keddy 1989). Our application of the same principle to patches of Avens growing in four habitats on Brier Island (BM1, BM3b, BM5 and CRS) failed to show significant differences in Avens cover after two years of treatments (cover of treated/control: 1.68/1.68, 2.37/1.90, 19.7/13.1, 15.8/18.6, p values single factor ANOVA = 1, 0.56, 0.3, 0.7, n = 8, 8, 7 and 6).

The lack of response of Avens to the removal of shrub and the field observation that Avens leaves in the shrub removed treatments were often redder that the Avens in the controls, led to a test of whether there were differences in the photosynthetic efficiency of the leaves in the treated and control plots. There was a substantial and significant decrease in the photosynthetic efficiency as measured by a reduction in quantum yield of photosystem II (Fig. 4.17). That the removal of shrubs and their shading canopy negatively affected the photosynthetic efficiency of the Avens leaves suggests that their removal resulted in a change in light intensity that saturated and inhibited the photosynthetic machinery of the more shade-adapted leaves. Alternatively, the cutting back of shrubs could reduce the performance of Avens if the roots of Avens were interconnected with mycorrhizae of the shrubs that were supplying photosynthate or minerals to Avens. The shrub removal experiments suggest that interactions between the rare plant and its community in this peatland ecosystem are quite different from the more naturally disturbed lake shore.

(b)Facilitating Reestablishment of Avens

Since the hydrological restoration, it has become clear that there is a major flow toward the northeast of the Meadow and rewetting has been most pronounced along the eastern lagg. Despite this rewetting of what most likely was Avens habitat prior to ditching (Fig. 4.15), there remains thick shrub and moss layers that established in the lagg fen in the decades following the drawdown of the water table due to the ditching. Since this residual biomass will prevent the natural establishment of Avens seedlings in lagg that has been returned to pre-ditching water table levels, we have conducted trials of scarifying lagg areas and sowing quadrat areas with Avens seeds. Tests were made of sowings of Avens seeds in hummock and hollow sites with and without scarification that removed the top 10 cm layer of *Sphagnum*. The seedling density of sown sites in the following season was ten-fold greater in hollows than in hummocks, and seedling density was greatest in scarified hollows (Fig. 4.18). Scarification removed the competition for light with *Sphagnum* mosses that engulf most of the germinating seedlings; a scatter plot of moss capitulum vertical growth (measured after Clymo 1970) over one season for non-scarified and scarified hollows shows how 50% of the variation in seedling density may be accounted for by the rate of *Sphagnum* growth (Fig. 4.19). Paradoxically, recruitment of seedlings occurs preferentially in hollows despite the fact that *Sphagnum* growth over a year, measured

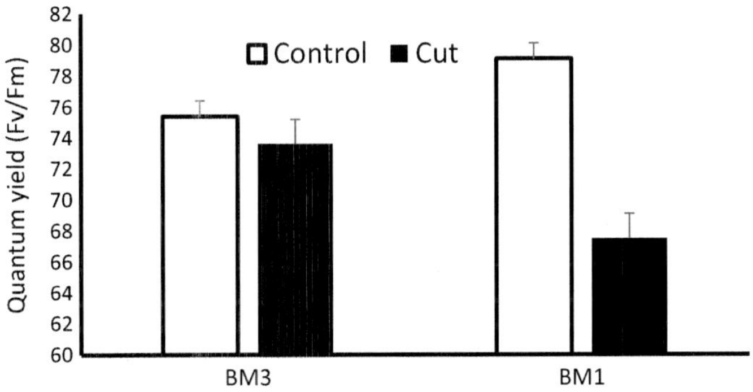

Fig. 4.17 Quantum yield (a measure related to electron flow between PSII and PSI) in shrub clipped and unclipped quadrats. Means ± s of all (2 good and 2 bad leaves per station), leaves on 9 plants per treatments at each site, $n = 36$ leaves). ANOVA Two Factor with replication showed the significance ($p < 10^{-5}$) of site, treatment (clipped and unclipped) and leaf quality (green versus discolored)

at two fens, was greatest in hollows (Brads Fen = 2.4 cm, Coastal Fen = 1.8 cm) and not in hummocks (Brad's Fen = 1.5 cm, Coastal Fen = 0.6 cm; $n = 11$ and 14 per treatment at respective fens, $p < 0.05$ and < 0.001 at Brad's and at Coastal fens). This suggests that for both *Sphagnum* and seedlings, hummocks are a harsher habitat for growth.

Our sowing experiment showed that despite the greater competition from *Sphagnum sp.* in the hollows, seedling establishment and survival (over one summer) is greater in hollows than in hummocks. Seedling establishment and survival is greater when a seed bed is made by cutting back the top 15 cm layer of *Sphagnum* capitula. The scarification experiment, the measurement of moss growth and Avens survival, and the demonstration of stratification-enhanced germination all show how Avens can be re-established in the field from seeds. Avens patches have now been re-established along the lagg areas of Big Meadow Bog in four areas in thirty 1 m^2 quadrat areas where the above ground shoots of shrubs and the layer of loose Sphagnum peat were removed and then the contents of 12 seed heads (containing on average 40 seeds/head) were sowed on this scarified fen surface. Seedling establishment occurs preferentially in peat that is just above the water table (Fig. 4.20).

Furthermore, seedling growth (rosette diameter) is greatest in scarified peats 4 cm above the water table (Fig. 4.21). We have not observed a natural scarification that allows for such seedling establishment in the field although the fire in the northwest lagg of Big Meadow Bog, BM1 (COSEWIC 1986). Seedlings are observed along deer paths that become a continuous hollow area. Genetic analysis (Chap. 3) demonstrates that successful sexual reproduction has occurred in both mesic and wet fens as well as in former sheep uplands since 25 m^2 quadrat areas contain from 3 to 7 unique genetic individuals, dispelling any conception that we are trying to save a population that is maintained through ramet production and fragmentation.

Natural examples of scarification in the Avens fens is rarely seen, however, the botanists who found Avens on Brier Island, Albert Roland, and his colleague, E.C.

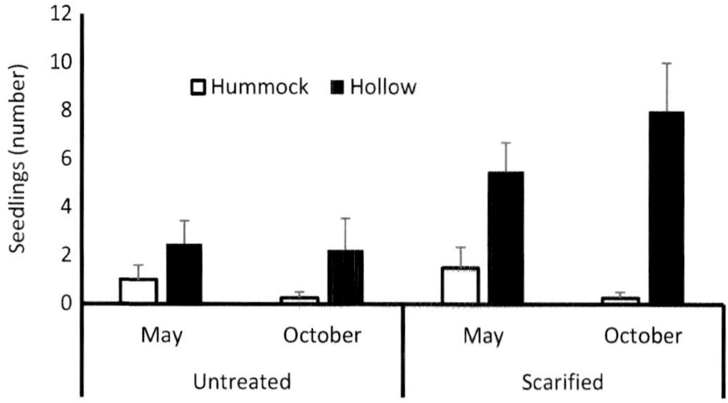

Fig. 4.18 Seedlings established in 0.025 m^2 quadrat areas after November 2016 sowings on untreated and scarified hummocks and at mesic fen, Brad's Bog (GH6), observed in May and October 2017

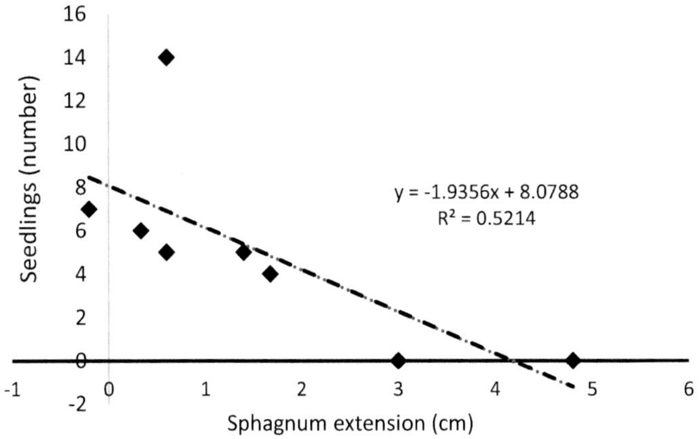

Fig. 4.19 Numbers of Avens seedlings in hollows one year after being sown. The graph includes results from 4 scarified plots (< 1 cm *Sphagnum* extension) and 4 untreated plots (extension rates > 1 cm)

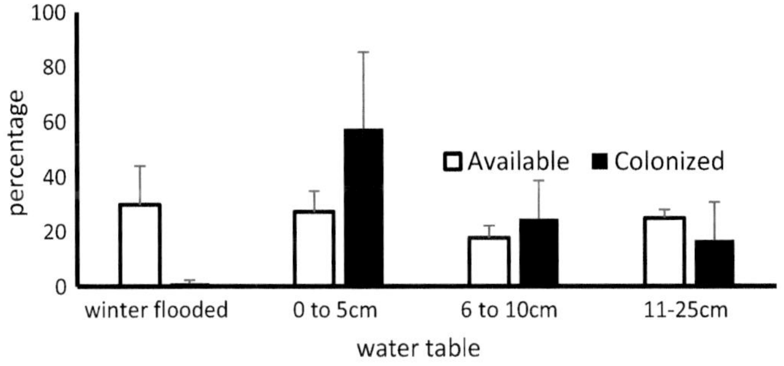

Fig. 4.20 Percent availability of four water table microsite classes and the percent of the established seedlings at microsites (mean ± s for percentages over 12 sown, scarified quadrats)

Smith, wrote "*Geum peckii* Pursh: Abundant and scattered over an area of several acres in a bog at the roadside and on burned areas between Westport and Big Cove on Brier Island" (Roland and Smith 1969). A fire at the Western Light bog, documented to be one of the oldest peatlands in Nova Scotia (Stea and Mott 1998), burned the only patch of *Geum peckii* in that bog in April of 2018. The rhizomes survived the fire and the reproductive effort from this patch in 2018 was greater than in any previous year. In recent times, two other fens on Brier Island that supported *Geum peckii* were burned: the northwest lagg (BM1) burned in the late 1980s (COSEWIC 1986) and Coastal Fen (GR3) at the south end of Gull Rock Road burned in 2012. In both latter cases, the resident Avens populations increased after the fires. Prescribed burns can allow for the periodic recruitment of rare species into plant communities (Schuyler and Stasz 1985; Snyder et al. 1992; Martinez-Hernandez and Rodriguez-Trejo 2008;

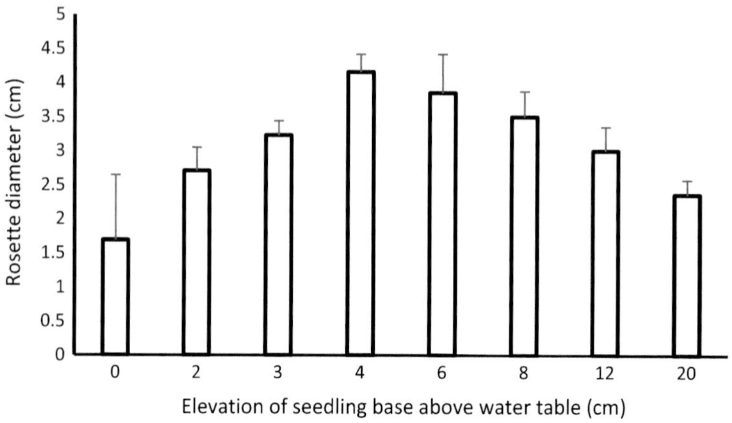

Fig. 4.21 Rosette sizes of seedlings established at different elevations in scarified quadrats

Landau et al. 2023). Fires are a main disturbance agency in boreal ecosystems (Elliott-Fisk 2000; McLaren and Turkington 2013) and these burns affect peatland succession (Bourgeau-Chavez et al. 2020; Guene-Nanchen et al. 2021; Kousmanen et al. 2023). Periodic fires can temporarily reduce Sphagnum cover and while this regrows a few years after a burn (Kousmanen et al. 2023) the low competition arena after a fire may provide a regeneration niche.

According to the results of our field interventions and to the genetic fingerprinting results, regeneration from seed has maintained the population of the Eastern Mountain Avens. We have a dim understanding only of the age of the genetic individuals—their vegetative ramets and offshoots—that grow beside one another in these fens or even in lands that were once rough pasture for sheep. These genets start in a scarification, in a patch from a seed of a plant whose rhizome survived a fire or in a deer trail or a former ATV trail in many of the fens. Once the seedling has developed into a shoot-rhizome storage system, the shoot produces leaves on petioles that place the leaf in an optimum light regime.

4.3.6 Understanding and Planning Conservation in a Changing Climate

Avens has been regarded as an arctic-alpine species (Roland and Smith 1969) which fits its White Mountain habitats lying above 5000 feet elevation but not the fens at sea-level on Brier Island. As Garbary and Hill elaborate in Chapter 5, there are major differences in the two climates, Brier and the White Mountains. The Brier Island habitat is one of the windiest on mainland Nova Scotia and has historically been kept cool through upwelling of deep waters of the Bay of Fundy but as we noted in the introduction, Avens is exposed to arctic level temperatures in the White

Mountain winters and climatic stresses (greater winds and short growing seasons, Bliss 1963) that make one question how this stress-tolerant alpine plant has survived in Maritime Canada. Over the past three decades, Brier Island has warmed more than any other location in Nova Scotia, a finding first recognized in 2021 when the temperature data over two climate normal was analyzed using corrections to account for sporadic weather records with much missing data (Chap. 5). "Planning for and mitigating impacts of climate change on Eastern Mountain Avens" had been ranked as priority (Environment and Climate Change Canada 2018) but after the above analyses revealed a near 2 °C increase in average temperature on Brier between the most recent climate normals (1961–1990 to 1991–2020), there are on-going projects both to predict the impact of warming on the plant community and understand whether warming may already have contributed to Avens declines.

Rather than wait for extirpation or maintain a one-track in situ approach to conservation, we have two studies that try to get ahead of climate change driven extirpation and give us an ability to predict. The first introduces open top chambers into wet (Runway Bog, BM3) and mesic (Brad's Bog, GH6) Avens fens and into the drier raised Reference Bog to mimic climate change. Open top chambers are used in both arctic and alpine habitats and can identify the indicators of change (Walker et al. 2006; Bjorkman et al. 2018). Avens is relegated to the low shrub areas of fens (Fig. 4.4 and Table 4.2) and population declines can be linked to an increase in shrub overgrowth (Table 4.7). Overgrowth by tall deciduous shrubs, "shrubification", is occurring throughout the arctic and is linked to climate warming (Mekonnen et al. 2021).

A second experiment is retrospective. Monitoring Avens over the past decade contains cases where patches of Avens have survived or died out. We have inserted temperature recorders at ground level to test whether survival versus extirpation patches ($n = 12$ pairs) differ in local temperature. In addition to providing insights to help separate local threats to Avens from those driven by climate change, these experiments can help us better forecast the impact and speed of the climate change associated shifts and what they augur for the survival of Avens. Such insights will inform local adaptive management of Big Meadow Bog.

The field of conservation needs to understand how to bridge the efforts restore and protect habitat and species in place with the need to facilitate the migration of species unable to keep up with the speed of climate change (Brodie et al. 2021). The seriousness of the Avens plight on Brier Island called together a multi-day workshop on ex situ conservation at Acadia University (September 2023). We systematically assessed in detail the feasibility, risks and potential benefits of the ex situ conservation and translocation of Avens. The workshop benefited from systematic approach and guidelines developed by the IUCN and the One Plan Approach, the formal facilitation, compilation and reporting by Canada Species Initiative staff and the attendance, participation and dedication of many of the key Avens conservation community members. This work provided a sound and transparent foundation and a comprehensive knowledge base for Avens conservation actions moving forward. There was strong support to employ ex situ conservation methods for Avens recovery with appropriate checks and balances. In fact, we and others, now view ex situ conservation methods as an essential insurance plan for the medium- to long-term survival of Eastern Mountain Avens.

4.4 Conclusions

The global distribution of the Eastern Mountain Avens shows a plant that is variously adapted to stress: to high winds and extreme over-winter cold in alpine New Hampshire and to waterlogging in fens of Nova Scotia. At the microsite level, Avens occurs in areas where the importance of competition from shrubs is low as we show here in the Realized Niche section. Stress-tolerance and competitive ability are evolutionary strategy trade-offs according to many ecologists (Haegeman et al. 2014) and we have reported how disturbances that have increased the productivity of Avens habitats, notably the ditching lowering water table or the Herring Gulls increasing nutrient inputs, have led to major population losses. Climate warming can also be predicted to increase community biomass through various mechanisms (Henry et al. 2022; Mekonnen et al. 2021). All of these threats are anthropogenic and all threaten Avens and other rare specialist organisms still occurring in (viz. *Betula michauxii, Blysmopsis rufa, Carex livida, Riccardia multifida*) or extirpated from (*Lophiola aurea, Schizaea pusilla, Selaginella selaginoides*) Big Meadow Bog.

The dire warning in the last assessment of the conservation status of Avens (COSEWIC 2010) led to the restoration program of Big Meadow Bog, the subject of this book. The restoration aimed to return the bog and the lagg ecosystems to their original hydrological regimes. This was carefully done through surveying and planning (Chap. 10) and the changes in water table after the ditch blockings are documented by Hiscock et al. (Chap. 7). Restoration has been called the acid test of ecological understanding (Bradshaw 1987) and yet there were considerations that in retrospect were obvious but that at the time were overshadowed by the anticipation of the restoration itself. The peatland hydrology expert, Dr. Jonathan Price, advised us in 2013, prior to embarking on the restoration: "you are nudging this ecosystem with a sledgehammer". Fifty years had elapsed since the ditching of the Meadow but the blocking of the ditches was done over a summer and restoration of hydrological regime to levels close to the original occurred in a few years. There have been surprises such as the large reduction in the Avens population at Beauty Bog (BM6) but there is a continued commitment to ensure the in situ recovery of the population as we report here using a scarification and sowing technique to reestablish Avens in the lagg of Big Meadow Bog.

John Harper, the pioneer of plant population biology, once surmised that the conservation of rare plants was an ideal that was motivated by nostalgia (Harper 1992). Nostalgia is a longing for a return to home, a desire for a former state. In this restoration project, all levels of government determined to make a positive change to bring a degraded ecosystem back and to recover an endangered species. While there are compelling genetic reasons not to lose this disjunct Canadian population of Avens (Chap. 3), there is a strong ecological, sociological and economic argument to restore Big Meadow and recover Avens. Rare specialist organisms are indicators of ecological processes and ecological integrity (Hill et al. 2018). The restoration of the Meadow was guided by what we knew about the requirements of this specialist: it required a high water table (to reduce shrub competition) and it was eradicated by the disturbance and enrichment attendant with thousands of nesting gulls. A restoration that continues to be guided by the requirements of Avens will ultimately lead to a recovery of *Sphagnum* moss in the center of the bog, carbon sequestration as greenhouse gas emissions decline, and clean, filtered water from the bog to the coastal waters. All of these goals are ecosystem services and all have sociological and economic benefits. In this view, conservation is much more than nostalgia; it is an optimism but one that is based on action and making things happen.

Acknowledgements This work would not have been possible without the support of the people of Brier Island and a group of dedicated summer students: May Buckland Nicks, Jackie Bradbury, Megan Fraser, Michelle, Jose Islas, Kasey Chinner and Emil Jurga. This work was partially funded by NSERC Discovery Grants to DJG and NSERC scholarships to JP and EM. Major funding and support came from Nova Scotia Department Natural Resources and Renewables and Environment Climate Change Canada and the restoration depended upon the generosity and vision of the Municipality of the District of Digby.

Appendix

See Table 4.8

Table 4.8 Woody species in Eastern Mountain Avens habitats on Brier island and on Mount Washington

Group	Brier	Mt Washington
Dwarf Shrubs (< 30 cm height)		
Andromeda glaucophylla	X	
Diapensia lapponica		X
Empetrum nigrum	X	
Juniperus horizontalis	X	
Kalmia procumbens		X
Rhododendron lapponicum		X
Rubus flagellaris	X	
Salix uva-ursi		X
Sibbaldiopsis tridentata		X
Vaccinium angustifolium	X	
Vaccinium macrocarpon	X	
Vaccinium oxycoccus	X	X
Vaccinium vitis-idaeea		X
Shrubs (30–60 cm height)		
Aronia melanocarpa	X	
Betula michauxii	X	
Chamaedaphne calyculata	X	
Dasiphora fruticosa	X	
Gaylussacia baccata	X	
Gaylussacia bigeloviana	X	
Juniperus communis	X	
Kalmia angustifolia	X	
Kalmia polifolia	X	
Lonicera villosa	X	
Morella pensylvanica	X	
Myrica gale	X	
Rhododendron canadense	X	
Rhododendron groenlandicum	X	X
Rosa nitida	X	
Rosa virginiana	X	
Spiraea alba	X	
Tall woody		
Alnus incana	X	
Betula glandulosa		X
Betula minor		X
Picea glauca	X	
Picea mariana	X	

(continued)

Table 4.8 (continued)

Group	Brier	Mt Washington
Salix planifolia		X
Viburnum nudum	X	

References

Anonymous (2022) Flowers of the red list of endangered species have been helped by grazing with wild horses and European bison, scientific research has shown. European Wildlife. https://www.eurowildlife.org/news/flowers-from-the-red-list-of-endangered-species-have-been-helped-by-the-grazing-of-wild-horses-and-european-bison-scientific-research-has-shown/. Accessed 28 June 2024

Baskin C, Baskin JM (2014) Seed eclology, Biogeography, and evolution of dormancy and germination. 2nd edition. Elsevier

Bjorkman AD et al (2018) Plant functional trait change across a warming tundra biome. Nature 562(7725):57–62. https://doi.org/10.1038/s41586-018-0563-7

Bliss LC (1963) Alpine communities of the presidential range, new hampshire. Ecology 44:678–697. https://doi.org/10.2307/1933014

Bourgeau-Chavez LL, Grelik SL, Bilmire M, Jenkins LK, Kakischike ES, Turtetsky MR (2020) Assessing boreal peat fire severity and vulnerability of peatlands to early season wildland fire. Front Glob Change 3:20. https://doi.org/10.3389/ffgc.2020.00020

Bradshaw AD (1987) Restoration: an acid test for ecology. In: Jordan WR, Gilpin ME, Aber JD (eds) Restoration ecology: a synthetic approach for ecological research. Cambridge University Press, Cambridge, pp 23–30

Brodie JF et al (2021) Global policy for assisted colonization of species. Science 372:456–458

Brylinsky M, Sollows J (2014) Results of the 2013 water quality survey of eleven lakes located in the Carleton River watershed area of Digby and Yarmouth Counties, Nova Scotia. Report to Nova Scotia Environment

Chislock MF, Doster E, Zitomer RA, Wilson AE (2013) Eutrophication: causes, consequences, and controls in aquatic ecosystems. Nat Educ Knowledge 4(4):10

Clymo RS (1970) The growth of *Sphagnum*: Methods of measurement. J Ecol 58(1):13–49

COSEWIC (1986) Assessment and status report on the Eastern Mountain Avens, *Geum peckii*, in Canada. Committee on the Status of Endangered Wildlife in Canada

COSEWIC (2010) Assessment and status report on the Eastern Mountain Avens, *Geum peckii*, In: Canada. Committee on the Status of Endangered Wildlife in Canada. p 33

Edwards AL, Weakely AJ (2001) Population biology and management of rare plants in depressional wetlands of southeastern coastal plain, U.S.A. Nat Areas J 21:12–35

Elliott-Fisk DL (2000) The taiga and boreal forest. In: M.G. Barbour and W.D. Billings, editors. North American Terrestrial Vegetation, 2nd edition. Cambridge University Press, Cambridge, U.K.

Environment and Climate Change Canada (2018) Action plan for the Eastern Mountain Avens (*Geum peckii*) in Canada. Species at Risk Act Action Plan Series. Environment and Climate Change Canada, Ottawa. v + 23

Eriksson O, Froberg H (1996) Windows of opportunity for recruitment of long-lived plants: experimental studies of seedling establishment in *Vaccinium* shrubs. Can J Bot 74:1369–1374

Espelland EK, Emam TM (2011) The value of structuring rarity: the seven types and links to reproductive ecology. Biol Conserv 20:963–985

Flynn PT, Garbary DJ, Novaczek I, Miller A, Quijón PA (2018) The unique giant Irish moss (*Chondrus crispus*) from Basin Head: health assessment in relation to reference sites on Prince Edward Island. Botany 96(11):805–811

Gajewski W (1959) Evolution in the genus *Geum*. Evolution 13:378–388

Garbary DJ, Hill NM, Miller A (2013) Invasion of *Rosa rugosa* (Rugosa Rose) into coastal plant communities of Brier Island. Nova Scotia Can Field Nat 127(4):319–331
Gaudet CS, Keddy PA (1988) Predicting competitive ability from plant traits: a comparative approach. Nature 334:242–243
Greene DF, Johnson EA (1997) Secondary dispersal of tree seeds on snow. J Ecol 85:329–340
Griggs RF (1940) The ecology of rare plants. Bull Torrey Bot Club 67:575–594
Grime JP (1979) Plant strategies and vegetation processes. Wiley, Toronto
Guene-Nanchen M, Leblanc M, Rochefort L (2021) Post-fire vegetation recovery: a case study in open fens of the Canadian boreal forest. Botany 100:435–447. https://doi.org/10.1139/cjb-2021-0194
Haegeman B, Sari T, Etienne RS (2014) Predicting coexistence of plants subject to a tolerance: competition trade-off. J Math Biol 68:1815–1847. https://doi.org/10.1007/s00285-013-0692-4
Harper JL (1992) Foreword In: PL Fiedler and SK Jain (eds) The Theory and Practice of Nature Conservation Preservation and management. Chapman Hall, NY
Henry GRH, Hollister RD, Klanderud K, Bjork RG, Bjorkman AD, Elphinstone C, Jonsdottir SI, Molau U, Petraglia A, Oberbaeur SF, Rixen C, Wookey PA (2022) The International Tundra Experiment (ITEX): 30 years of research on tundra ecosystems. Arctic Sci. https://doi.org/10.1139/as-2022-0041
Hill NM, Keddy PS (1992) Prediction of rarity from habitat variables: coastal plain plants on Nova Scotian lakeshores. Ecology 73(5):1852–1859
Hill NM, Chinner K, Toms B, Garbary DJ (2019) Recovery succession of native plant communities at big meadow bog after disturbance and eutrophication from Herring Gulls. Wetlands 39:533–548. https://doi.org/10.1007/s13157-018-1099-1
Hill NM, Crowell M, Lapaix R, Hicks S (2018) The rare southern twayblade (*Neottia bifolia*): sentinel of ecological integrity in sphagnous swamps. Rhodora 120(982):117–142. https://doi.org/10.3119/17-11
Hill NM, Belliveau A (2014) Delineation of wetland areas associated with the Eastern Mountain Avens (*Geum peckii* Pursh). Report on contract to Canadian Wildlife Service, Sackville, New Brunswick
Hill NM (2015) Benchmarks and goalposts to guide restoration of the endangered Avens, Big Meadow. Final report to species at risk conservation fund Grant. Wildlife Division, NS Department Natural Resources. p 62
Howie SA, Van Meerveld TI (2011) The essential role of the lagg in raised bog function and restoration: a review. Wetlands 31:613–622. https://doi.org/10.1007/s13157-011-0168-5
Hwang SJ (2020) Eutrophication and the ecological health risk. Int J Environ Res Public Health 17(17):6332. https://doi.org/10.3390/ijerph17176332
Keddy PA (1989) Effect of competition from shrubs on herbaceous wetland plants: a 4-year field experiment. Can J Bot 67:708–716
Keddy PA (2010) Wetland ecology. Cambridge University Press
Keddy PA, Reznicek AA (1986) Great Lakes vegetation dynamics: the role of fluctuating water levels and buried seeds. J Great Lakes Res 12:25–36
Keddy PA, Wisheu IC (1989) Ecology, biogeography, and coastal plain plants: some general principles from the study of Nova Scotian wetlands. Rhodora 91:72–94
Kousmanen N, Valiranta M, Piilo S, Tuittila E, Oksanen P, Wallenius T (2023) Repeated fires in forested peatlands in sporadic permafrost zone in Western Canada. Environ Res Lett 18:094051. https://doi.org/10.1088/1748-9326/acf05b
Kozlowski TT (1997) Responses of woody plants to flooding and salinity. Tree Phys 1:1–29
Kulbaba MW, Tardif JC, Staniforth RJ (2009) Morphological and ecological relationships between burrs and furs. Amer mid Nat 161(2):380–391
Landau D, Cahalan GD, Batley PN (2023) Causal impact of fire on a globally rare wetland plant: a 40-year Bayesian timeseries analysis. Fire Ecol 19:2. https://doi.org/10.1186/s42408-022-00154-3

LaRue D (2016a) In situ and ex situ propagation of the globally rare *Geum peckii* (Rosaceae). Rhodora 118(975):315–319
LaRue D (2016b) Natural regeneration of the globally rare G*eum peckii* (Rosaceae) on Brier Island. Nova Scotia Rhodora 118(975):310–314. https://doi.org/10.3119/15-39
Ma X, Geng Q, Zhang H, Bian C, Chen HYH, Djang D, Xu X (2021) Global negative effects of nutrient enrichment on arbuscular mycorrhizal fungi and on plant diversity and ecosystem multifunctionality. New Phytol 229:2957–2969. https://doi.org/10.1111/nph.17077
Manning E (2018) Seed germination of the globally rare flowering plant. Directed Studies, St. Francis Xavier University, Antigonish, Nova Scotia, Geum peckii
Martinez-Hernandez HC, Rodriguez-Trejo DA (2008) Species diversity after prescribed burns at different intensities and seasons in a high altitude *Pinus hartwegii* forest. Interciencia 33:337–344
Mauracher S, Wagner J (2021) Flower preformation in the nival plant *Ranunculus glacialis* L: Shoot architecture and impact of the growing season length on floral morphogenesis and developmental dynamics. Alp Botany 131:1–12. https://doi.org/10.1007/s00035-021-00249-8
McLaren J, Turkington R (2013) Boreal forest ecosystems. In: SA Levin (ed). Encyclopedia of Biodiversity, 2nd Edition, Academic Press
Mekonnen ZA, Riley WJ, Berner LT, Bouskill NJ, Torn MS, Iwahana G et al (2021) Arctic tundra shrubification: a review of mechanisms and impacts on ecosystem carbon balance. Environ Res Lett 16(5):053001. https://doi.org/10.1088/1748-9326/abf28b
Myers JA, Vellend M, Gardescu S, Marks PL (2004) Seed dispersal by white-tailed deer: implications for long-distance dispersal, invasion and migration of plants into eastern North America. Oecologia 139:35–44. https://doi.org/10.1007/s00442-003-1474-2
Owczarek A, Rozalski M, Krajewska U, Olszeweska MA (2017) Rare ellagic acid sulfate derivatives from the rhizome of *Geum rivale* L—Structure, cytotoxicity, and validated HPLC-PDA assay. Appl Sci 7(4):400. https://doi.org/10.3390/app7040400
Paradis E, Rochefort L, Langlois M (2015) The lagg ecotone: an integrative part of bog ecosystems in North America. Plant Ecol 216:999–1018. https://doi.org/10.1007/s11258-015-0485-5
Petersen TK, Bruun HH (2019) Can plant traits predict seed dispersal probability via guts, fur and hooves? Ecol Evol 9:9768–9781. https://doi.org/10.1002/ece3.5512
Poirier J (2016) Experimental transplantation of Eastern Mountain Avens (*Geum peckii*) over a a range of peatland moisture conditions on Brier Island, Nova Scotia. Honours thesis, St Francis Xavier University, Antigonish, Nova Scotia
Read DJ (1996) The structure and function of the ericoid mycorrhizal root. Annals Bot 77:365–374. https://doi.org/10.1006/anbo.1996.0044
Reed PB Jr (1988) National list of plant species that occur in wetlands: 1988 national summary. Biol Rep 88(24). Washington, DC: U.S. Fish and Wildlife Service. http://www.usace.army.mil/CECW/Documents/cecwo/reg/plants/list88.pdf
Roland AE, Smith EC (1969) The Flora of Nova Scotia. Nova Scotia Museum, Halifax, Nova Scotia.
Rysiak A, Chabuz W, Sawicka-Zujag W, Zdulski J, Grzywaczewski G, Kulik M (2021) Comparative impacts of grazing and mowing on the floristics of grasslands in the buffer zone of Polesie National Park, eastern Poland. Global Ecol Conserv 27:e01612. https://doi.org/10.1016/j.gecco.2021.e01612
Schuyler AE, Statz JL (1985) Influence of fire on the reproduction of *Scirpus longii*. Bart 51:105–107
Schwinzer C, Tiepkema J (1983) Seasonal pattern of energy use, respiration, and nitrogenase activity in root nodules of Myrica gale. Candian J Botany 61(11):2937–2942. https://doi.org/10.1139/b83-328
Nature Serve 2024 https://explorer.natureserve.org/Taxon/ELEMENT_GLOBAL.2.146728/Geum_peckii accessed January 11, 2024)
Shlepr KR, Ronconi RA, Hayden B, Allard KA, Diamond AW (2021) Estimating the relative use of anthropogenic resources by Herring Gull (*Larus argentatus*) in the Bay of Fundy Canada. Avian Conserv Ecol 16(1):2. https://doi.org/10.5751/ACE-01739-160102

Skene KR, Sprent JI, Raven JA, Herdman L (2001) Myrica gale L. J Ecol 88(6):1079–1094. https://doi.org/10.1046/j.1365-2745.2000.00522.x
Snyder JR, Herndon A, Robertson WB (1992) South Florida Rockland. In: Myers RL, Ewel JJ (eds) Ecosystems of Florida. University of Central Florida Press, Orlando, FL, USA, pp 230–276
Stea R, Mott R (1998) Deglaciation of Nova Scotia: stratigraphy and chronology of lake sediment cores and buried organic sections. Geograph Phys Quat 52:1–19
Toms B, LaRue D (2012) Building for the future: Establishing a foundation for the recovery of Eastern Mountain Avens, Kempt, NS, CA: Mersey Tobeatic Research Institute
Toms B (2015) Analysis of vegetation cover in Big Meadow Bog on Brier Island, Nova Scotia (1955–2015). Report to environment climate change Canada, Mersey Tobeatic Research Institute, Kempt, Nova Scotia. http://www.merseytobeatic.ca/userfiles/file/Analysis%20of%20vegetation%20cover%20in%20Big%20Meadow%20Bog%201955-2015.pdf
Toms B (2023) Current status of Eastern Mountain Avens in Nova Scotia in 2022 and analysis of a decade of monitoring on Brier Island. Mersey Tobeatic Research Institute Report to Environment Climate Change Canada
VanderKloet SP, Hill NM (2000) *Bacca quo vadis*: Regeneration niche differences among seven sympatric *Vaccinium* species on headlands of Newfoundland. Seed Sci Res 10:89–97
Vandvik V, Klanderug K, Meineri E, Maren IE, Topper J (2016) Seed banks are biodiversity reservoirs: species-area relationships above versus below ground. Oikos 125:118–128
Verhoeven JTA, Arheimer B, Yin C, Hefting MM (2006) Regional and global concerns over wetlands and water quality. TRENDS Ecol Evol 21:96–103. https://doi.org/10.1016/j.tree.2005.11.015
Vitt D (2006) Functional characteristics and indicators of boreal peatlands. In: Weider RK, Vitt DH (eds) Boreal Peatland Ecosystems. Springer, pp 9–24
Vittoz P, Engler R (2007) Seed dispersal distances: a typology based on dispersal modes and plant traits. Bot Helvet 117:109–124. https://doi.org/10.1007/s100035-007-0797-8
Voss EG, Reznicek AA (2012) Field Manual of Michigan flora. University of Michigan Press, Ann Arbor, MI
Walker MD et al (2006) Plant community responses to experimental warming across the tundra biome. Proc Natl Acad Sci USA 103:1342–1346
Wells ED, Pollett FC (1983) Peatlands, In: G.R. South (ed), Biogeography and Ecology of the Island of Newfoundland. pp 207–266
Wisheu IC, Keddy PA (1992) Competition and centrifugal organization of plant communities: theories and tests. J Veg Sci 3:147–156
Zinck MC (1996) Numerical evaluation of the taxonomy of *Geum radiatum* and preliminary studies of the pollination biology of its Nova Scotian population. BSc Thesis, Acadia University, Wolfville, Nova Scotia. https://scholar.acadiau.ca/islandora/object/theses:1058

Chapter 5
Climate Change on Brier Island: Prospects for Long-Term Survival of *Geum Peckii*

David J. Garbary and Nicholas M. Hill

Abstract The endangered plant *Geum peckii* has found refuge on Brier Island, but is likely being put in peril by significant climate change. We evaluated climate change since the 1960s in the context of regional atmospheric and oceanic influences. On the outer southern edge of the Bay of Fundy, Brier Island sits in an ocean hotspot that has warmed at least 2 °C since the 1990s. This 2 °C rise in ocean temperature is reflected in the elevated mean air temperature of 1.9 °C on Brier Island in the same period. This change was double the change of 1 °C over the last 60 years at the nearby sites of Grand Manan, on the north side of the Bay of Fundy, and Yarmouth near the southwestern extreme of mainland Nova Scotia. Temperature changes are presented along with an analysis of precipitation and wind. The underlying causes of these changes can be attributed to both local and global climatic influences, especially in light of ocean warming. These results suggest a changing climate regime for *G. peckii,* in which higher temperatures may be decreasing its competitive abilities through the more vigorous growth of shrubs in its primary microhabitats on Brier Island. These climate changes, in addition to changes in hydrology and eutrophication of its primary habitat in the Big Meadow Bog, add further risk to the long-term survival of *G. peckii* on Brier Island.

5.1 Introduction

Surrounded by the cold waters that upwell from the Bay of Fundy, the boreal ecosystems of Brier Island at the very end of the Digby Peninsula, have developed in relative isolation from the rest of Nova Scotia. This isolation provided the refuge for Eastern

D. J. Garbary (✉)
Department of Biology, St. Francis Xavier University, Antigonish, NS B2G 2W5, Canada
e-mail: dgarbary@stfx.ca; dgarbary@gmail.com

N. M. Hill
UNESCO Southwest Biosphere Association, 399 Wyman Road, Sand Beach, NS B5A 5G4, Canada
e-mail: nick@swnovabiosphere.ca

N. M. Hill et al. (eds.), *Wetland Restoration for Endangered Species Recovery*, Environmental Contamination Remediation and Management,
https://doi.org/10.1007/978-3-031-71344-6_5

Mountain Avens, *Geum peckii* (hereafter *Geum* or Avens) an imperiled rare plant for over 10,000 years. The climate of Brier Island (hereafter, 'Brier') is as much dependent on local oceanographic conditions as it is on regional atmospheric conditions. While the weather record since the 1960s is a shallow dip into the climate record that has formed the boreal ecosystems on Brier, these changes are concurrent with anthropogenic changes whose impacts have led to major population declines. Regardless, the rate of change we describe here over the last 60 years is cause for alarm to the Recovery Team responsible for the conservation management of the Eastern Mountain Avens (Hill and Garbary 2021; Nova Scotia Department of Lands and Forests 2021). In this chapter we describe the changing climate, starting from the early 1960s for Brier, and compare it to changes at two nearby sites, Grand Manan Island (New Brunswick, hereafter Grand Manan) and Yarmouth (on the southwestern corner of Nova Scotia's mainland). This time period covers major environmental changes in the primary habitat of *Geum*, Big Meadow Bog, that include: (a) the ditching of Big Meadow Bog (BMB) in the 1950s, (b) the arrival in the 1980s of up to 5100 breeding pairs of (mostly) *Larus argentatus* (Herring gulls) with consequent eutrophication and trampling that led to the decline of *Geum*, and (c) the damming of draining ditches in 2018 (Blaney 2010; Toms 2015, 2023; East Coast Aquatics 2018; Schlepr et al. 2021). Previous accounts of climate change described temperature changes over a similar period for Prince Edward Island (Garbary 2018) and Nova Scotia (Garbary and Hill 2021). These accounts provide a pattern of temperature change in which two periods were distinguished: an early period (1960–1997) with no significant changes, and a recent period (1998–2023) with an elevated temperature regime. We use these studies with their apparent break between 1997 and 1998 to evaluate temperature changes on Brier along with changes in precipitation and wind over the same period.

It should not be surprising that climate changes would affect the persistence of an alpine herb restricted mainly to elevations above 1500 m in the White Mountains (New Hampshire, USA) and to the boreal Brier Island that projects into the Bay of Fundy. When the COSEWIC (2010) report and the Recovery and Action Plans (Environment Canada 2010) were written, the main threats to populations of *Geum* included competitive exclusion resulting from ditching of Big Meadow Bog as well as changing land use (e.g., abandonment of livestock grazing), the eutrophication and disturbance impacts of nesting gulls, and disturbance by all-terrain vehicless. Climate change was mentioned, but only as it applied populations in the White Mountains (COSEWIC 2010). In the same vein, neither an assessment of the history of the peatland of Big Meadow Bog since glaciation (Spooner et al. 2017) nor the ability of its plant community to recover by succession after gull eutrophication (Hill et al. 2019) mentioned present day climate change. We address these issues by focusing on climate change for Brier Island in the last 60 years and consider how oceanographic factors may be the source of climatic changes independent of changes on mainland Nova Scotia.

5.2 Methods

For our analyses, climate data were extracted manually from the Environment and Natural Resources Canada 'Historical Data' portal (Government of Canada webpage: climate.weather.gc.ca). At least two weather stations were required for each of our three sites of interest: (1) for Brier, stations Westport 8206260 and Brier Island 8200606, (2) for Grand Manan, stations Grand Manan 8101920, Grand Manan SAR CS 8101015, and Seal Cove, 810M937, and (3) for Yarmouth, stations Yarmouth Airport 8206500, and 8206495. Data gaps for climate metrics are indicated in the figures and were more extensive for precipitation and wind than for temperature. Changes for Brier were benchmarked against data associated with temperatures previously published for Nova Scotia (Garbary and Hill 2021) and newly analyzed data up to 2023. Climate change is normally considered over 30-year periods to determine the climate normals for a particular site. Thus, the period since 1960 provides two such periods (1961–1990 and 1991–2023) for consideration. However, instead of this arbitrary division of time periods we used an apparent step change in temperature for eastern Canada that occurred between 1997 and 1998 (Garbary 2018; Garbary and Hill 2021) that is also apparent in the other Atlantic Provinces. This break also makes for a more even distribution of temperature data given the extensive missing data for Brier and Grand Manan with only 30 and 33 years, respectively, of useful temperature data between 1961 and 2023. This allowed us to fit the data that does exist with the pattern from Yarmouth that is consistent with the rest of Nova Scotia (Garbary and Hill 2021). We used Grand Manan and Yarmouth as our comparison sites because of their proximity to Brier (both ~ 50 km away) and the common marine influence in their climate regimes. The data gaps for Brier and Grand Manan were even greater for other climate metrics, i.e., precipitation and wind. Statistical analysis was carried out using Excel for descriptive statistics, and GraphPad (https://www.graphpad.com/quickcalcs/linear1/) was used for linear regressions.

The temperature regime for Brier and our two comparison sites were considered from a variety of perspectives. We used mean annual temperature as our initial metric to evaluate change and to compare Brier with the rest of Nova Scotia where ~ 1 °C elevation has occurred (Garbary and Hill 2021, Fig. 5.1). We then considered if these temperature changes were associated with particular months, or if the changes were uniformly distributed throughout the year. In addition, we examined intra- and inter-annual temperature extremes and how these reflected the overall temperature regime. Accordingly, we counted the number of days in which temperatures exceeded 20, 25, and 30 °C. We also examined the length of the growing season by counting the number of days between the last frost (i.e., less than 0 °C) of spring and the first frost of autumn. A final metric associated with temperature was the number of frost days from autumn to spring; this evaluated temperature amelioration in winter associated with the elevation of mean annual temperature. The underlying assumption behind these measures is that *Geum*, with its origin as an arctic–alpine species would be physiologically less able to perform under new warming conditions, and that such

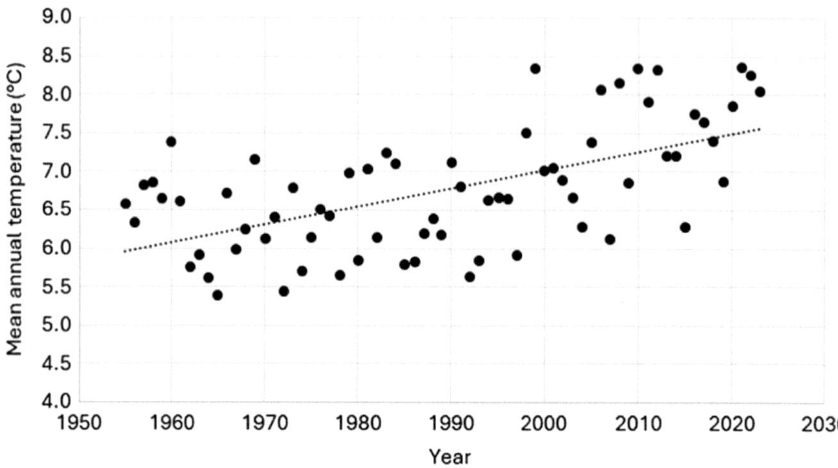

Fig. 5.1 Mean annual temperature for five Nova Scotia domestic and military airports (Yarmouth, Greenwood, Halifax, Shearwater, Sydney), 1955–2023. Linear regression is significant at $P < 0.0001$ and indicates an overall temperature increase of 0.02 °C/y. See Table 5.1 for details of linear regressions for the total period (1955–2023), early period (1955–1997) and recent period (1998–2023)

climate changes would disadvantage the species because of lowered competitiveness with native shrubs, for example, *Myrica pensylvanica* (Bayberry; see Hill et al. 2024).

Precipitation was evaluated based on yearly total precipitation and how this was reflected in seasonal changes based on monthly totals. We further evaluated precipitation by asking: is precipitation occurring as more extreme rainfall events? This was done by calculating the highest percentage of monthly precipitation that occurred on a single day. Accordingly:

$$\text{Intensity} = \frac{\text{maximum daily precipitation in a given month}}{\text{total precipitation for given month}} \times 100$$

with values for an entire year averaged, i.e., Intensity/12.

With its high wave exposure in the outer Bay of Fundy, Brier has been exposed to major storm events. The most dramatic of these was the Groundhog Day gale in February 1976 (Parkes et al. 1997) in which much of the waterfront of Westport was washed away and ocean waves breached Big Meadow Bog. Coming at a time when ocean temperatures are increasing and resulting in storms of greater number and intensity, we asked the question: have the frequency and intensity of storms changed over the last 60 years? Thus, for each month, we recorded the maximum wind gust and the number of storm days. We defined a storm day as a day in which the strongest wind gust was at least 75 km/h, i.e., a strong gale and level 9 on the Beaufort Scale.

5.3 Changes in Climate Metrics

5.3.1 Mean Annual Temperature

Overall changes in mean annual temperature for Nova Scotia (Fig. 5.1) serve as a benchmark for interpretation of the more local changes to our three test sites of Grand Manan, Yarmouth and Brier (Fig. 5.2). The primary pattern of mean annual temperature change for Nova Scotia was a 1 °C increase over the period 1955–2023, i.e., from 6.4 ± 0.5 °C (mean ± s) during 1955–1997 to 7.4 ± 0.7 °C during 1998–2023 (Fig. 5.1 and Table 5.1). Across five Nova Scotia airports (Yarmouth, Greenwood, Shearwater, Halifax, Sydney) temperature increases varied from 0.9 °C at Sydney to 1.5 °C at Greenwood. In addition, there was no significant change in mean annual temperature during 1960–1997 and then a period from 1998 to 2022 in which there were elevated temperatures (Fig. 5.1 and Table 5.1). The overall significance of this temperature increase is demonstrated in the highly significant linear regressions for the overall period ($R^2 = 0.4695$; $P < 0.0001$), but with no significant change in either the early or late periods (early: $R^2 = 0.0409$, $P = 0.2299$; recent: $R^2 = 0.0271$, $P = 0.1383$). Consistent with the non-significance of the linear regressions, the changes during the early and recent periods for Brier were 0.02 °C/y and – 0.05 °C/y, respectively; values for Yarmouth with a complete dataset were 0.01 °C/y and 0.3 °C/y for early and recent periods, respectively. The extent of these differences is such that, in the recent period, only in 2015 was the annual mean below the mean for the previous period. Similarly, only in 1983 did the annual mean of 7.2 in the early period reach the overall mean of the recent period. Thus, climate warming in the measured area has occurred not as a gradual change but in a single step-like manner beginning in 1998. This pattern of change was observed for Prince Edward Island over 1960–2015 (Garbary 2018) and for Nova Scotia over 1960–2020 (Garbary and Hill 2021).

We use temperatures for the province in the early and recent periods, i.e., 1960–1997 and 1998–2022, respectively, to highlight changes on Brier, Grand Manan, and Yarmouth. In the early period, mean annual temperatures for Brier Island were all below those of Yarmouth, whereas after 1998, roughly half of the years had temperatures substantially above Yarmouth. The remaining years were either slightly lower or data was missing. Thus, despite the distance of only 50 km between the two sites, the temperature on Brier has been rising faster; this was confirmed by the more positive overall slope and higher R^2 values in the linear regressions (Table 5.1).

The warming trend shown in Figs. 5.1 and 5.2 for mean annual temperatures for Nova Scotia and Brier is shown in greater detail in Fig. 5.3 with mean monthly temperatures for early and recent periods. In every month, temperature in the recent period was elevated by 1–3 °C (all significant at $P < 0.001$). Differences were less in spring to early summer, and greater in late fall and early winter. A striking change in these mean monthly temperatures is that in the early period four months of the year (December, January, February, March) mean temperatures were below 0 °C, whereas in the recent period, only January and February had means below 0 °C, with

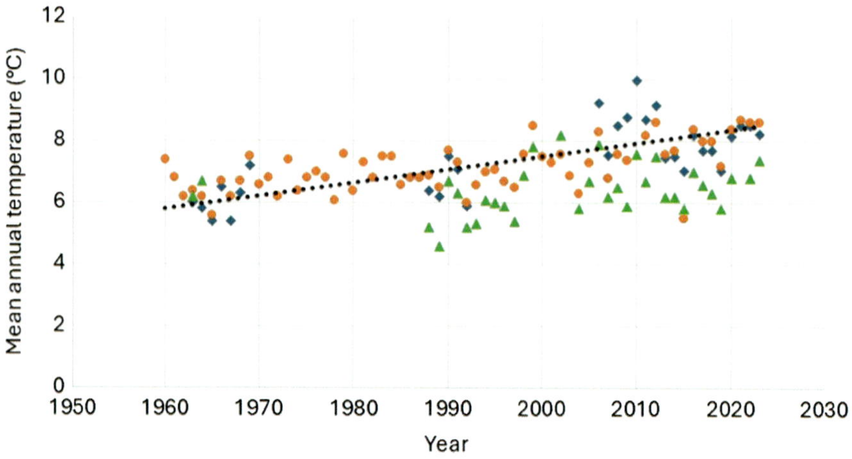

Fig. 5.2 Mean annual temperature for Yarmouth (orange circles), Grand Manan (green triangles), and Brier Island (blue diamonds); trend line shown for Yarmouth; see Table 5.1 for details of linear regressions and differences between the early period (1960–1997) and recent period (1998–2023)

Table 5.1 Changes in mean annual temperature for Nova Scotia (combined data from Yarmouth, Greenwood, Shearwater, Halifax, Sydney), Brier, Grand Manan, and Yarmouth with results from linear regressions of the overall time period (1960–2023), early period (1960–1997), and recent period (1998–2023)

Site	Time period (number of years)	Mean annual temperature (°C)	Slope	R^2	P
Nova Scotia	1955–2023 (68)	6.8 ± 0.8	0.0238	0.3505	**< 0.0001**
	1955–1997 (42)	6.4 ± 0.5	– 0.0019	0.0020	0.7769
	1998–2023 (26)	7.4 ± 0.8	0.0140	0.0274	0.429
Brier	1963–2023 (30)	7.4 ± 1.2	0.0424	0.5535	**< 0.0001**
	1963–1992 (12)	6.3 ± 0.7	0.0243	0.2016	0.1431
	2006–2023 (18)	8.2 ± 0.8	– 0.0501	0.1144	0.1698
Grand Manan	1963–2023 (33)	6.4 ± 0.8	0.0316	0.1583	0.0241
	1963–1997 (12)	5.8 ± 0.6	– 0.0229	0.1643	0.1922
	1998–2023 (21)	6.8 ± 0.7	– 0.0277	0.0760	0.2142
Yarmouth	1960–2023 (63)	7.1 ± 0.8	0.0254	0.3736	**< 0.0001**
	1960–1997 (38)	6.8 ± 0.50	0.0098	0.0478	0.1874
	1998–2023 (25)	7.7 ± 0.8	0.0350	0.1211	0.0882

Bold indicates highly significant regressions
Note negative slopes for some places and time periods

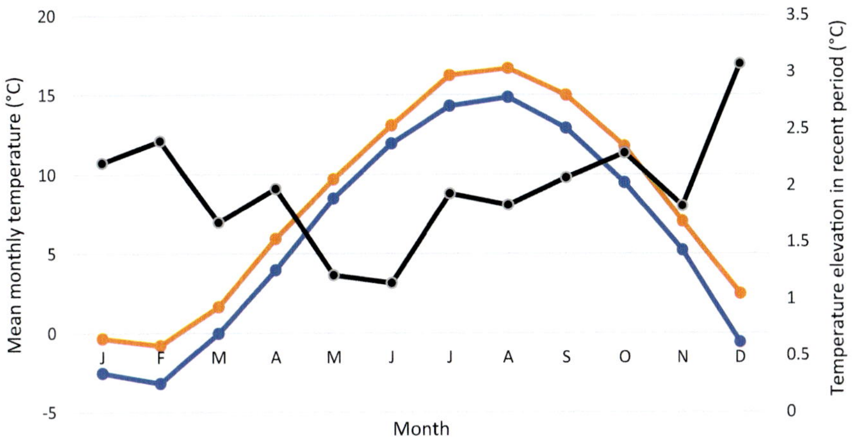

Fig. 5.3 Mean monthly temperature for Brier (left axis) during the early period (blue, 1960–1997, $n = 14$) and recent period (orange, 1998–2023, $n = 18$); black line (right axis) indicates differences between early and recent periods (all significant at $P < 0.001$)

eight of the last 18 years having mean January temperatures above 0 °C. Thus, even a slight moderation of temperatures in January would result in above zero values on a regular basis.

We divided the 60 plus years of temperature change into early and recent periods based on a major jump of temperature between 1997 and 1998. Around this break, we evaluated linear regressions in periods 1960 to present, 1960–1997 and 1998–2022. The linear regressions for the overall periods were significant for all sites, with differences in means ranging from 0.9 °C for Yarmouth, 1.0 °C for Grand Manan and 1.9 °C for Brier. It is of interest that neither of the within-period regressions, i.e., early or recent, were significant, and that some of the within-period regressions had negative slopes, even though none of these were significant. This regression analysis is consistent with a step change in temperature regime rather than a change in the rate of temperature increase post-1998.

5.3.2 *Extreme Warm and Cold Temperatures*

In addition to an average increase of 1.9 °C on Brier and the monthly changes, we recorded the occurrence of elevated temperatures, i.e., the number of days per year in which maximum temperature reached the thresholds of 20, 25 and 30 °C in the early and late periods relative to Grand Manan and Yarmouth (Table 5.2). Brier had a 60% increase in the number of days over 20 °C, well above the equivalent increases of 22% and 36% for Grand Manan and Yarmouth, respectively. The total number of days above 20 °C, i.e., 40, was about half the number for the other sites. Despite the > 150% increase in days per year above 25 °C, post-1998 Brier experienced less

Table 5.2 Number of days per year in early (1960–1997) and recent (1998–2023) periods where temperatures were at least 20, 25, and 30 °C

Site	Period (*n*)	> 20 °C	> 25 °C	> 30 °C
Brier	Early (12)	25.6 ± 17.0	0.8 ± 1.2	0
	Recent (16)	40.3 ± 11.5	2.2 ± 2.1	0
Grand Manan	Early (2)	61.5 ± 17.7	12.5 ± 2.1	0
	Recent (21)	79.4 ± 7.1	19.0 ± 6.8	1.2 ± 1.5
Yarmouth	Early (37)	52.2 ± 10.6	3.6 ± 2.0	2 days in 37 years
	Recent (26)	71.0 ± 12.0	7.1 ± 4.4	1 day in 25 years

n number of years

than a third that of Yarmouth and about 12% that of Grand Manan. Brier has not yet recorded temperatures above 30 °C and these were rare at Grand Manan and Yarmouth.

Extreme cold temperatures have also changed at the three sites (Table 5.3) with the most extreme elevation of the maximum annual cold temperature occurring at Brier (3.6 °C) with Grand Manan and Yarmouth having less change, 3.0 °C and 1.6 °C, respectively. Equivalent changes of extreme annual temperature maxima also occurred, although the elevation change for Yarmouth of 0.6 °C was not significant. These changes in annual temperature extremes were consistent with the monthly changes in temperature (Fig. 5.2) in which the greatest changes were for December, January, and February, with these months ranking first, fourth and second, respectively in terms of greatest mean change. The extreme cold and warm temperatures and the overall temperature range for the three sites provided a reduced temperature range between early and recent periods, but the change was significant only for Brier. This is most apparent for Brier and least apparent for Yarmouth (Table 5.3). This reduction of the annual range between early and recent periods was caused by the greater elevation in extreme low temperatures for Brier (3.6 °C) relative to Grand Manan (3.0 °C) and Yarmouth (1.9 °C).

5.3.3 *Frost Days and Growing Season*

The increase in annual mean temperature and the rising extreme annual minimum and maximum temperatures provide the basis for two other temperature metrics: the number of continuous frost-free days over the summer (i.e., the growing season) and the number of days with sub-zero temperatures over the cold season (October to May; i.e., the number of frost days). Brier, with its greater increase in annual temperature, had larger changes both in terms of absolute numbers and as percent changes (Table 5.4). In addition, the decline in frost days was significant in all months from November to April for Brier but only for December, January, April, and May for Yarmouth (data not shown). The magnitude of the reduction in days with frost

Table 5.3 Comparison of extreme minimum, extreme maximum and extreme temperature range for Brier, Grand Manan, and Yarmouth in early period (up to 1997) and recent period (1998 to present)

Site	Period (*n*)	Extreme annual minimum (°C)	Extreme annual maximum (°C)	Range (°C)
Brier	Early (15)	– 17.7 ± 2.4	24.1 ± 2.0	41.8 ± 2.5
	Recent (18)	– 14.1 ± 2.6	25.8 ± 1.1	39.9 ± 2.3
	P-value	**< 0.001**	**< 0.005**	**< 0.05**
Grand Manan	Early (12)	– 23.3 ± 2.4	28.3 ± 1.8	51.5 ± 2.6
	Recent (23)	– 20.3 ± 3.5	29.7 ± 1.8	50.2 ± 3.2
	P-value	**< 0.05**	**< 0.05**	NS
Yarmouth	Early (27)	– 17.5 ± 2.0	26.9 ± 1.5	44.3 ± 2.6
	Recent (22)	– 15.6 ± 2.4	27.5 ± 1.6	43.1 ± 2.7
	P-value	**< 0.005**	NS	NS

Bold indicates significant differences
Significance based on two-tail Student's *t*-test at $P < 0.05$
n number of data-points, *NS* not significant

reflects the almost 2 °C increase in annual temperature for Brier, with the largest increases occurring December to February.

Table 5.4 Changes in the number of continuous frost-free days per year and number of days per year with sub-zero temperatures in earlier period (1961–1997) and recent period (1998–2022); significance based on Student's *t*-test

Site / Temperature metric	Early period (days ± s)	Recent period (days ± s)	Difference (days, %)	Significance
Brier				
Continuous frost-free days	206 ± 15	227 ± 13	+ 21, 10.1	< 0.001
Number of frost days	111 ± 7	85 ± 17	– 26, 23.4	< 0.001
Yarmouth				
Continuous frost-free days*	182 ± 16	197	+ 15, 9.8	
Number of frost days	130 ± 7	116 ± 12	– 14, 10.8	< 0.001

The number of reported frost days is limited to the months from October to May
*Data from Garbary and Hill (2021)

5.3.4 *Precipitation*

There was no pattern of change in total precipitation during the > 60 years we evaluated for either Brier or Yarmouth. For Yarmouth, the means for the early and recent period were 1267 ± 171 and 1235 ± 193 mm/y, respectively, whereas the means for Brier Island over the same periods were 1125 ± 104 and 1137 ± 147 mm/y. Linear regression for the overall period, the early period, and recent periods were all non-significant. A conspicuous feature of the overall period was the extent of year-to-year variation producing a wide scatter of points (Fig. 5.3).

Despite the lack of change in annual precipitation, we examined the possibility that precipitation events were becoming more extreme, i.e., that in a given month that the day with the most precipitation consisted of a greater proportion of the monthly total (Fig. 5.4). While the overall linear regression for the combined sites has a positive slope (Table 5.5), the regressions for the early period are negative and significant for the combined data and for Brier. There is a decline in the early period (Fig. 5.5, Table 5.5) and then an increasing trend for Brier and Yarmouth. The resulting change gives Brier a change of 7.3% from the lows in 1988–1993 (25.8 ± 2.3%) to the post-2006 values (33.1 ± 4.0%). This represents an overall increase of over 25% in mean intensity between the lows of 1988–1993 and the mean of the recent period. For Brier, in only one year since 2009 was the mean of this index below the mean of the low period (1988–1993). Missing data within years and several extreme intensity values in months with low precipitation, e.g., August, may make this 'trend' more apparent than real.

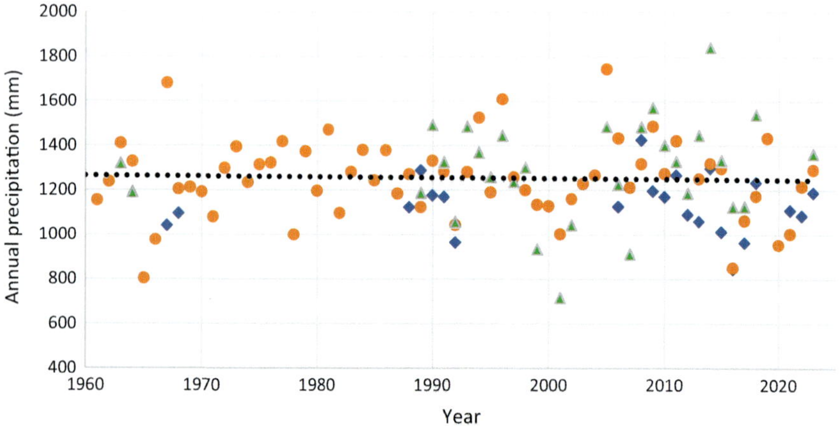

Fig. 5.4 Annual precipitation on Brier (blue diamonds), Yarmouth (orange circles), and Grand Manan (green triangles), 1961–2023; regression line is for the mean of the three sites

Table 5.5 Percent of monthly precipitation occurring on the day with the highest rainfall

Site and years covered	Mean ± s (*n*) %	Slope	*R*2	*P*
Brier				
1962–2023	32.5 ± 4.2 (34)	0.0052	0.0007	0.8833
1962–1993	30.7 ± 4.5 (15)	− 0.2601	0.5624	0.0013
2006–2023	33.1 ± 4.8 (19)	0.4166	0.2146	0.0528
Grand Manan				
1962–2023	33.7 ± 5.7 (40)	0.1218	0.1155	0.0319
1962–1997	29.6 ± 2.8 (14)	− 0.1104	0.2736	0.0540
1998–2023	36.0 ± 5.6 (26)	− 0.1104	0.0002	0.9433
Yarmouth				
1961–2022	32.5 ± 3.5 (62)	0.0503	0.0669	0.0423
1961–1997	31.5 ± 2.7 (37)	− 0.0762	0.0922	0.0678
1998–2023	33.9 ± 4.1 (25)	0.2138	0.1457	0.0597
Combined sites				
1961–2023	32.6 ± 3.6 (63)	0.0586	0.0878	0.0184
1961–1997	31.2 ± 2.9 (37)	− 0.01171	0.1860	0.0077
1998–2023	34.3 ± 3.6 (26)	0.1151	0.0580	0.2357

Linear regressions for Brier, Grand Manan, Yarmouth, and means of the three combined sites. Except for Brier, the means between the early and recent periods were significantly different (Student's *t*-test)
n number of years

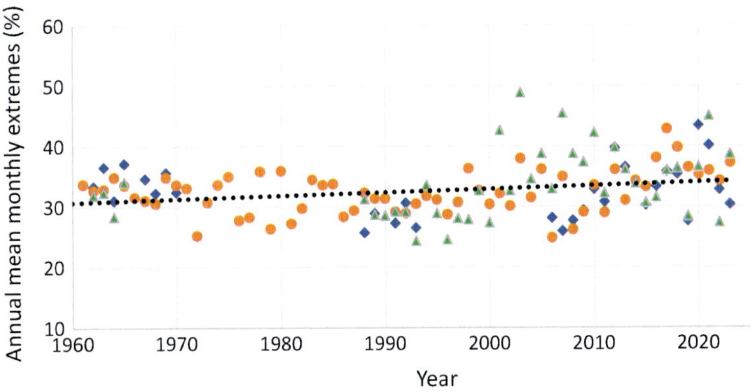

Fig. 5.5 Percent of monthly precipitation from the most extreme daily value; each point is the average of available months of the year for which data was available (orange circles, Yarmouth; blue diamonds, Brier; green triangles, Grand Manan). See Table 5.5 for results of linear regressions at different sites

5.3.5 *Wind and Storms*

Brier is very windy with numerous gales (Figs. 5.6 and 5.7). Indeed, it may be the second windiest region of the province after the northwest coast of Cape Breton Island where the locally termed "Les Suêtes" have winds that regularly exceed 150 km/h (McIldoon and Pilon 2008; The Nature of Things 2021). Indeed, since 2008, only in 2018 was the annual mean of monthly maxima for Brier less than 80 km/h, and in most years this value exceeded 85 km/h (Fig. 5.6). The two comparison sites had 10–15 km/h weaker winds. Wind has strong seasonal variation with the weakest winds during summer and the stronger winds from November to March. Brier has the strongest winds in every month, and these are significantly different from the mean monthly maxima for Yarmouth. The declining wind intensity on Brier from 2008 to 2023 was statistically significant ($R^2 = 0.2908$, $P = 0.031$) but contrary to the pattern from Yarmouth where an increasing trend occurred over the same period ($R^2 = 0.4287$, $P = 0.0059$). The latter pattern is consistent with our analyses of four other sites in Nova Scotia (Fig. 5.6).

In addition to the wind intensity, Brier far exceeds the other sites in terms of number of storm days per year, i.e., days in which wind gusts were $\geq$ 75 km/h. From 2008 to the present Brier had five to ten times the number of gales as Grand Manan, and at least double the number as Yarmouth. The stability of the number of storm days for Brier and Grand Manan contrasts with the increasing number of wind events at Yarmouth. The increasing number of storm days for Yarmouth is consistent with values across mainland Nova Scotia and Cape Breton. There was no significant change for Brier over this period whereas Yarmouth showed a highly significant increase ($R^2 = 0.4562$, $P = 0.004$). The trend for Yarmouth (Fig. 5.8) was consistent

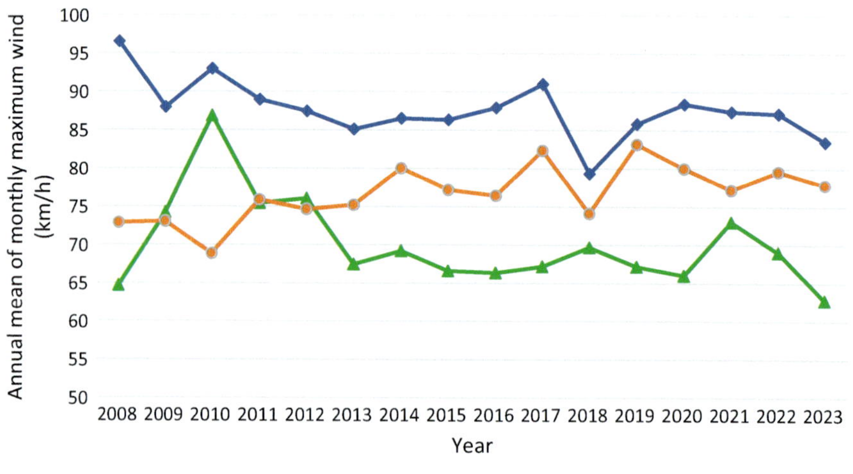

Fig. 5.6 Annual mean of monthly wind maxima for Brier (blue diamonds), Yarmouth (orange circles), and Grand Manan (green triangles)

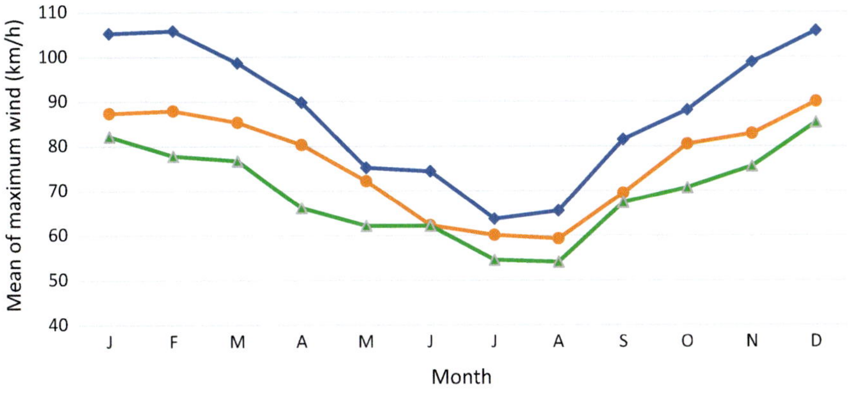

Fig. 5.7 Mean of monthly wind gust maxima from 2008 to present for Brier (blue diamonds), Grand Manan (green triangles), and Yarmouth (orange circles); data prior to 2008 unavailable for Brier or Grand Manan

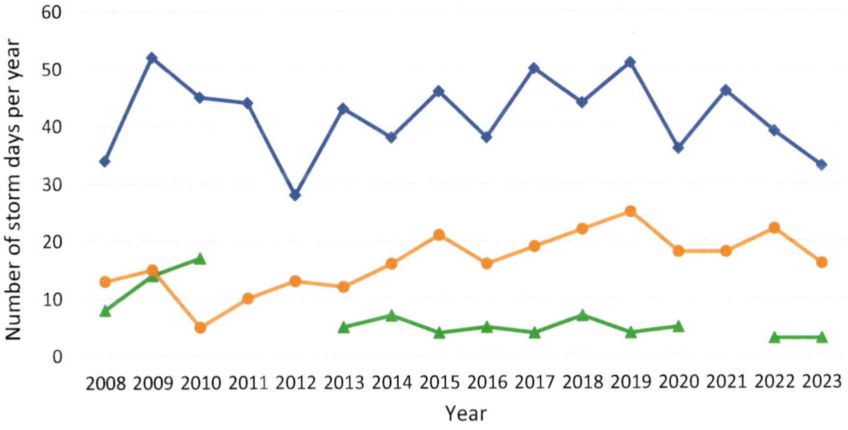

Fig. 5.8 Number of storm days per year for Brier (blue diamonds), Yarmouth (orange circles), and Grand Manan (green triangles); data prior to 2008 not available for Brier Island or Grand Manan

with the trend for four other sites in Nova Scotia and the long-term trend in which there was greater storm activity in the 1960s and 1970s (not shown).

5.4 Sixty Years of Climate Change

The hypothesis that the climate has changed sufficiently on Brier to affect the survival of the Eastern Mountain Avens was not posed during the lead-up to the active restoration of Big Meadow Bog (Recovery and Action Plans, Environment Canada

2010; COSEWIC Report 2010). When it was privately discussed in conversation, we pointed to the incomplete temperature record and left Brier and other stations out of the temperature change analysis (Garbary and Hill 2021). These missing years (1970–1987, 1993–2005), however, "masked climate change concern for endangered plants in Nova Scotia" (Hill and Garbary 2021). Here we use data from Grand Manan on the north side of the Bay of Fundy as a comparison site and Yarmouth to the south to present an analysis of how temperature, precipitation, and wind have changed over the past 60 years.

The temperature regime for Brier has warmed significantly since the 1960s with an increase of 1.9 °C based on mean monthly increases. These temperature changes confirm a regional break in temperature regime from 1997 to 1998 in which the temperature increased and has remained largely stable but with year-to-year variation. This pattern provides a basis for defining an early period up to 1997, and a recent period, beginning in 1998; we used this as the basis for our analysis rather than the arbitrary 30-year periods used for defining climate normals (for e.g., Environment Canada, https://climate.weather.gc.ca/climate_normals/).

At Brier, we found an increased mean annual temperature, increase in number of days with temperatures over 20 and 25 °C, decline in number of days below 0 °C, and increase in the number of continuous frost-free days (i.e., growing season). The basic trends for Brier and our two comparison sites are similar to the rest of the Maritime provinces (Garbary 2018; Garbary and Hill 2021). However, it is now apparent that Brier is a local hotspot, with possibly the warmest temperature regime in Atlantic Canada with the greatest increase in mean annual temperature since the 1960s. As pointed out in Hill et al. (2024), the COSEWIC report and the Recovery and Action Plans, the main threats to populations of *Geum* included competitive exclusion resulting from changes in land use and the impact of gulls. Climate change was mentioned only as it might apply to the White Mountains populations (COSEWIC 2010. We can now conclude that significant climate warming has occurred on Brier and that these changes are impacting the long-term potential for survival of *Geum* on Brier (Hill et al. 2024).

There were no changes in total annual precipitation and all three sites had about 1200 mm/y (Fig. 5.3). In addition, there were no significant changes in the monthly distribution of precipitation in early and recent periods. A more subtle change was apparent in the intensity of precipitation in which a greater proportion of the monthly precipitation occurred in single daily events (Fig. 5.5, Table 5.5).

5.5 Importance of Ocean Warming

The scale of changes on Brier relative to the rest of the Maritime provinces suggests that factors may be operating to exacerbate the regional temperature increases associated with mainland Nova Scotia and nearby Grand Manan (55 km away). First, while surrounded by the similar tidal currents of the Bay of Fundy, Grand Manan is close enough to the mainland to experience a more continental climate with colder winters and warmer summers than Brier Island (Table 5.3). At only 50 km from Brier, Yarmouth experiences a major upwelling of colder water (Tee et al. 1993), that maintains a more stable temperature regime. It is of interest that pre-1997 Yarmouth was one of the warmest areas of the province (i.e., almost 7 °C annual average temperature), but it has experienced a relatively small increase (i.e., 0.7 °C, Garbary and Hill 2021) and some other sites have higher annual temperatures and are experiencing greater increases in the post-1998 climate regime (e.g., Shearwater with a 1.3 °C increase and post-1998 mean annual temperature of 7.8 °C). The differences between Brier and Yarmouth can be attributed to changing oceanographic conditions in which the Bay of Fundy and the Gulf of Maine are undergoing rapid temperature increases resulting from an extensive ocean warming hotspot that has been highly disruptive of the ecosystem (Pershing et al. 2015, 2021; Saba et al. 2015; Siedov et al. 2021). This has reversed a 900-year cooling trend in the Gulf of Maine (Whitney et al. 2022). These warmer surface water temperatures have been confirmed in the outer Bay of Fundy midway between Brier and Grand Manan with a more than 2 °C increase since the early 1990s (Atlantic Zonal Monitoring Program (AZMP), Hebert et al. 2023; https://www.cbc.ca/news/canada/new-brunswick/grand-manan-dulse-worst-season-1.6951770). However, temperature increases in the Yarmouth area have been moderated by localized upwelling (Tee et al. 1993).

The rapid warming of the Gulf of Maine and Bay of Fundy may be associated with more global oceanographic fluctuations including the Atlantic Multidecadal Oscillation (AMO). AMO is a long-term oscillation in the North Atlantic Ocean of 0.4 °C in which complete cycles of elevated and lowered water temperatures occur over periods of 50–80 years (Dijkstra et al. 2006; Knight et al. 2006; Li et al. 2009; Wyatt et al. 2012; McCarthy et al. 2015). This cycle switched into a positive phase in the 1990s and was coincident with the increased ocean temperatures associated with the development of the ocean hotspot described in the previous paragraph. If the AMO was part of the trigger for initiating the jump in regional marine temperatures, there is a possibility that the local ocean hotspot may revert to its previous cooler state along with temperatures conditions on Brier Island.

Changing weather conditions on Brier Island and regionally may also be associated with ENSO events (NOAA 2023). The break in regional air temperatures that occurred over 1997–1998, when the annual temperature rose 2 °C, was coincident with a period of La Niña that began in summer of 1998 and extended through to the summer of 2001 (https://origin.cpc.ncep.noaa.gov/products/analysis_monitoring/ensostuff/ONI_v5.php). This period marked two of the warmest years for Brier Island and Nova Scotia (i.e., mean temperatures over 8 °C). Similarly, the warmest year on Brier at 9.9 °C in 2010 was also coincident with a period of La Niña. The latest period of La Niña began in the summer of 2020 and extended to the fall of 2023. This period has been characterized by near maximum annual temperatures in 2021, 2022, and 2023 on Brier Island and Nova Scotia generally (Fig. 5.1). Conversely, the lowest annual temperatures regionally in the last 25 years occurred in 2015 in a year-long episode of El Niño.

A final oceanographic phenomenon that may become critical for local, regional and global climate change is the Atlantic meridional overturning circulation

(AMOC). This system controls the global transport of oceanic heat and salt via the Gulf Stream and the formation of the deep ocean circulation beginning in the northeast Atlantic. This circulation is currently at its weakest point in the last millennium (Caesar et al. 2021), and there are concerns that it may be on the verge of a dramatic change. Modeling has suggested that AMOC is at a tipping point based on increased freshwater inputs from melting of the Greenland ice sheet (van Westen et al. 2024). Should this tipping come about, displacement of the Gulf Stream will likely lead to substantive changes to AMO and ENSO and result climate cooling especially in northwestern Europe, change that would include Brier (van Westen et al. 2024).

5.6 Wind and Dispersal of Geum

Nova Scotia is a windy province (Keys et al. 2017). Apart from the mountain-associated, adiabatic winds of western Cape Breton Island (McIldoon and Pilon 2008; The Nature of Things 2021), Brier may be the place in the Maritime provinces with the strongest winds and the most frequent storms. Our dataset for wind on Brier Island starts only in 2008. Thus, it is unclear if longer term wind patterns are consistent with continent-wide patterns of wind stilling (e.g., Klink 1999; McVicar et al. 2008; Pryor et al. 2009) and their apparent reversal since early this century that is apparent in the overall wind record for Nova Scotia. Given the importance of wind for pollen and seed dispersal, and the overall wind conditions on Brier, it would not be surprising if this played a key role in the historical dispersal of *Geum* (see Chap. 4).

Modeling studies have pointed to the multiple factors that affect a seed's migration ability via wind and have highlighted seed dispersal, seed structure and weight, seed release time, seed head height above surface, wind speed, maximum wind speeds and turbulence, and landscape structure (Nathan et al. 2002; Heydel et al. 2014). Empirical studies are more difficult, but Zhang et al. (2022) stained thousands of Canada goldenrod seeds with safranin and recorded wind dispersal of these small achenes with their pappi (parachute-like hairs at top of seed) of up to 2 km downwind. We can only speculate as to whether the *Geum* population at the mouth of the Bay of Fundy was founded by a long distance dispersal or whether it and the White Mountain populations are relic survivors among once continuous post-glacial populations along latitude 44°N. Geological studies suggest that Fundy shore (Stea and Mott 1998) deglaciated before the White Mountains (Thompson et al. 1999); however, genetic data would suggest that the genetically diverse New Hampshire populations are the source population (see Chap. 3). Both of the landscapes where *Geum* now survives are among the windiest in their areas—Bliss (1966) reported a maximum wind speed of 370 km/h in the White Mountains, and Brier Island (as was shown above) is the windiest spot in the mouth of the Bay of Fundy. While *Geum* (see Chap. 4) has no structural adaptations at the seed level for either wind or animal dispersal, its seeds are borne up above the low shrub vegetation and the contents of their seed heads (hypanthia or cups holding 40 seeds that tend to cling together by dense ovary hairs) are gradually released from November on into the winter by wind rattling the stems. The range of local seed dispersal by wind will be increased in winters when fens are covered with snow and the seeds are blown from the open seed heads. The near zero temperatures in December through February (Fig. 5.2) post-1998 suggest limited snow cover during this period to facilitate dispersal.

The changing temperature regime on Brier Island may be important for the long-term survival of *Geum*. The high elevation, arctic–alpine climate regime of Mt. Washington is quite different from the temperate conditions of Brier (Bliss 1966). The decline of *Geum* on Brier Island can be primarily attributed to the Herring gulls with their associated eutrophication and trampling. It also appears that the reclamation process on Big Meadow Bog of damming the ditches (East Coast Aquatics 2018) has resulted in extensive flooding of the lagg zone that has impacted *Geum* populations already in decline. Thus, the declines in *Geum* at higher elevations on Brier, e.g., Brad's Bog, may be associated with competition with overgrowing shrubbery brought on by climate warming. Regardless, the continuing declines in *Geum* suggest that the conservation of the species in Canada may depend on ex situ transplantation to the highlands of Cape Breton, or possibly to Newfoundland, where climate conditions may be more compatible with the environment of Mt. Washington where the species is not endangered.

5.7 Sea-Level Rise and Long-Term Prospects for the Survival of Geum

In addition to climate change effects, low-lying areas of Brier including Big Meadow Bog (BMB) may be threatened by storm-driven landscape changes and sea-level rise. BMB developed from a sea lagoon through many progressive wetland stages—saltmarsh, freshwater marsh, swamp, fen and bog. In 2015, the two southern most sub-populations in the lagg fen next to the Big Pond Cove barrachois (pond above sea beach) were flooded when the outflow of the barrachois became blocked by storm-driven cobbles. This resulted in losses of 50% and 66%, respectively in these two sub-populations of *Geum* (Toms 2023). This shows the vulnerability of the low-lying Big Meadow plants to flooding. The decline in freshwater drainage from BMB resulting from damming of the drainage ditches may also allow greater saltwater intrusion into Big Pond Cove. Since much of BMB is within 1.5–2.0 m of the high tide water mark in Big Pond Cove, storm surges and sea level rise are major concerns, particularly for the lagg fen.

With global climate change and resulting sea level rise, the Bay of Fundy is predicted to have a 0.5 m increase in sea level in the next 50 years and an increase of about 1.0 m by the end of the current century (Greenberg et al. 2012). This would exacerbate flooding of large areas of BMB by marine waters during storm surges associated with spring tides. Such flooding occurred during the Groundhog Day gale of 1976 with a tidal surge of 1.6 m (Parkes et al. 1997) in which waves traversed

BMG. Whereas storm activity on Brier in terms of both frequency and intensity has been stable over the last 15 years, future storms associated with higher sea levels will endanger the primary habitat of *Geum* at both the northern end of BMB at Westport and at the southern end of BMB at Big Pond Cove. The long-term trend of increasing sea level and regional subsidence following the end of isostatic rebound (Stea et al. 2001) is the gradual transformation of BMB from freshwater peat ecosystems to a salt marsh, a reversal of the states observed between marine muds (start date 9000 y BP) and the beginning of *Sphagnum* peats (2500 y BP) documented in a dated core of BMB (Spooner et al. 2017). Major breaches of the barrier beach in Big Pond Cove will hasten these changes and restrict *Geum* populations to progressively more elevated peatlands on Brier, the elevations available for peatland establishment upon deglaciation of Nova Scotia (Stea and Mott 1998).

5.8 Conclusions

With a 1.9 °C increase over the last 15 years, temperature has risen faster on Brier than comparable sites in Nova Scotia; it now has the warmest annual temperature in the province of 8.2 ± 0.8 °C with 2020–2023 all averaging over 8 °C. The increased temperatures are greatest in the fall and winter with December increases of 3 °C, and with the smallest change of 1.1 °C during summer. While these changes may not directly impact the growth and survival of *Geum*, they likely will have a negative impact on its ability to compete with shrub species in its primary habitat. We conclude that based on the impacts of climate warming and rising sea level that the long-term prospects for survival of *Geum* on Brier are poor.

References

Blaney SC (2010) COSEWIC assessment and status report on the Eastern Mountain Avens, *Geum peckii*, in Canada. Committee on the Status of Endangered Wildlife in Canada. https://publications.gc.ca/site/eng/372122/publication.html

Bliss LC (1966) Productivity in alpine microenvironments on Mt. Washington, New Hampshire. Ecol Monogr 36:125–155. https://doi.org/10.2307/1942152

Caesar L, McCarthy D, Thornally DJR, Cahill N, Rahmstorf S (2021) Current Atlantic meridional overturning circulation weakest in last millennium. Nat Geosci 14:118–120. https://doi.org/10.1038/s41561-021-00699-z

COSEWIC (2010) COSEWIC assessment and status report on the Eastern Mountain Avens *Geum peckii* in Canada. Committee on the Status of Wildlife in Canada. Ottawa. https://publications.gc.ca/site/eng/372122/publication.html

Dijkstra HA, te Raa L, Schmeits M, Gerrits J (2006) On the physics of the Atlantic multidecadal oscillation. Ocean Dyn 56:36–50. https://doi.org/10.1007/s10236-005-0043-0

East Coast Aquatics (2018) Big Meadow Bog hydrology restoration: as built report. March 16, 2018. Final. Prepared for Fernhill Institute for Plan Conservation. South Berwick, Nova Scotia. 42 pp

Environment Canada (2010) Recovery strategy for the Eastern Mountain Avens (*Geum peckii*) in Canada. *Species at Risk Act* recovery strategy series. Environment Canada, Ottawa. https://www.registrelep-sararegistry.gc.ca/virtual_sara/files/plans/rs_eastern_mountain_avens_final_2010_e.pdf

Garbary DJ (2018) Demonstrating climate change in Prince Edward Island—a procedure using climate normals and weather data suitable for classroom use. Proc Nova Scotian Inst Sci 49(2):293–311. https://doi.org/10.15273/pnsis.v49i2.8165

Garbary DJ, Hill NM (2021) Climate change in Nova Scotia: temperature increases from 1961 to 2020. Proc Nova Scotian Inst Sci 51(2):411–442. https://doi.org/10.15273/pnsis.v51i2.11174

Greenberg DA, Blanchard W, Smith B, Barrow E (2012) Climate change, mean sea level and high tides in the Bay of Fundy. Atmos-Ocean 50(3):261–276. https://doi.org/10.1080/07055900.2012.668670

Hebert D, Layton C, Brickman D, Galbraith PS (2023) Physical oceanographic conditions on the Scotian Shelf and in the Gulf of Maine during 2022. Can Tech Rep Hydrogr Ocean Sci 359. https://publications.gc.ca/site/eng/9.925351/publication.html

Heydel F, Cunze S, Bernhardt-Römermann M, Tackenberg O (2014) Long-distance seed dispersal by wind: disentangling the effects of species traits, vegetation types, vertical turbulence and wind speed. Ecol Res 29:641–651. https://doi.org/10.1007/s11284-014-1142-5

Hill NM, Chinner K, Toms B, Garbary DJ (2019) Recovery succession of native plants communities at Big Meadow Bog after disturbance and eutrophication from Herring gulls. Wetlands 39:533–548. https://doi.org/10.1007/s13157-018-1099-1

Hill NM, Garbary DJ (2021) The missing years: how the inadequacy of the weather record masked climate change of conservation concern for endangered plants in Nova Scotia. Presentation to the annual meeting of the Canadian Botanical Society.

Hill NM, Boates JS, Johnston MO, Toms BE, Manning E, Poirier J, Garbary DJ (2024) Ecology and conservationof the Eastern Mountain Avens in Nova Scotia. (Chapter 4, this book)

Keys K, Garron E, Oikle D, Quigley E, Neily P, McGrath T, Bruce J (2017) Digital wind exposure map of Nova Scotia. Nova Scotia Department of Natural Resources. Forest Research Report 2017–15, No. 99. https://novascotia.ca/natr/library/forestry/reports/Wind_map_FRR_99.pdf

Klink K (1999) Trends in mean monthly maximum and minimum surface wind speeds in the coterminous United States, 1961 to 1990. Clim Res 13:193–205. https://doi.org/10.3354/cr013193

Knight JR, Folland CK, Scaife AA (2006) Climate impacts of the Atlantic multidecadal oscillation. Geophys Res Lett 33:L17706. https://doi.org/10.1029/2006GL026242

Li SL, Wang YM, Gao YQ (2009) A review of the researches on the Atlantic multidecadal oscillation (AMO) and its climate influence. Trans Atmos Sci 32:458–465

McCarthy GD, Haigh ID, Hirschii JJ-M, Grist JP, Smeed DA (2015) Ocean impact on decadal Atlantic climate variability revealed by sea-level observation. Nature 521:508–510. https://doi.org/10.1038/nature14491

McIldoon T, Pilon M (2008) Examples of mountain induced winds in Atlantic Canada. In: Presentation to the American meteorological society, 13th conference on Mountain meteorology. Session 12B.4. https://ams.confex.com/ams/13MontMet17AP/webprogram/Paper141099.html

McVicar TR, Van Niel TG, Li LT, Roderick ML, Rayner DP, Ricciardulli L, Donohue RJ (2008) Wind speed climatology and trends for Australia, 1975–2006: capturing the stilling phenomenon and comparison with near surface reanalysis output. Geophys Res Lett 35:L20403. https://doi.org/10.1029/2008GL035627

Nathan R, Katul GG, Horn HS, Thomas SM, Oren R, Avissar R et al (2002) Mechanisms of long-distance dispersal of seeds by wind. Nature 418:409–413. https://doi.org/10.1038/nature00844

NOAA (2023) ENSO: recent evolution, current status and predictions. https://www.cpc.ncep.noaa.gov/products/analysis_monitoring/lanina/enso_evolution-status-fcsts-web.pdf

Nova Scotia Department of Lands and Forests (2021) Recovery plan for the Eastern Mountain Avens (*Geum peckii*) in Nova Scotia [Final]. Nova Scotia Endangered Species Act Recovery Plant

Series. https://novascotia.ca/natr/wildlife/species-at-risk/docs/RECOVERY_PLAN_Adopted_Eastern_Mountain_Avens.pdf
Parkes GS, Ketch L, O'Reilly CT (1997) Storm surge events in the Maritimes. In: Skafel MG (ed) Proceedings of the 1997 Canadian Coastal Conference, 21–24 May 1997. Guelph, Ontario; Canadian Coastal Science and Engineering Association, pp 115–129
Pershing AJ, Alexander MA, Brady DC, Brickman D, Curchitse EN, Diamond AW et al (2021) Climate impacts on the Gulf of Maine ecosystem: a review of observed and expected changes in 2050 from rising temperatures. Elementa Sci Anthropoc 9(1):00076. https://doi.org/10.1525/elementa.2020.00076
Pershing AJ, Hernandez CM, Kerr LA, Le Bris A, Mills KE, Nye JA et al (2015) Slow adaptation in the face of rapid warming leads to collapse of the Gulf of Maine cod fishery. Science 350(6262):809–812. https://doi.org/10.1126/science.aac9819
Pryor SC, Barthelimie RJ, Young DT, Takle ES, Arritt RW et al (2009) Wind speed trends over the contiguous United States. J Geophys Res 114:D14105. https://doi.org/10.1029/2008JD011416/doi:10.1029/2008JD011416
Saba VS, Griffies SM, Anderson WG, Winton M, Alexander MA, Delworth TL et al (2015) Enhanced warming of the Northwest Atlantic Ocean under climate change. JGR Oceans 121(1):118–132. https://doi.org/10.1002/2015JC011346
Schlepr KR, Ronconi RA, Hayden B, Allard KA, Diamond AW (2021) Estimating the relative use of anthropogenic resources by Herring gull (*Larus argentatus*) in the Bay of Fundy, Canada. Avian Conser Ecol 16(1):2. https://doi.org/10.5751/ACE-01739-160102
Seidov D, Mishonov A, Parsons R (2021) Recent warming and decadal variability of Gulf of Maine and slope water. Limnol Oceanog 66(9):3472–3488. https://doi.org/10.1002/lno.11892
Spooner I, Principato S, Hill N, White H, Dunnington D, Neily T, Stolze S (2017) Late Holocene records of changing moisture regime from wetlands from southwestern Nova Scotia, Canada: Implications for wetland conservation and restoration. Northeastern Natural 24:331–348. https://doi.org/10.1656/045.024.0310
Stea R, Fader GBJ, Scott D.B., Wu P (2001) Glaciation and relative sea-level change in Maritime Canada. In: TK Weddle, MJ Ratelle (eds) Deglacial history and relative sea-level changes, northern New England and adjacent Canada. Geol Soc Am Special Paper 351:35–49. https://doi.org/10.1130/0-8137-2351-5.35
Stea R, Mott R (1998) Deglaciation of Nova Scotia: Stratigraphy and chronology of lake sediment cores and buried organic sections. Geograph Phys Quatern 52:1–19. https://doi.org/10.7202/004871ar
Tee KT, Smith PC, LeFaivre D (1993) Topographic upwelling off southwest Nova Scotia. J Phys Oceanog 23:1703–1726. https://doi.org/10.1175/1520-0485(1993)023%3c1703:TUOSNS%3e2.0.CO;2
The Nature of Things (2021) Cape Bretoners ride out Canada's strongest winds with epic storm parties. Canadian Broadcasting Corporation. https://www.cbc.ca/documentaries/the-nature-of-things/cape-bretoners-ride-out-canada-s-strongest-winds-with-epic-storm-parties-1.5880823
Thompson WB, Fowler BK, Dorion CC (1999) Deglaciation of the northwestern White Mountains, New Hampshire. Geograph Phys Quatern 53:59–77. https://doi.org/10.7202/004882ar
Toms B (2015) Analysis of vegetation cover in Big Meadow Bog on Brier Island, Nova Scotia (1955–2015). Mersey Tobeatic Research Institute, Kempt, Nova Scotia. https://novascotia.ca/natr/wildlife/habfund/final15/NSHCF15_15_Hill_BigMeadow.pdf
Toms B (2023) Current status of Eastern Mountain Avens in Nova Scotia in 2022 and analysis of a decade of monitoring on Brier Island. Mersey Tobeatic Research Institute Report to Environment Climate Change Canada.
Van Westen RM, Kliphuis M, Dijkstra HA (2024) Physics-based early warning signal shows that AMOC is on tipping course. Sci Adv 10:eadk1189. https://doi.org/10.1126/sciadv.adk1189
Whitney N, Wanamaker AD, Ummenhofer CC, Johnson BJ, Cresswell-Clay N, Kreutz KJ (2022) Rapid 20th century warming reverses 900-year cooling in the Gulf of Maine. Nat Commun Earth Environ 3:article 179. https://doi.org/10.1038/s43247-022-00504-8

Wyatt MG, Kravtsov S, Tsonis AA (2012) Atlantic multidecadal oscillation and Northern Hemisphere's climate variability. Climate Dynam 38:929–949. https://doi.org/10.1007/s00382-011-1071-8//doi:10.1007/s00382-011-1071-8

Zhang Z, Wen G, Bu D, Sun G, Quang S (2022) Long-distance wind dispersal drives range expansion of *Solidago canadensis*. Plants 11(20):2734. https://doi.org/10.3390/plants11202734

Chapter 6
Geological History and Ecosystems Evolution at Big Meadow Bog, Brier Island, Nova Scotia

Ian Spooner, Deanne van Rooyen, Mark L. Mallory, Nelson J. O'Driscoll, Nicholas M. Hill, and Dewey D. Dunnington

Abstract An understanding of the geological history and ecosystems evolution of Big Meadow Bog peatland at Brier Island, Nova Scotia is an important element of informing the conservation management of critical habitats for a complex of nationally listed, rare, disjunct wetland species. The ecology of the peatland is influenced by both chemistry and morphology of the bedrock and glacial sediments within the watershed. The Big Meadow Bog peatland is bordered to the north and south along its long axis by basaltic ridges of the North Mountain. Drainage from these watershed ridges flows to the peatland, which has both a northern and southern outflow to the sea. Paleogeographic reconstruction suggests that Brier Island was deglaciated sooner than the rest of southwestern Nova Scotia. Big Meadow Bog peatland contains a stratigraphic record of a late Holocene transition to a moister climate shortly after 3000 cal. BP. However, several periods of late Holocene drying took place from

I. Spooner (✉) · D. van Rooyen · N. J. O'Driscoll
Department of Earth and Environmental Science, Acadia University, Wolfville, NS B4P 2R6, Canada
e-mail: ian.spooner@acadiau.ca

D. van Rooyen
e-mail: deanne.vanrooyen@acadiau.ca

N. J. O'Driscoll
e-mail: nelson.odriscoll@acadiau.ca

M. L. Mallory
Department of Biology, Acadia University, Wolfville, NS B4P 2R6, Canada
e-mail: mark.mallory@acadiau.ca

N. M. Hill
UNESCO Southwest Biosphere Association, 399 Wyman Road, Sand Beach, NS B5A 5G4, Canada
e-mail: nick@swnovabiosphere.ca

D. D. Dunnington
Voltron, 650 Castro St., Ste 120 PMB 96571, Mountain View, CA 94041, USA
e-mail: dewey.dunnington@gmail.com

N. M. Hill et al. (eds.), *Wetland Restoration for Endangered Species Recovery*, Environmental Contamination Remediation and Management,
https://doi.org/10.1007/978-3-031-71344-6_6

1950 cal. BP until present, one of which is recorded stratigraphically by the presence of a conifer forest indicating significant lowering of the local water table. Under these conditions, both wetland productivity and water table varied considerably. When designing potential habitat preservation and restoration strategies, these site-specific records of moisture and habitat structure variability need to be considered. Specific sites in the Brier Island peatlands may respond very differently to particular restoration strategies aimed at establishing long-term residency of threatened and endangered species at wetland sites.

6.1 Introduction

Brier Island is the westernmost land of the Digby Peninsula (44°15′N, 66°21′W, Fig. 6.1) in southwestern Nova Scotia and is located at the mouth of the Bay of Fundy. Big Meadow peatland (commonly called Big Meadow Bog peatland, BMB in this manuscript) is a 2 km long, oval-shaped peatland with a central, raised bog that formed between basalt ridges of the North Mountain at Brier Island which dip gently toward the Bay of Fundy. Glaciation has played a significant role in modifying the topography on Brier Island and has resulted in the deposition of the relatively thin veneer of variable surficial sediments that overlie the basalt (Webster et al. 2006). The post-glacial geological and environmental history of Brier Island is complex and was influenced by fluctuating relative sea level, changing climate and variable hydrogeological conditions. Though Nova Scotia has one of the densest archives of regional paleoenvironmental data in Eastern North America because of the excellent preservation of lake sediments of Late Wisconsinan and Holocene age (Stea and Mott 1998), relatively few records exist for wetland systems. In this chapter, we summarize both the geological and the environmental history of Brier Island, commenting on the impact on environmental history and the ecology of BMB. Paleoecological studies of peatlands provide details of past wet and dry periods, and the record from specific sites over time can help to indicate whether different sites are prone to periodic drying episodes, and whether sites are equally affected by such events. Knowing which areas are more likely to experience drying events or be affected by rising or lowering of the local water table can help to inform the design of specific habitat restoration or preservation strategies.

6.2 Geology

Brier Island lies along the southern shore of the mouth of the Bay of Fundy, a large macrotidal embayment situated on the east coast of Canada between the provinces of New Brunswick and Nova Scotia (Fig. 6.1). Two main branches extend off the central Bay: Minas Basin to the east, which is dominated by large accumulations of sand; and Chignecto Bay to the northwest, which is characterized by large accumulations

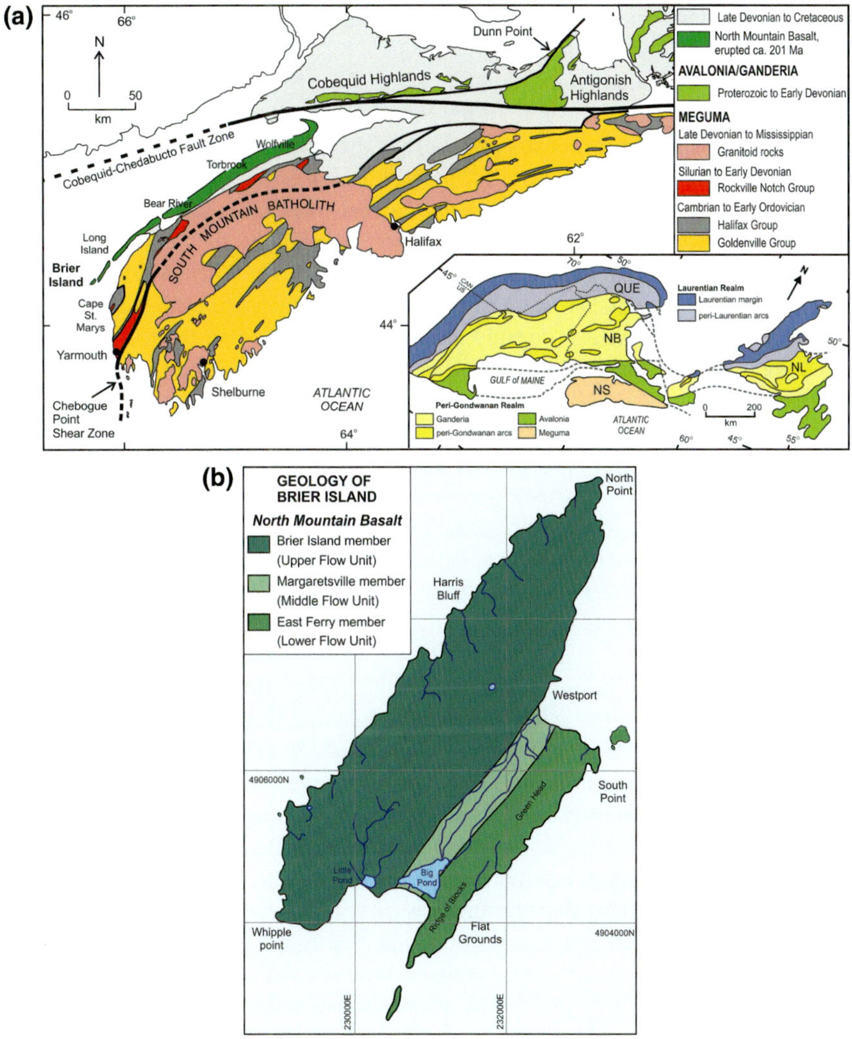

Fig. 6.1 **a** Geology of Nova Scotia with inset map for the Appalachian orogen (modified from Barr et al. 2022 after Hibbard 2006). **b** Geology of Brier Island showing the three major divisions of the North Mountain Basalt (modified from White and Horne 2012a, b, after Kontak and Webster 2010). Big Meadow peatland overlies the Margaretsville Member of the North Mountain Basalt.

of silt. The Bay of Fundy is in the Appalachian geological region (Amos et al. 1991), an area characterized primarily by deformed sedimentary and volcanic rocks. There is great variability in local relief because of glacial and fluvial erosion on bedrock of differing resistance. Chignecto Bay is a glacially excavated, elongated macro-tidal estuary (Amos et al. 1991) with blankets of transgressive, silty, glacial and post-glacial unconsolidated sediments in the upper reaches. The Minas Basin, with Brier

Island at its outermost edge, is a large, triangular basin at the head of the Bay of Fundy.

Brier Island is the southernmost section of the North Mountain, a prominent cuesta that extends along the south shore of the Fundy Basin (Kontak and Webster 2010). Brier Island consists entirely of basaltic rocks of the North Mountain Formation, also known as the North Mountain Basalt (Fig. 6.1), which is part of the Fundy Group, a Triassic to Jurassic succession of rocks that represents continental rift environments that evolved into volcanic rifts shortly before the breakup of the supercontinent Pangaea (Wade et al. 1996; Hodych and Dunning 1992). The North Mountain Basalt was erupted on continental crust, mostly as sub-aerial tholeiitic basalt but some parts of it were erupted in shallow oceanic basins that formed along the rifting margins (Greenough and Dostal 1992). This volcanic activity is part of the Central Atlantic Magmatic Province (CAMP), remnants of which are located on the eastern margin of North America from Nova Scotia to Florida, in northern South America, West Africa, and southern Europe (Cirilli et al. 2009; Marzoli et al. 2018). The major peak of volcanic activity in CAMP occurred at ca. 201 Ma, and the event only lasted for about 1 million years. High-precision U–Pb zircon geochronology of the North Mountain Basalt shows it erupted at 201.38 ± 0.31 Ma and this and other high precision zircon dates have been used to link the eruption of CAMP volcanism and resultant global cooling and glaciation to the Triassic mass extinction (Schoene et al. 2010).

The North Mountain Basalt is about 400 m thick (Kontak 2006) and is made up of three major units as summarized in maps by White and Horne (2012a, b). The East Ferry Member is the lowermost flow unit and is about 180 m thick. It consists of gray-green, fine- to medium-grained, massive basalt with microcrysts of plagioclase and pyroxene and in places has well-developed columnar jointing. This unit also contains coarse-grained gabbroic pegmatite, and the upper few meters of the unit is vesicular (Kontak 2006). The Margaretsville Member is the middle flow unit (about 150 m thick) and consists of many thin basaltic flows, each between 10 and 15 m thick with abundant vesicles which commonly contain zeolite minerals (Pe-Piper and Miller 2002; Kontak 2006). This unit contains sparse layers of sediments deposited between flows. It is dark gray to gray-green, fine-grained basalt and contains abundant zeolite minerals that form in vesicles and fractures (Kontak 2006). The Brier Island Member is the uppermost flow unit (at least 100 m thick) and consists of dark gray to gray-green, medium-grained, ophitic basalt with plagioclase and pyroxene microcrysts (Kontak 2006).

The structures and textures of the basalt flows influence how local weathering and drainage affects the rocks. The North Mountain Basalt dips gently toward the Bay of Fundy and forms two prominent ridges seen in the topography of the Island. These ridges consist of the upper (Brier Island Member) and lower (East Ferry Member) flow units (Kontak 2006). These two units are massive and do not have individual flow layers, so they are generally strong and resistant to weathering. In contrast, the middle flow unit (Margaretsville Member) consists of thin, repeated lava flows (Kontak 2006) which allows more weathering and water flow through the unit, forming a swampy valley between the two ridges. The upper and lower flow units both have distinct columnar jointing characteristic of thick basalt flows, and zeolite minerals formed by hydrothermal circulation are common in vesicles and fractures (Kontak 2006; Pe-Piper and Miller 2002). The North Mountain Basalt is a popular quarry stone for building and is used as armor stone in several harbors along the Funday shoreline (Prime 2008). It is also being investigated as a possible soil amendment to encourage sequestration of CO_2 in soils (Jauer 2022).

6.3 Glacial History

Models for Late Wisconsinan glaciation in western Nova Scotia are complex. King (1993, 1996), Stea et al. (1992, 2011) and others have shown that a series of glacial advances from shifting ice centers and the development of offshore ice divides resulted in complex striation records and the presence of stacked till sheets. These data have been used to reconstruct at least four distinct ice flow phases in southwestern Nova Scotia, each with at least one recognizable till sheet (Stea et al. 1992).

Bedrock at Brier Island is variably overlain by till deposited during the Wisconsinan glaciation. The Lawrencetown Till is the most common glacial unit and consists of muddy till with 20–30% gravel, 30–40% sand and 30–50% silt and clay with low permeability (Stea et al. 1992). At some locales the till forms a > 3 m thick, near continuous blanket over the basalt surface, whereas at other sites a < 3 m thick till veneer exists with common bedrock outcrops (Spooner et al. 2017). Ice last flowed northwestward over the North Mountain and into the Bay of Fundy from the Scotian Ice divide which formed across the axis of Nova Scotia (Stea et al. 1992). Glaciation on Brier Island likely accentuated the existing local topography in which the middle flow unit of the North Mountain basalt (Margaretsville Member) is flanked by the more resistant upper and lower members. The difference in competence of these three members is responsible for the ridge and valley topography on Brier Island.

The post-glacial isostatic and eustatic history of the Bay of Fundy region is poorly resolved. Ongoing landscape evolution combined with the development of a macro-tidal shoreline environment complicates both the identification and resolution of indicators of past sea level. Research in ice sheet modeling, tidal power development, Mi'kmaq habitation patterns, and coastal erosion dynamics requires a better understanding of relative sea level change in the region (Taylor 2020). Paleo-shoreline features have not been recognized on Brier Island.

During the final stages of deglaciation, remnant ice caps developed in highland areas of central Nova Scotia; some of these ice caps may have been rejuvenated during a short-lived but severe cool period called the Younger Dryas (Stea and Mott 1998). The basal ages of lake cores have been used to reconstruct the timing of deglaciation; pollen, diatom, and macrofossil records have been used to reconstruct regional climate conditions.

6.4 Post-glacial Environmental Change

Dates from a raised marine deposit suggest that Brier Island was likely the earliest land on Nova Scotia to deglaciate (15.9–14.7 cal. BP, Stea et al. 2011) and may have deglaciated 2000–2500 years before alpine New Hampshire (cf. values in Stea et al. 2011 and Thompson et al. 1999). The impact of late glacial and early Holocene (~11,000–8000 cal yr BP) environmental change on ecosystems at Brier Island is unclear. Cwynar and Levesque (1995) described Younger Dryas-equivalent gravel and sand deposits in lakes in western New Brunswick and Maine which have been used to infer dramatic cooling and landscape instability. Terrestrial records of both deglaciation and Younger Dryas cooling and reinvigoration of ice caps have also been recognized in mainland Nova Scotia (Mott and Stea 1993; Spooner et al. 2005). Stea et al. (2011) suggested that it is likely that a corridor for animals and plants was available along the edges of the Bay of Fundy in the early stages of deglaciation. This thin maritime band of boreal climate along Nova Scotia's shoreline suggests a potential pathway for the continued migration of the boreal organisms on Brier Island.

Post-glacial landscape evolution at Brier Island is strongly linked to both isostatic recovery, relative sea level change, and the development of a macrotidal shoreline environment. Relative sea levels have been rising in most parts of the region since deglaciation and continue at present (Shaw et al. 1993). Mean sea level in the Bay of Fundy has lowered from + 40 m at 13 ka to − 25 m at 8 ka, and has been rising since (Amos and Zaitlin 1985). To date, few paleo-shoreline features associated with relative sea level change have been recognized and those that are known to exist exhibit little continuity (Webster et al. 2006). Paleo beach and terrane features appear to be most common along the eastern shoreline of the upper Bay of Fundy and are found at 19 m and more above current sea level (Taylor 2020). Areas to the southwest, such as Digby Neck (which includes Brier Island) show minimal evidence of raised beaches (Davis 1983a, b).

Conventional post glacial paleoclimate models from Nova Scotia postulate that post-glacial warming was abruptly interrupted by Younger Dryas cooling (ca. 12,500 cal. BP; Stea and Mott 1998) during which a shrub and herb dominated landscape developed. At the same time, the basin dimensions of the Bay of Fundy changed primarily due to sea level rise though sedimentation and erosion, tectonics, and compaction of the subsurface sediments were also likely contributors to morphometric change. Abiotic proxy records indicate that the middle Holocene thermal maximum was followed by a transition to cooler and moister climate, though the timing of this transition is poorly constrained (Lennox et al. 2010; Spooner et al. 2014). Similarly, most pollen records from lake and wetland archives in Nova Scotia indicate a muted response to this transition (Davis 1983a, b; Green 1987; Ogden 1987). Most high-resolution paleoclimate records from southwestern Nova Scotia indicate that the late Holocene was a time of increasing precipitation and cooler air temperatures (Lennox et al. 2010; Spooner et al. 2014).

Wetlands have been the subject of far fewer paleoecological studies than lake sites in Nova Scotia though they have the potential to preserve long (> 9000 years) and continuous records of hydrological variability, a proxy for moisture regime (Robichaud and Bégin 2009; Spooner et al. 2014; Suarez-Gonzalez et al. 2015). Stea and Mott (1998) obtained a core from a Brier Island bog that contained a basal unit of sand containing marine shells in which radiocarbon dating suggest deglaciation before 13.6 ka. (13.2–10.3 thousand years BP). Lennox et al. (2010) in a paleolimnological study at Canoran Lake, located about 80 km to the southeast of BMB, noted both fluctuating and higher than normal lake levels in the late Holocene. Spooner et al. (2014) found that a wetland record from Pleasant River Fen (located in south-central Nova Scotia about 120 km northwest of BMB) indicated an increase in precipitation and water table fluctuation in the late Holocene from 5250 cal. BP onwards. This record coincides broadly with other records of the development of cooler and wetter conditions during the late Holocene (Spooner et al. 2014).

A study at Baltzer Bog in the eastern Annapolis Valley (120 km from Brier Island) recognized two woodland—wetland moisture transitions sometime after 1640 cal. BP and 1045 cal. BP (Spooner et al. 2017). A disparity in timing of precipitation increase between the Baltzer Bog record and the Pleasant River Fen record was noted but was attributed to local geomorphological conditions at Baltzer Bog that resulted in a lag in the ecological response to the initial increase in precipitation and flooding.

At BMB, groundwater flows from swamps on bounding basalt ridges, resulting in a marginal fen bounded and influenced by both the central raised bog and the surrounding swamps. These conditions have produced a classic "lagg" ecotone (Howie and Tromp-van Meerveld 2011) which originally consisted of a system of pools and streams comprising habitat for a complex of rare boreal organisms (Spooner et al. 2017). Construction of ditches in the lagg zones and the central raised bog in 1958 accelerated outflow and lowered the water table (Fig. 6.2). A biostratigraphic record from BMB obtained by trenching and vibracoring the central zone of the raised bog (Spooner et al. 2017) indicated that BMB is underlain by marine sand and lacustrine mud, and wood at this transition produced a date of 9435 cal. BP (Fig. 6.3). The lacustrine mud was overlain by mats of various *Sphagnum* types; a wood layer within this succession was dated at 1760 cal. BP. Spooner et al. (2017) developed a stratigraphic profile of the *Sphagnum* moss community for the BMB section to explore whether shifts in the relative abundances of *Sphagnum* species might reflect changes in peatland conditions. The degree of decomposition (humification index-bog surface wetness index; BSW) of the *Sphagnum* peat in the BMB section was used as an indicator of changing bog surface moisture levels.

The *Sphagnum* successions were interpreted to indicate substantially fluctuating water table (Fig. 6.3). The presence and preservation of the wood mat and conifer needles were interpreted to indicate development of a conifer forest at the site, an indication of significant drying at 1760 cal. BP, likely associated with significantly reduced local water table. The abrupt transition above the wood mat to *Sphagnum* indicates that shortly after 1760 BP, a rapid increase in water table elevation took place (Spooner et al. 2017).

Fig. 6.2 Sampling the extruded vibracore sediment at BMB (Photo credit Dewey D. Dunnington)

The BMB record along with records from Baltzer Bog and Pleasant River Fen represent independent wetland records that collectively suggest the occurrence of a series of Holocene wetting periods as far back as 5250 cal. BP and as recent as 1045 cal. BP. The data also indicate an apparent overlap between the timing of water table rise (increased moisture), particularly at Baltzer Bog and BMB. Spooner et al. (2017) suggested significant wetting events at 1045 cal. BP and 1640–1760 cal. BP, though Payette (1988) suggested dry conditions elsewhere in eastern North America at this time. However, paleoecological records from northeastern Canada indicate significant climate variability in the last millennium rather than sustained cooling, a condition also likely in Nova Scotia (Lemus-Lauzon et al. 2016).

Though the water table rise is a local phenomenon, the apparent persistent increase in moisture recorded at BMB, Baltzer Bog and Pleasant River Fen likely indicates a significant climate regime shift. Other paleoenvironmental records indicate similar but much subtler late Holocene trends (Lennox et al. 2010; Stea and Mott 1998). Historical records show that over the past 100 + years, wetlands in southwestern Nova Scotia have been in transition from wetland to woodland. These data may indicate coincident declining water tables but also may be influenced by anthropogenic activity at almost all sites, in particular drainage modification and peat harvesting (Environment Canada 2010).

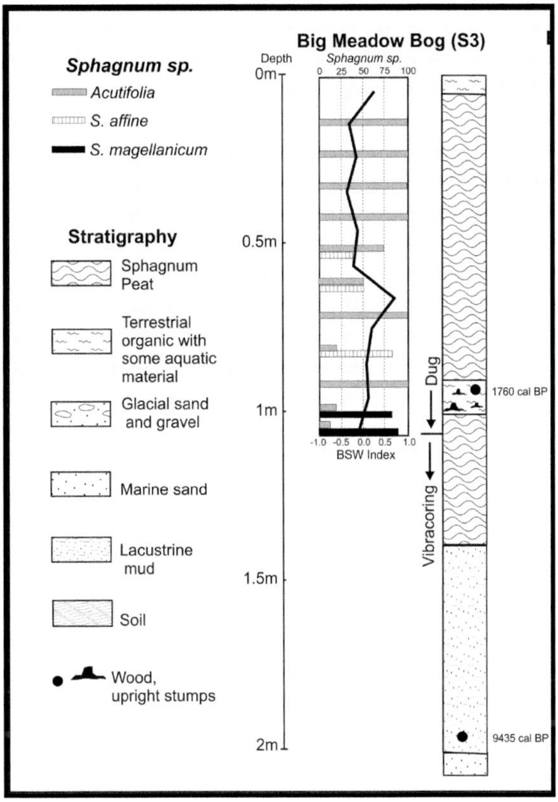

Fig. 6.3 Section stratigraphy at Big Meadow Bog showing the age of the wood mats, *Sphagnum* species composition (Big Meadow Bog), and Bog Surface wetness (BSW index). The BSW index indicates variable water table elevation from 1760 cal. BP onwards. Drying at about 1760 cal. BP is indicated by the presence of a substantial wood mat with a spruce forest inferred at that time. Figure modified from Spooner et al. (2017)

6.5 Implications of Geology and Environmental Change on the Ecology of Big Meadow Bog

Wetland paleoenvironmental records provide perspective on moisture variability that may offer insight into species resilience to environmental change and habitat suitability for wetland restoration and/or species re-introduction (Spooner et al. 2014). BMB is now a highly disturbed wetland as it has been ditched for agriculture, then abandoned. It is now the site of an extensive wetland restoration driven by Recovery Plan for the Endangered Species, *Geum peckii,* the Eastern Mountain Avens, which thrives only in the mountains of New Hampshire and wetlands on Brier Island (Munro et al. 2014). Recent studies on the suitability of wetlands in Nova Scotia for restoration and species re-introduction have focused on reversing the impact of anthropogenic alteration of local hydrology (e.g., *Geum peckii*, Big Meadow Bog) and determining optimum forest and wetland structure (e.g., *Emydoidea blandingii* at Pleasant River Fen; Spooner et al. 2014). The paleorecord from BMB suggests that fens to the north of Brier Island, also located between the same basalt ridges that formed Big Meadow Bog, may in future become raised bog with lagg habitats essential to *Geum peckii.* As well, with an understanding of the paleohydrological change at BMB and regional geology, it may be possible to locate additional sites which may serve as suitable future habitat. Therefore, the landscape features at BMB that serve, in part, to support *Geum peckii* may be a critical component to the success of this rare plant.

It is apparent that an understanding of moisture regime in Nova Scotia and its impact on wetland environments is required to understand the scope of conditions in wetland environments that provide habitat for rare and specialist species such as *Geum peckii* (Hill and Keddy 1992; Munro et al. 2014). For example, this approach was used at Pleasant River Fen to provide a long-term baseline record on habitat availability for Blanding's turtles (*Emydoidea blandingii*) which exist in Nova Scotia as a disjunct population complex at the extreme northeastern area of the species range (Spooner et al. 2014). As well, recent studies in Southwestern Nova Scotia suggest the long-term residency of disjunct wetland species regardless of rapid variability in water table observed in the wetland records and the coincident change in wetland habitat structure (Newton and Herman 2009). These observations may be used to better understand the resilience of these species considering both anthropogenic and natural environmental change. Therefore, an understanding of how these wetlands responded to past environmental change is required to provide context for habitat management and restoration efforts given predicted future changes.

Acknowledgements The authors acknowledge the Natural Science and Engineering Research Council (NSERC), and Acadia University for financial support of this research project. We are also grateful for the permission from Northeastern Naturalist to refer to text and a figure from Spooner et al. (2017).

References

Amos C, Zaitlin B (1985) The effects of change in tidal range on a sublittoral macrotidal sequence, Bay of Fundy, Canada. Geo-Marine Lett 4:161–169

Amos C, Tee KT, Zaitlin B (1991) The post-glacial evolution of Chignecto Bay, Bay of Fundy, and its modern environment of deposition

Barr SM, Kuiper Y, van Rooyen D, White CE (2022) Geological comparisons and correlations among crustal blocks in Eastern North America, Northwest Africa, and Western Europe. Field trip part of the International Geoscience Programme (IGCP) project 683 (https://igcp683.org/), held during GAC-MAC-IAH-CNC-CSPG Joint Meeting, May 15–18, Halifax NS

Cirilli S, Marzoli A, Tanner L, Bertrand H, Buratti N, Jourdan F, Bellieni G, Kontak D, Renne P (2009) Latest triassic onset of the central Atlantic magmatic province (CAMP) volcanism in the Fundy basin (Nova Scotia): new stratigraphic constraints. Earth Planet Sci Lett 286:514–525

Cwynar LC, Levesque AJ (1995) Chironomid evidence for late-glacial climatic reversals in Maine. Quatern Res 43:405–413

Davis MB (1983) Holocene vegetation history of the eastern United States. In: Wright HE (ed) Late-Quaternary environments of the United States, vol 2. The Holocene. University of Minnesota Press, Minneapolis, pp 166–181

Davis SA (1983b) Rising sea levels threaten archaeological sites. Can Geogr J 103:40–46

Environment Canada (2010) Recovery strategy for the Eastern Mountain Avens (*Geum peckii*), species at risk act recovery strategy series. Environment Canada, Ottawa, 31 pp

Green DG (1987) Pollen evidence for the postglacial origins of Nova Scotia's forests. Can J Bot 65:1163–1179

Greenough JD, Dostal J (1992) Geochemistry and petrogenesis of the early Mesozoic North Mountain Basalts of Nova Scotia, Canada. Geol Soc Amer Spec Paper 268:149–159

Hibbard JP, van Staal CR, Rankin DW, Williams H (2006) Lithotectonic map of the Appalachian Orogen, Canada-United States of America; Geological Survey of Canada, Map 2096A, scale 1:1 500 000

Hill NM, Keddy PA (1992) The prediction of rarities from habitat variables: coastal plain plants on Nova Scotian lakeshores. Ecology 73:1852–1859

Hodych JP, Dunning GR (1992) Did the Manicouagan impact trigger end-of-the-Triassic mass extinction? Geology 20:51–54

Howie SA, Tromp-van Meerveld I (2011) The essential role of the lagg in raised bog formation and restoration. Wetlands 31:613–622

Jauer C (2022) Nova Scotia basalt for CO_2 capture and enhanced crop yield. GAC-MAC-IAH-CNC-CSPG 2022 Halifax meeting abstracts. Geosci Canada 49:59–226. https://doi.org/10.12789/geocanj.2022.49.188

King LH (1993) Proposed younger Dryas glaciation of the eastern Scotian shelf. Can J Earth Sci 31:401–417

King LH (1996) Late Wisconsinan ice retreat from the Scotian shelf. GSA Bull 108:1056–1067

Kontak DK (2006) Geological map of the North Mountain Basalt from Cape Split to Brier Island, with comments on its resource potential. Nova Scotia Dept Nat Res Mineral Res Branch Report ME 1:39–66

Kontak DJ, Webster TL (2010) Overview map for the bedrock geology maps of basaltic rocks of the North Mountain Formation from Brier Island to Cape Split, Nova Scotia. Nova Scotia Dept Nat Res Mineral Res Branch Open File Map ME 7(scale 1):200000

Lemus-Lauzon I, Bhiry N, Woollent J (2016) Assessing the effects of climate change and land use on Northern Labrador forest stands based on palaeoecological data. Quatern Res 86:260–270

Lennox B, Spooner I, Jull T (2010) Post-glacial climate change and its effects on a shallow dimictic lake in Nova Scotia, Canada. J Paleolimnol 43:15–27

Marzoli A, Callegaro S, Dal Corso J, Davies JH, Chiaradia M, Youbi N, Bertrand H, Reisberg L, Merle R, Jourdan F (2018) The Central Atlantic magmatic province (CAMP): a review. In: Tanner LH (ed) The Late Triassic world: Earth in a time of transition. Springer-Verlag, New York, pp 91–125

Mott RJ, Stea RR (1993) Late-glacial (Allerod/Younger Dryas) Buried Organic Deposits, Nova Scotia, Canada. Quat Sci Rev 12:645–657

Munro MC, Newell RE, Hill NM (2014) Nova Scotia plants. Museum of Nova Scotia Publication, Halifax, NS, p 25

Newton EJ, Herman TB (2009) Habitat, movements, and behaviour of overwintering Blanding's turtles (*Emydoidea blandingii*) in Nova Scotia. Can J Zool 87:299–311

Ogden JG III (1987) Vegetational and climatic history of Nova Scotia. I. Radiocarbon-dated pollen profiles from Halifax, Nova Scotia. Can J Bot 65:1482–1490

Payette S (1988) Late Holocene development of sub-arctic ombrotrophic peatlands: allogenic and autogenic succession. Ecology 69:516–531

Pe-Piper G, Miller L (2002) Zeolite minerals of the north shore of the Minas Basin, Nova Scotia. Atlantic Geol 38:11–28

Prime G (2008) An overview of the industrial mineral potential of the North Mountain Basalt. Nova Scotia Dept Nat Res Mineral Res Branch Report ME 1:81–104

Robichaud A, Bégin Y (2009) Development of a raised bog over 9000 years in Atlantic Canada. Mires Peat 5:1–19

Schoene B, Guex J, Bartolini A, Schaltegger U, Blackburn TJ (2010) Correlating the end-Triassic mass extinction and flood basalt volcanism at the 100-ka level. Geology 38:387–390

Shaw J, Taylor RB, Forbes DL (1993) Impact of the Holocene transgression on the Atlantic coastline of Nova Scotia. Géograph Phys Quatern 47:221–238

Spooner IS, MacDonald I, Beierle B, Jull T (2005) A multi-proxy lithostratigraphic record of Late-Glacial and Holocene climate change from Piper Lake, Nova Scotia. Can J Earth Sci 42:2039–2049

Spooner IS, Stolze S, Martin B, Robichaud A, Herman T, Mockford S, Caverhill B, Mazzucchi D, White H (2014) A 10 000-year record of environmental change from Pleasant River Fen, Nova Scotia, Canada: impact on Blanding's turtle habitat. Wetlands 34:1145–1158

Spooner IS, Principato S, Hill N, White H, Dunnington D, Neily T, Stolze S (2017) Late Holocene records of changing moisture regime from wetlands in southwestern Nova Scotia, Canada: implications for wetland conservation and restoration. North Nat 24:331–348

Stea RR, Mott RJ (1998) Deglaciation of Nova Scotia: stratigraphy and chronology of lake sediment cores and buried organic sections. Géograph Phys Quatern 52:1–19

Stea RR, Forbes DL, Mott RJ (1992) Quaternary geology and coastal evolution of Nova Scotia. Geological Association of Canada/Mineralogical Association of Canada, Joint Annual Meeting, Wolfville 1992, Field Excursion A-6, Guidebook, 125 p

Stea RR, Seaman AA, Pronk T, Parkhill GA, Allard S, Utting D (2011) The Appalachian glacier complex in maritime Canada. In: Ehlers J, Gibbard PL, Hughes PD (eds) Quaternary glaciations—extent and chronology (Develop Quatern Sci 15:631–659)

Suarez-Gonzalez A, Sutton JT, Trant AJ, Zamlynny E, Good S (2015) Rethinking refugia: tree topology, divergence dates, and demographic history trace the distribution of the endangered Plymouth Gentian (*Sabatia kennedyana*) from the Pleistocene glaciation to present day. Am J Bot 102:609–620

Taylor CA (2020) Paleogeography, sea-level change, and the peopling of the Maritimes: an archaeological perspective. Ph.D. Thesis, Dalhousie University, Halifax, Nova Scotia, 274 pp

Thompson WB, Fowler BK, Dorion CC (1999) Deglaciation of northwestern White Mountains, NH. Geograph Phys Quatern 53:59–77

Wade JA, Brown DE, Traverse A, Fensome RA (1996) The Triassic-Jurassic Fundy Basin, eastern Canada: regional setting, stratigraphy and hydrocarbon potential. Atlantic Geol 32:189–231

Webster TL, Murphy JB, Gosse J, Spooner IS (2006) The application of lidar-derived digital elevation model analysis to geological mapping: an example from the Fundy Basin, Nova Scotia, Canada. Can J Remote Sens 32:173–193

White CE, Horne RJ (2012a) Bedrock geology map of the Church Point area, NTS sheet 21B/08, Digby County, Nova Scotia. Nova Scotia Dept Nat Res Mineral Res Branch, Open File Map ME 081(scale 1):50000

White CE, Horne RJ (2012b) Bedrock geology map of the Meteghan area, NTS sheet 21B/01, Digby and Yarmouth counties, Nova Scotia. Nova Scotia Dept Nat Res Mineral Res Branch Open File Map ME 86(scale 1):50000

Chapter 7
Pre and Post Restoration Hydrology of Big Meadow Bog

Audrey Hiscock, Gavin Kennedy, John C. Brazner, Craig Lake, Lindsay Johnston, and Rob Jamieson

Abstract The historical drainage activities in Big Meadow Bog (BMB) on Brier Island, Nova Scotia have altered the hydrologic regime and physical characteristics of the wetland complex. The effect of restoration activities on water levels throughout the wetland complex was investigated on two different spatial scales: sitewide and within a focused study plot. The spatial variability of water level recovery was investigated and comparisons were made to a reference wetland. Growing season water levels were closer to ground surface for longer periods post-restoration, indicating that the restoration was successful at raising water levels across BMB. However, some areas of the bog showed poorer water level recovery. The prolonged drying and decomposition of the peat could be affecting its ability to retain water. Future research should include monitoring other comparable peat wetland systems as references, characterizing the health of the peat material, and monitoring the community of vegetation in the bog and potential re-colonization of bog moss species.

Keywords Restoration · Hydrology · Wetland · Peat · Water level

7.1 Introduction

The Big Meadow Bog (BMB) wetland complex on Brier Island, Nova Scotia (NS) has been impacted by drainage activities since the late 1950s. Many peatlands across Nova Scotia have historically been altered, mainly by lowering their water table, in order to increase the quality of land for agricultural purposes. In the case of BMB,

A. Hiscock · C. Lake · L. Johnston · R. Jamieson (✉)
Dalhousie University, Halifax, Nova Scotia, Canada
e-mail: jamiesrc@dal.ca

G. Kennedy
Nova Scotia Environment and Climate Change, Halifax, Nova Scotia, Canada

J. C. Brazner
Nova Scotia Natural Resources and Renewables, Kentville, Nova Scotia, Canada

N. M. Hill et al. (eds.), *Wetland Restoration for Endangered Species Recovery*, Environmental Contamination Remediation and Management,
https://doi.org/10.1007/978-3-031-71344-6_7

drainage ditches were dug and modifications to the natural drainage channels were made in order to create desirable lands for farming (see Chap. 1). These drainage activities have altered the hydrology of the system; since then water levels have decreased and the upper layers of peat have become unsaturated and more decomposed. Other impacts to BMB over the past ~80 years include the construction and subsequent abandonment of a secondary road and culvert near the northern outlet of the wetland complex, the construction and maintenance of Gull Rock Road along the eastern margin, the construction of a private runway and associated vegetation removal and site drainage modifications, and historical off-highway vehicle usage in parts of the peatland. These impacts have facilitated the establishment of woody vegetation within BMB over the years, which has accelerated drying of the peat material due to increased evapotranspiration. The woody vegetation has also created competition within the vegetation community and made it more difficult for bog vegetation to be dominant.

The peatland has also been impacted by the emergence of herring and great black-backed gulls. Since the 1970s, the gulls have taken over many portions of the bog to use it as a nesting habitat. The presence of gulls has increased nutrient levels in the bog, impacting the survival of native bog vegetation. The combined effects of the historical drainage and the gulls has threatened the habitat of the Eastern Mountain Avens (EMA), *Geum peckii*, a globally rare plant that gives the bog its high ecological value.

In this chapter the effects of restoration activities on the hydrology of Big Meadow Bog are presented. The chapter will describe the baseline 'disturbed' hydrology of the wetland complex, the effect the restoration activities had on water levels in Big Meadow Bog, and the spatial variability in water level recovery.

7.2 Background

7.2.1 Peatland Hydrology

Peatlands are defined as wetland areas with an accumulation of organic matter in excess of 40 cm (National Wetland Working Group 1997). They typically develop according to a successional gradient from mesotrophic (nutrient rich) fens to *Sphagnum* dominated poor fens to oligotrophic bogs (Kuhry et al. 1993). Peatlands are characterized by a diplotelmic soil structure with the upper (< 0.5 m) acrotelm consisting of surface layers of permeable, living and poorly decomposed organic matter, and the underlying catotelm consisting of more decomposed, less permeable saturated peat. The diplotelmic structure of the peat system regulates peatland hydrological responses, limiting water table and soil moisture variability (e.g., Price and Schlotzhauer 1999). During a period of drought, the acrotelm's high permeability and low capillarity limits evaporative losses as water is retained in the underlying catotelmic layers, which are associated with lower hydraulic conductivity and increased water retention due to the smaller average pore size diameter of the peat matrix (Kennedy and Price 2005). Shrinkage (contraction of peat matrix with increasing soil water pressure in unsaturated zone) and compression (decrease in pore volume in saturated zone due to primary consolidation and secondary compression under loading) of the deformable peat matrix during a period of water deficit allows the peatland to maintain wetter soil moisture conditions and a higher water table relative to the peat surface, which is critical in providing non-vascular Sphagnum access to an adequate supply of soil moisture (Price and Schlotzhauer 1999). The reduction in peat pore volume also reduces hydraulic conductivity while increasing water retention (Kennedy and Price 2005). During a rain event, precipitation rapidly infiltrates through the permeable acrotelm, limiting water losses due to runoff but allowing the system to shed water through shallow lateral groundwater flow while groundwater levels are high (Bay 1969). The acrotelm's high storativity and swelling of the peat allows the system to rapidly accommodate significant increases in water content.

Peatland drainage, typically accomplished by digging a drainage ditch, causes a lowered and more variable water table. Water table variability tends to increase with time since the drainage was established (Kennedy and Price 2005). Poulin et al. (1999) showed that drainage can impact vegetation up to 60 m from drainage ditches in peatlands located in Quebec and New Brunswick. Drainage is associated with changes to the pore structure of peat in both acrotelmic and catotelmic layers (Price et al. 2003). Lowering of the water table due to drainage increases the effective stress on underlying compressible peat in excess of pre-consolidation pressures (Lang 2002) and promotes peat oxidation due to a thicker zone of aeration (humification and conversion of carbonaceous material via mineralization to water and CO_2) (Waddington et al. 2002), changing the peatland's morphology and impairing the system's self-regulating function. Van Seters and Price (2001) measured a total subsidence of 80 cm for a peatland in the Bas-Saint-Laurent, Quebec that had been drained for 57 years.

7.2.2 Physical and Hydrological Characteristics of Big Meadow Bog

Big Meadow Bog is a 40 ha wetland complex located on Brier Island, a small (1700 ha) island off the westernmost tip of Nova Scotia where the Bay of Fundy meets the Gulf of Maine. The peatland has an elongate shape, stretching approximately 1800 m from the community of Westport in the northeast, to Big Pond in the southwest. The southwest outlet, Big Pond, is a barrachois with episodic barrier breaches draining ponded brackish water into the Gulf of Maine. The width of the peatland varies from about 350–450 m.

The wetland type at BMB varies according to an elevation gradient from upland slope swamp and fen to a central basin bog (National Wetland Working Group 1997).

The peatland is underlain by Triassic North Mountain Formation basalt bedrock (Kontak and Webster 2010) and the surrounding area is mantled by stony till plain deposits (Stea et al. 1992). The peat depth within the peatland varies but is greatest toward the center of the bog (Kennedy et al. 2015), exceeding 4 m in the central portion according to a peat depth survey in 2014, and at least 2 m, based on measurements at 30 random locations by Brown (2003). An 8 m core collected from a bog approximately 1 km west of BMB on Brier Island showed 5 m of peat underlain by 2.3 m of gyttja (organic lake sediments), overlying sand and gravel with marine shells (Wilson et al. 1993).

The hydrologic characteristics of the disturbed BMB have been previously assessed in studies dating back to 2003, but the most thorough assessment was completed in 2013–2014 when 16 monitoring wells were installed throughout the wetland complex. These investigations revealed localized lowering of the water table (> 70 cm observed in north-central area of the bog in 2014 during dry conditions) in the central area of BMB due to the effect of the drainage ditching, and it was suspected that this was also associated with higher rates of peat subsidence. The raised bog character noted in historical descriptions of the bog was no longer apparent in most areas. A conceptual model illustrating the BMB system and how it was impacted by the ditching is presented in Fig. 7.1. These previous studies demonstrated significantly more water table variability compared to an undisturbed system and water level change was exaggerated toward the center of the wetland complex, near the central drainage ditch. The components of the site's water balance have not been well quantified, and seepage losses or groundwater interaction with underlying sediments is not known. However, it is expected that the principal mechanisms for water losses from the central basin bog include evapotranspiration and groundwater discharge to the central ditch and peripheral streams (Kennedy et al. 2015). Discharges from the central basin bog to the northern and southern outflows of the central ditch are estimated to be of similar magnitude, however the central drainage ditch did not appear to efficiently convey water during the growing season, displaying a muted discharge response to rain events compared to the peripheral streams (Webster et al. 2015).

7.2.3 *Hydrologic Restoration of Big Meadow Bog*

The restoration of BMB aimed to restore an ombrotrophic state in the domed portion of the bog and re-establishment of the lagg (bog margins) zone. The restoration objectives included (Chap. 10; East Coast Aquatics Inc. 2018):

- Restoring summer low flow period shallow groundwater elevation to within 0–20 cm of the ground surface in lagg areas.
- Re-create natural ground saturation conditions unsuitable for nesting gulls.
- Re-create natural rate of discharge on the site.
- Re-establish bog/fen hydrology with little to no future maintenance.

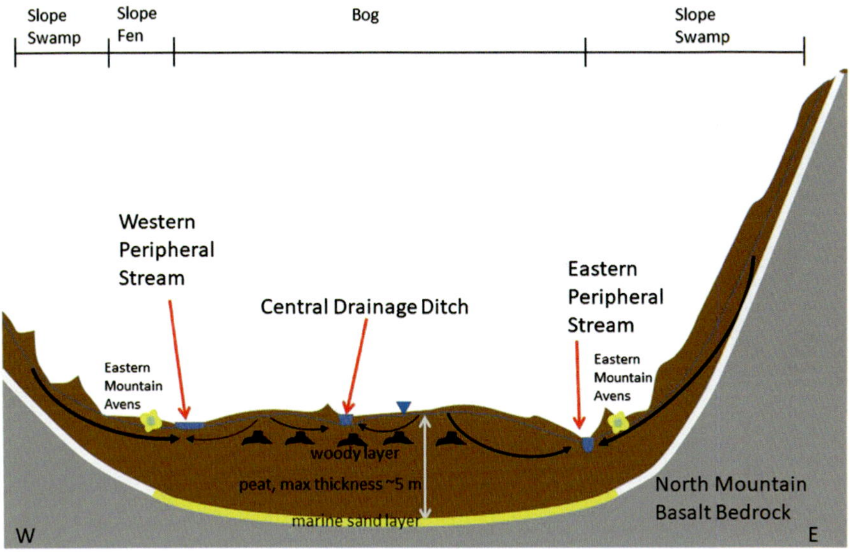

Fig. 7.1 Conceptual model of the big meadow bog wetland complex and disturbances

The restoration of BMB involved the two primary activities of (i) constructing ditch blocks and reprofiling the ditches, and (ii) removing woody vegetation. Ditch blocks were constructed to block any flow of water in the existing ditches and redirect surface water to natural channels. The ditch blocks were built using saturated, catotelmic peat from nearby borrow pits. A total of 123 ditch blocks were installed throughout the BMB site, with spacing determined by the slope of the ground surface. The lengths of the drainage ditch between ditch blocks were re-profiled to create a more gradual slope and allow for native vegetation re-colonization. Lastly, the removal of woody vegetation in the central raised portion of the bog was completed to reduce evapotranspiration and promote recovery of water levels and native vegetation. The restoration of the southern end of BMB commenced in the spring of 2016 and continued into the spring of 2017. The northern portion of the bog underwent restoration during the fall of 2017. Figure 7.2 illustrates the sequence of restoration activities (East Coast Aquatics Inc. 2018). Readers can refer to Chap. 10 for more details.

7.3 Methods

Following the restoration work, research was conducted to understand the effect of restoration activities on the hydrology of the system. Water levels across the site were studied from June 2017 to September 2019. Fourteen water level loggers were installed in monitoring wells located throughout the wetland complex to provide

Fig. 7.2 Sequence of restoration activities completed by east coast aquatics in big meadow bog, illustrating ditch block installation locations (East Coast Aquatics Inc. 2018)

an overview of the sitewide water level response and pattern of spatial variability. A more focused study plot was established and instrumented to investigate water levels and spatial patterns at a smaller scale in BMB (Fig. 7.3). This study plot was a 50 m × 50 m plot in the northern portion of the bog, with the existing central ditch running through the middle of the plot. This area of the bog was chosen because of its characteristics; increased drying and peat decomposition from lower water levels had occurred in this area, which had been accelerated by the presence of gulls and the associated nutrient enrichment. The monitoring equipment in the study plot was installed prior to the restoration of the northern portion of BMB in the fall of 2017.

A reference wetland was also monitored during this post-restoration study period to provide a baseline comparison to an adjacent unimpacted wetland to BMB. The reference wetland is located northwest of BMB at a latitude of 44° 15′ 22″ N and a longitude of 66° 22′ 12″ W (Fig. 7.4) It is characterized as an undisturbed fen and bog peatland system. Water levels in the reference wetland were monitored from 2017 to 2019. The three well locations across the reference wetland were chosen to capture hydrologically different regions, which are characteristic in most domed bogs. As shown in Fig. 7.4, the raised center portion of the reference bog is represented by the 'Dry' water level location and the margins, or lagg, of the bog would be represented by the 'Wet' water level location.

The position and elevation of wells were determined with a Real-Time Kinematic (RTK) topographic survey. The monitoring wells were constructed out of 25 mm diameter PVC pipe and were 1.25–1.50 m in length and were perforated or screened. The wells were capped at the bottom and covered in a well filter sock to prevent in-filling with peat. The monitoring wells were instrumented with pressure transducers which logged pressure and temperature on a 30 min time step. Manual water level readings were recorded at regular intervals when downloading and deploying the transducers. Pressure readings were corrected using barometric pressure data from an Environment Canada climate station located on Brier Island (Climate ID: 8200604). Continuous water level measurements were summarized as average daily water levels and filtered for the growing season study period (May–September).

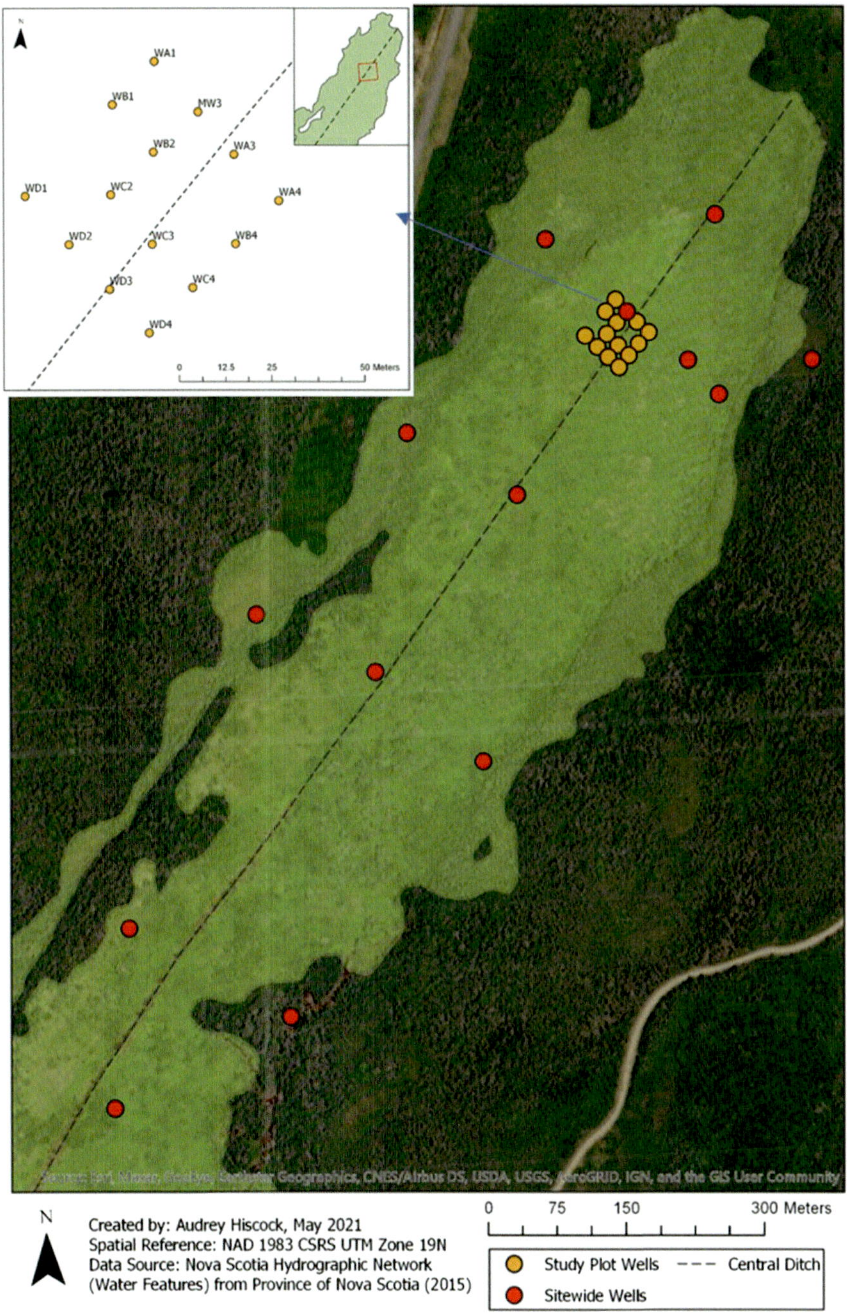

Fig. 7.3 Big meadow bog with sitewide and focused study plot monitoring well locations

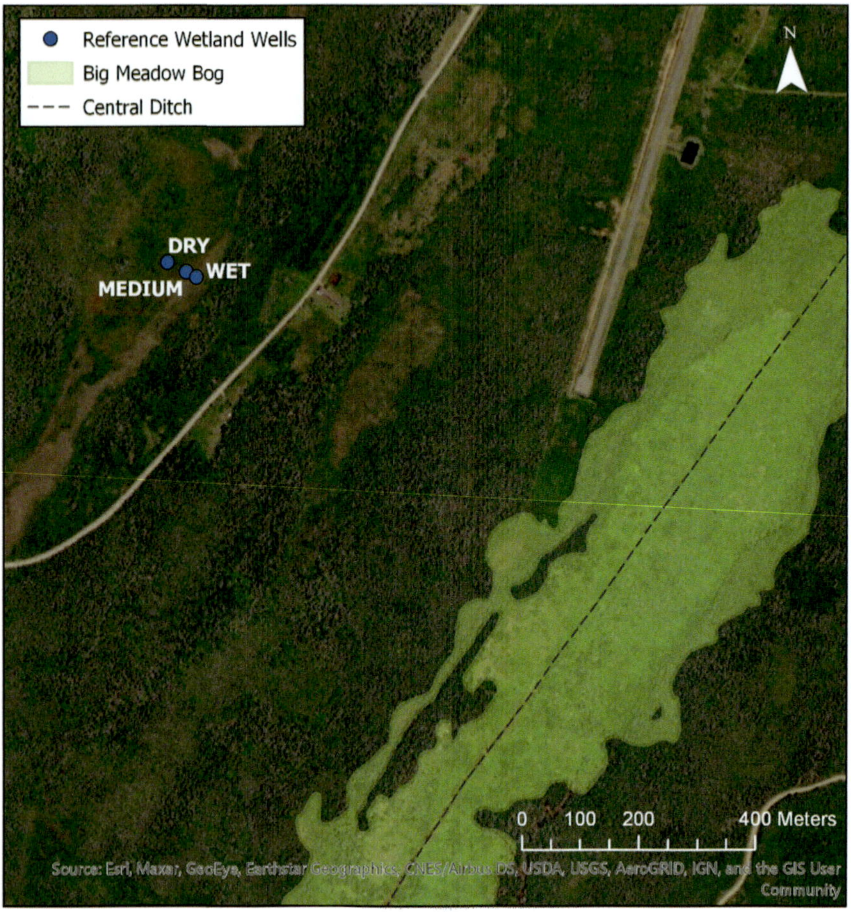

Fig. 7.4 Reference wetland, relative to big meadow bog wetland complex, with monitoring well locations

Statistical analysis was performed to investigate the differences in water levels between study years and location of wells on a sitewide and focused study plot scale. Boxplots were created in R Software to show the distribution and visually assess the hydrologic characteristics at each monitoring well location (R Core Team 2013). Descriptive statistics including mean, minimum, and maximum water level for each well were calculated to supplement visual methods of comparison. Time series of water levels, also produced in R Software, were used to visualize trends at specific well locations throughout the duration of the growing season (R Core Team 2013).

Two-way analysis of variances (ANOVAs) was completed in IBM SPSS Statistics 27 to compare the mean differences of water levels between study year and well location (IBM Corp 2020). The levels of each factor tested are summarized in Table 7.1.

Table 7.1 Levels of the factors considered in two-way ANOVA analysis

Study period	Well location	
	Sitewide	Focused study plot
2017	North of transect 1	Near central ditch
2018	Between transect 1 and 2	Far from central ditch
2019	South of transect 2	

Tukey's (HSD) post-hoc tests were completed in IMB SPSS Statistics 27 with the two-way ANOVA analysis (IBM Corp 2020). The data was evaluated for homogeneity of variances prior to completing post-hoc tests and this assumption was met. Paired t-tests were also completed in IBM SPSS Statistics 27 to compare the mean differences of the percent of time water levels were within 0.30 m of the ground surface for pre- and post-restoration (IBM Corp 2020). A p-value less than 0.05 was considered significant for all statistical tests.

The spatial variability of water levels throughout the study period was investigated through interpolation using Kriging. Sitewide and focused study plot water level interpolation was performed in the Geostatistical Analyst toolbox in ArcGIS Pro. Raster inputs into the Kriging model included the well locations with associated average daily water level minimums. Simple Kriging with a spherical semivariogram was the chosen model.

7.4 Results and Discussion

7.4.1 Study Period Climate and Reference Wetland Hydrology

Climate data from Yarmouth RCS Station for 2017–2019 are summarized in Table 7.2. Climate normals (1981–2010) for the Yarmouth RCS station showed annual precipitation of 1293 mm and growing season precipitation and mean temperature of 463 mm and 14.3 °C, respectively. The 2017 pre-restoration year was slightly drier than these climate normal values. Comparing 2017 to 2018, it is evident that the 2018 post-restoration year was a particularly dry year. When looking at the growing season precipitation values, 2018 was drier than both 2017 and 2019. The 2019 post-restoration year was the wettest of the study period with annual precipitation exceeding the climate normals. Although growing season precipitation for 2019 was slightly below climate normals, it was still considerably greater than 2017 or 2018 growing season precipitation totals.

Figure 7.5 illustrates boxplots of the water levels for the three reference wetland locations during each study year. On an annual basis, water levels were consistent throughout the reference wetland from 2017 to 2019, despite variations in precipitation. Spatial differences in water level between each location in the reference wetland followed expected trends. The Wet, Medium, and Dry locations had average water

Table 7.2 Climate date for Yarmouth RCS station (Climate ID: 8206491)

Year	Annual precipitation (mm)	Growing season (May–September)	
		Precipitation (mm)	Mean temperature (°C)
2017	1098	338	16.4
2018	996	270	16.8
2019	1478	435	15.7

Water year taken as October 1st–September 30th

levels near ground surface, 0.20 m, and 0.40 m below ground surface, respectively. These water levels are typical of what is expected in an undisturbed bog wetland complex, especially if the central portion of the bog is raised. These results are consistent with findings from MacIntyre (2017), which showed that water levels in a relatively unaltered domed bog in this area of the province can range up to 50 cm below the ground surface in the center of the bog.

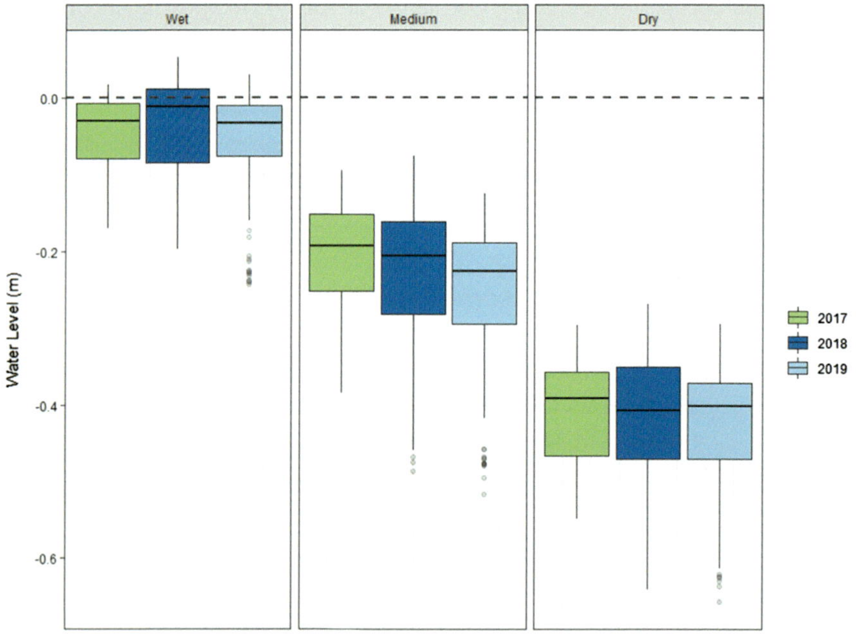

Fig. 7.5 Boxplots of mean daily water levels for wet, medium, and dry sites in reference wetland

7.4.2 *Sitewide Water Level Response*

Figure 7.6 illustrates the locations of the monitoring wells studied across the Big Meadow Bog site for 2017–2019. These include several wells installed in previous investigations that formed three distinct transects:

- Transect 1: MW38, MW3, MW5, and MW6.
- Transect 2: MW19, MW9, and MW12.
- Transect 3: MW14 and MW18.

The variation in mean daily water level for the sitewide monitoring wells during the 2017–2019 growing seasons is illustrated in Fig. 7.7. The boxplots are arranged to correspond with locations of the monitoring well as you travel south along the length of Big Meadow Bog. The first well, MW37, is located in the most northernly position of the wetland complex and the last well, MW34, is the most southernly monitoring

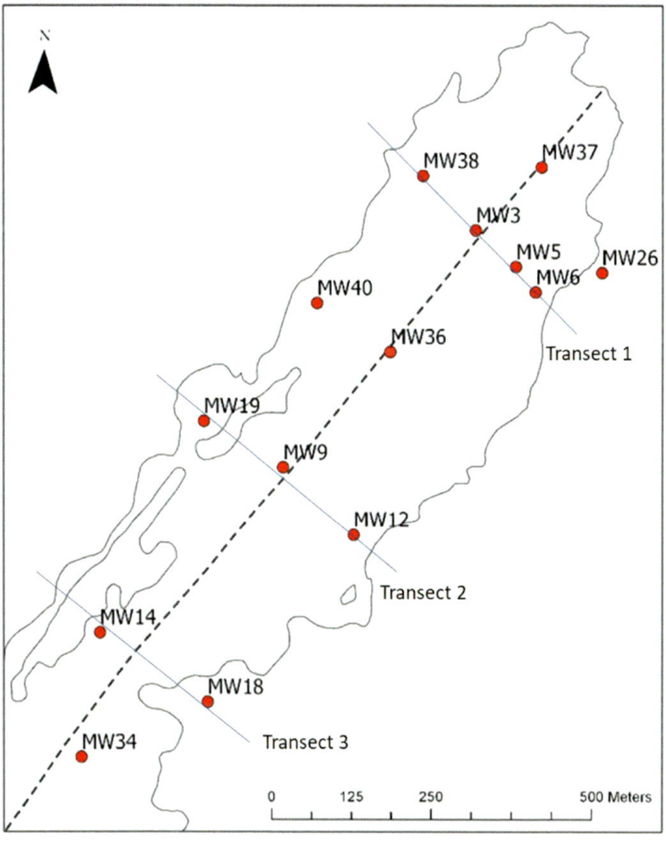

Fig. 7.6 Locations of sitewide monitoring wells in big meadow bog

location. Sitewide water levels showed increasing trends from pre-restoration in 2017 to post-restoration in 2018–2019. Larger increases in water level were seen between 2018 and 2019, most likely due to below average amounts of precipitation in 2018. The lack of precipitation in 2018 likely dampened the initial re-establishment of water levels during the first growing season post-restoration. The water level response, however, was not consistent throughout the wetland complex. Overall, the greatest change in water levels was observed in the northern portions of the bog. Water levels in several wells located in the central and southern part of the bog did not appear to change. A two-way ANOVA was applied to examine the effect of year (2017, 2018, and 2019) and well location (on or North of Transect 1 (T1), between T1 and Transect 2 (T2), South of T2) on mean daily water levels, with neither of the main effects or interaction terms found to be significant.

Many of the wells on or near the central ditch (MW37, MW3, MW36, MW34) experienced the greatest increase in water levels. These increases were expected as the central ditch was the source of substantial drawdown in areas of the bog adjacent to the ditch, between 0.5 and 0.7 m, for prolonged periods pre-restoration. The installation of the ditch blocks appears to have allowed water levels to return to the upper 0.5 m of the peat for a large portion of the growing season. The level of variation in growing season water levels in these wells also increased post-restoration, indicating a flashier hydrologic response. This could be partly attributable to the

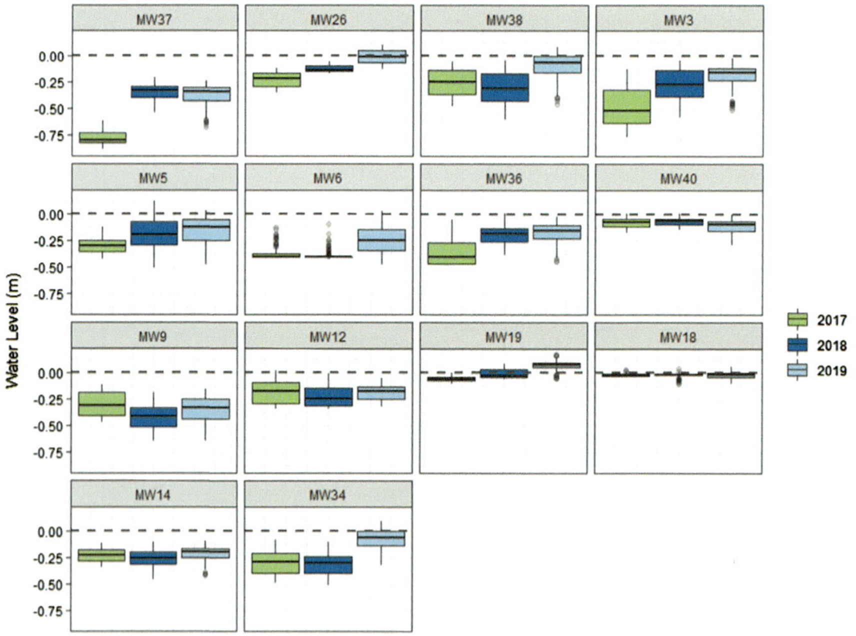

Fig. 7.7 Boxplots of mean daily water levels for sitewide wells relative to ground surface during the growing season (May–September)

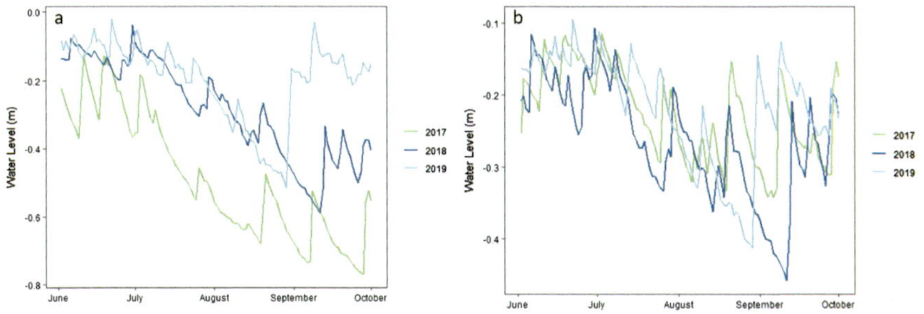

Fig. 7.8 Growing season water levels relative to ground surface for **a** MW3 and **b** MW14 from 2017 to 2019

physical state of the impacted peat, which has been subjected to excessive drying and decomposition. Figure 7.8a illustrates the growing season water levels in MW3 for 2017–2019. It is evident that average water levels have increased post-restoration, despite 2018 being drier. In 2019, water levels re-established to near surface due to precipitation events in late August–early September.

In general, wells along the margins of BMB (MW14, MW18, MW19, and MW40) demonstrated smaller increases in water level, if any. Figure 7.8b illustrates growing season water levels with similar trends for MW14 from 2017 to 2019. Variation in growing season levels only increased slightly in the two years post-restoration. However, MW26 and MW6, which are in the lagg area on the northeast boundary of the wetland complex, showed a marked increase in water level. These lagg areas are the preferred habitat of the *Geum peckii* and the re-establishment of near surface water levels in the lagg could limit vegetation competition for the *Geum peckii*.

Bog hydrology is characterized by a water table that is at or near the surface. A high water table is essential for *Sphagnum* moss, the primary peat forming vegetation, to access soil water and promote re-colonization (McCarter and Price 2013). Figure 7.9 shows the percent increase in time, from pre- (2017) to post-restoration (2018 and 2019), that the sitewide water levels were within 30 cm of the ground surface during the growing season (May–September). Overall, post-restoration water levels were within 30 cm of the ground surface for longer periods of time as compared to pre-restoration. Wells closer to the ditch (MW36, MW3, MW37, MW34, and MW5), experienced a greater than 20% increase in the time the water level was within 0.30 m of the ground surface. Wells located further away from the ditch generally did not show as great an increase. Pre-restoration water levels in several of these wells (MW18, MW19, and MW40) were already at or very near the ground surface and these levels were maintained post-restoration. When comparing means of all sitewide wells, there was no significant difference in this percentage pre- ($M = 60.29$, SD = 33.2) and post- ($M = 70.79$, SD = 26.12) restoration ($p = 0.065$).

There does appear to be a relationship between percent increase in the duration of shallow (< 0.30 m) water levels and distance from the ditch, with a few exceptions. Wells closer to the ditch experienced a larger percent increase in the amount of time water levels were within 0.30 m of the ground surface from pre- to post-restoration.

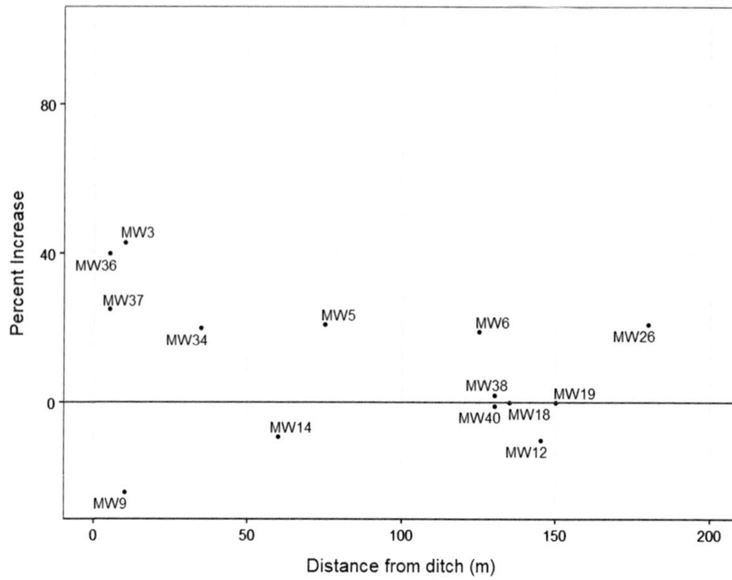

Fig. 7.9 Percent increase in time growing season (May–September) water levels were within 0.30 m of the ground surface for sitewide wells, pre- and post-restoration

Even so, it should be noted that MW9 showed the reverse trend, with a decrease in the duration of shallow water levels (−24%). Trends for MW9 mean daily water levels (Fig. 7.6) indicate that some of the hydrologic interventions have diverted water away from this area of the bog in the two years post-restoration. In 2018, MW9 daily water levels decreased compared to 2017, but 2019 water levels did recover somewhat and were more comparable to levels observed in 2017.

Interpolated contour plots of water levels were constructed to visualize the spatial variation of water levels throughout the site. Figure 7.10 illustrates the spatial variation in minimum (low) water levels from 2017 to 2019. Overall, there were increases in minimum daily water levels around the perimeter of the wetland. Ditch blocking the east and west ditches appeared to promote water retention in the lagg regions of the bog. A two-way ANOVA was conducted that examined the effect of year (2017, 2018, and 2019) and well locations (on or North of T1, between T1and T2, South of T2) on minimum daily water levels. There was a statistically significant effect of well location on the lowest (minimum) daily water levels observed [$F(2,33) = 3.965$, $p = 0.029$] but no significant effect of year on minimum daily level.

Visualizations of lowest daily water level provide a somewhat different understanding in the hydrology of the system as compared to previous analyses of the mean water levels (Fig. 7.6), which indicated an increase in water level across the site pre- and post-restoration. In Fig. 7.10, we see similar levels, and spatial trends, in minimum water levels across the bog in 2017 and 2018. However, 2019 minimum water levels produced a dominant gradient from the margins of the bog toward the

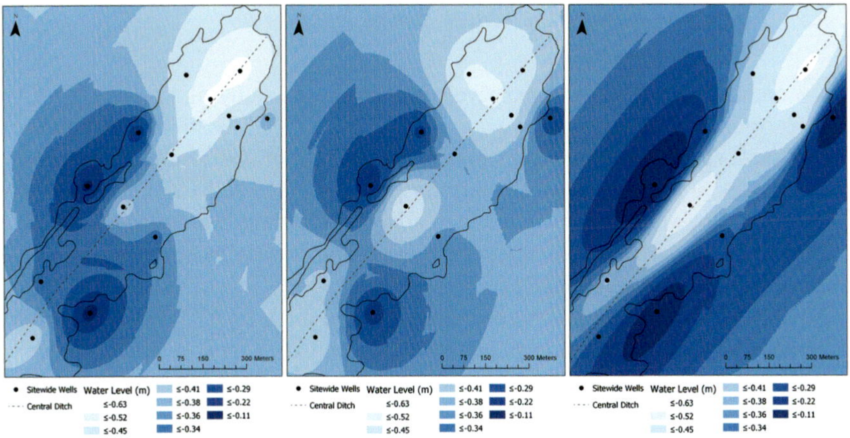

Fig. 7.10 Interpolated minimum daily average water levels for sitewide wells during the 2017 (left), 2018 (middle), and 2019 (right) growing seasons (May–September)

longitudinal middle of the bog. This could be attributed to changes in peat composition (more decomposed, humified) in the middle of the bog, as it was most affected by ditching and gull presence, or differing precipitation that summer. The gradient could indicate that the disturbed peat near the central ditch of the bog does not have the ability to retain water consistently throughout the growing season. At this early stage following the restoration, it is expected that the capacity of the peat to hold water during the growing season would be limited. This might improve as the bog stays rewetted for a longer period of time.

Studies have shown that ditch blocking can have an immediate effect on the water table in bog systems (McCarter and Price 2013; Howie et al. 2009; Price 1996). However, even with this immediate re-establishment of water levels, disturbed bog systems can take many years to re-start the peat forming process. The reversal of the impacts of ditch drainage will not immediately be achieved through raising the water table, although it is a critical first step. This is likely due to the continued structural impairment of the peat in the vicinity of the ditch due to compression and shrinkage from drying (Price and Schlotzhauer 1999). The lack of pore water pressure within the peat structure causes the peat matrix to shrink, which in turn alters the storage and retention capacity of the peat. Howie et al. (2009) explains that there are differences in the zone of influence of a drainage ditch before and after it has been restored using ditch blocking. We would expect to see a larger zone of influence pre-restoration and the disparity between these two would increase with the amount of time the peat was subjected to drying (Howie et al. 2009). It would be expected that it will take several years for the peat to recover and for the water level regime to stabilize, which is likely the case for Big Meadow Bog.

7.4.3 Focused Study Plot Water Level Response

The variation in mean daily water levels in the focused study plot for the growing seasons from 2017 to 2019 is shown in Fig. 7.11. The well locations are organized by transects, starting with the most northernly transect of wells, moving southwest (well name A to D), with the wells left to right (well number 1 to 4). The majority of the study plot wells demonstrated increases in mean daily water level from 2017 to 2019. Water levels were also closer to the ground surface more often during the growing season. Most water levels increased to be within 30 cm of the ground surface by 2019. As expected, well locations closest to the ditch (WA3, WC3, and WD3) had greater increases in water level between 2017 and 2018 (post-restoration) due to the immediate rewetting from the central ditch blocking. This increase was present even though 2018 was a comparatively dry year. A two-way ANOVA was applied to examine the effect of year (2017, 2018, and 2019) and position (near or far from the ditch) on mean daily water levels. There was a statistically significant effect of year on mean daily water level [$F(2,36) = 19.835$, $p = < 0.001$]. Closer examination of the main effects using a Tukey HSD post-hoc test showed that mean water levels in 2017 were lower than 2018 ($p = 0.001$) and 2019 ($p = 0.000$), while no significant differences were found between 2018 and 2019 ($p = 0.085$).

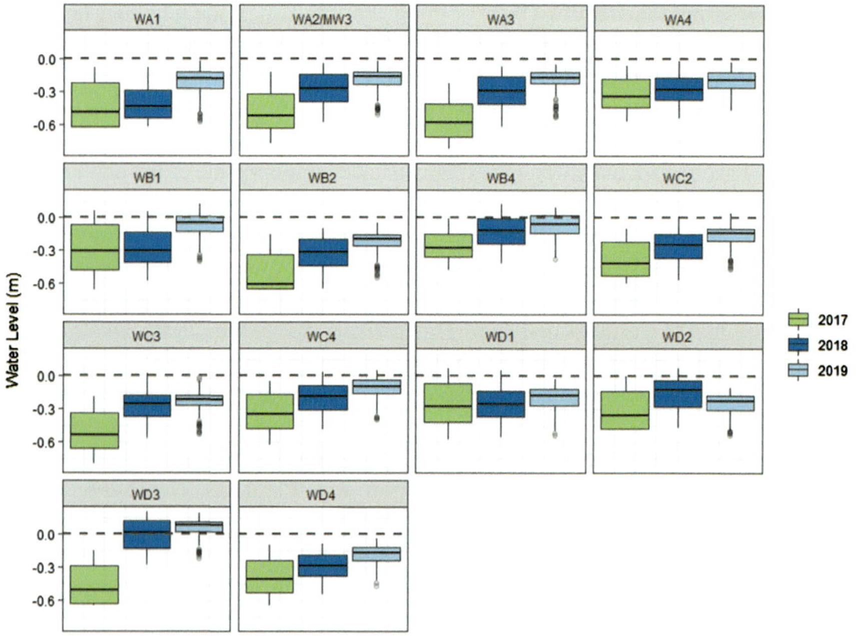

Fig. 7.11 Boxplots of mean daily water levels for focused study plot relative to ground surface during the growing season (May–September)

Well locations further away from the ditch (WA1, WB1, WC4, and WD4) continued to see increases in water levels through 2019. However, most wells near the ditch had mean water levels similar to 2018. On the study plot scale, water levels appeared to be more variable during the 2017 pre-restoration growing season compared to 2018 and even more so compared to 2019. This could be an indication that on a smaller scale, water levels are more consistent near the ditch throughout the growing season post-restoration.

Minimum (lowest) daily water levels were interpolated and contour plots produced to show the spatial variation in the study plot between 2017 and 2019 (Fig. 7.12). In 2017, daily minimum water levels were quite low near the central ditch, approximately 0.65–0.70 m below ground surface. This was expected as the central ditch was 0.70–1.0 m deep in the northern portion of the bog where the study plot was installed. Minimum water levels in 2017 ranged from 0.55 m below ground surface around the perimeter on the East and West edges to 0.70 m below ground surface in the center of the study plot. In 2018, minimum water levels ranged from 0.40 to 0.58 m below ground surface with the wetter conditions encountered around the southern half of the study plot. There was an area that was notably drier in the northern half of the plot in 2018 but water levels had still increased from 2017. In 2019, minimum water levels ranged from 0.40 to 0.50 m below ground surface. The steep gradient of water levels across the plot is much less prominent in 2019 compared to 2017, indicating less spatial variability in water level during the growing season.

A two-way ANOVA was conducted to examine the effect of year (2017, 2018, and 2019) and location (near or far from the ditch) on minimum daily water levels. There was a statistically significant effect of year on minimum daily average water level [$F(2,36) = 10.035$, $p = < 0.001$]. When applying a Tukey HSD post-hoc test it showed that mean water levels in 2017 were statistically lower than 2018 ($p = 0.025$) and 2019 ($p = 0.000$), while no significant differences were found between 2018 and 2019 ($p = 0.217$).

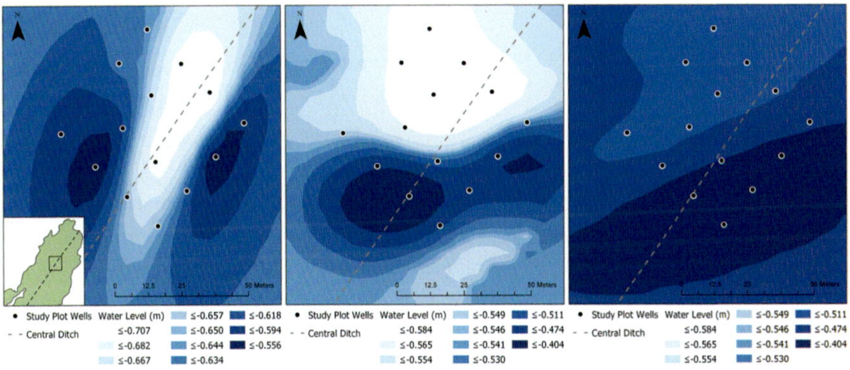

Fig. 7.12 Interpolated minimum daily average water levels for focused study plot wells during the 2017 (left), 2018 (middle), and 2019 (right) growing seasons

7.4.4 *Comparison to Reference Wetlands*

In Nova Scotia there are challenges with finding appropriate reference wetlands due to the limited availability of historical water level datasets of undisturbed peatlands. Comparisons of wetland types across Nova Scotia, in terms of hydrologic conditions, have been completed in recent years (Bell 2021; MacIntyre 2017). These studies aimed to establish hydrologic baselines that are relevant for wetland management, to ultimately aid in the development of restoration targets of disturbed sites. MacIntyre (2017) showed that domed bogs have differing hydrologic patterns (lower water tables) when compared to basin bogs, where water tables were in the shallower (0 – 20 cm) range. Bell (2021) had similar findings for wooded peatlands in Nova Scotia, which showed that these sites remained perennially saturated and mean water levels were maintained between 0.04–0.17 m below ground surface during the summer months.

Sitewide water level trends in Big Meadow Bog, post-restoration, followed similar spatial trends to the reference wetland. Water levels were near ground surface in the peripheral lagg area of the wetland, getting increasingly drier moving toward the middle of the bog. In the reference wetland, mean daily water levels in the middle, domed portion of the bog averaged around 0.40 m below ground surface. Comparatively, mean daily water levels in the raised middle of BMB averaged within 0.30 m of the ground surface, with greater fluctuations observed at BMB compared to the reference wetland.

The subsidence of the peat matrix in BMB from prolonged drying has altered the peat structure and domed shape of the wetland. The reference wetland also exhibits a hummocky peat surface, which due to drying and impacts from the gulls, is not as apparent in many areas of BMB. The findings from this study suggest that the restoration activities in BMB have raised the mean water table to levels slightly higher than those recorded in the reference wetland.

7.5 Conclusions

The findings of the study showed that the restoration activities were successful in raising water levels on both a sitewide and focused study plot scale. Many of the sitewide wells close to the central ditch had an increase, greater than 20%, in the amount of time mean daily water levels were within 30 cm of the ground surface throughout the growing season. The largest changes in water level occurred near the ditch. Water levels in the peripheral, lagg areas of the wetland also increased to be within 20 cm of the ground surface in a few focused areas across the site. This was evident in the areas of MW19, MW26, and MW38, which have known populations of Eastern Mountain Avens.

However, the effects of the restoration activities were variable across the wetland complex. Some of the impacted portions of the bog, mainly along the longitudinal middle, appear to have issues with water retention in the summer. This is most likely due to the disturbed nature of the peat, and continued monitoring will be required to determine if the peat can regain water holding capacity. At this time, the hydrology of BMB is now considered more comparable to other bogs in this region of Nova Scotia. However, there is uncertainty as to what should be used as a true 'reference' wetland. Overall, the findings from the study have shown that the hydrology of the bog is initially responding well to the restoration activities. The hydrology of a peatland is crucial to its health and functional value and therefore, the establishment of a higher, more consistent water table is an important first step for the bog to regain its natural functions.

References

Bay RR (1969) Runoff from small peatland watersheds. J Hydrol 9:90–102

Bell NM (2021) Establishing baseline hydrologic conditions of forested wetlands in Nova Scotia, Canada. Dissertation, Dalhousie University

Brown P (2003) Big Meadow Bog and Geum peckii: preliminary restoration plan. Nature Conservancy of Canada, Fredericton, New Brunswick

East Coast Aquatics Inc. (2018) Big Meadow Bog hydrology restoration: as built report. Available via http://eastcoastaquatics.ca/projects.html Accessed Jan 2020

Howie S, Whitfield P, Hebda R, Munson T, Dakin R, Jeglum J (2009) Water table and vegetation response to ditch blocking: restoration of a raised bog in southwestern British Columbia. Can Water Resour J 34:381–392

IBM Corp (2020) IBM SPSS statistics for windows, version 27.0. IBM Corp., Armonk, NY

Kennedy G, Drage J, Nixon C (2015) 2014 Update on baseline hydrological monitoring at Big Meadow Bog, Brier Island, Nova Scotia. Nova Scotia Department of Natural Resources Open File Report ME 2015-005. Available via https://novascotia.ca/natr/meb/data/pubs/15ofr05/ofr_me_2015-005.pdf. Accessed July 2018

Kennedy GW, Price JS (2005) A conceptual model of volume-change controls on the hydrology of cutover peats. J Hydrol 302:13–27

Kontak DJ, Webster TL, (2010) Bedrock geology map of basaltic rocks of the north mountain formation from Brier Island to Sandy Cove, Digby County, Nova Scotia. Nova Scotia Department of Natural Resources, Open File Map ME 2010-008, scale 1:50,000

Kuhry P, Nicholson BJ, Gignac LD, Vitt DH, Bayley SE (1993) Development of Sphagnum-dominated peatlands in boreal continental Canada. Can J Bot 71:10–22

Lang Y (2002) Influence of water table changes on saturated peat compressibility and permeability. Dissertation, University of Waterloo

MacIntyre MBJ (2017) Comparing shallow groundwater hydrodynamics in reference peatlands and the ditched Big Meadow Bog complex on Brier Island, Nova Scotia. Dissertation, Dalhousie University

McCarter CPR, Price JS (2013) The hydrology of the Bois-des-Bel bog peatland restoration: 10 years post-restoration. Ecol Eng 55:73–81

National Wetland Working Group (1997) In Warner B, Rubec C (ed) The Canadian wetland classification system, 2nd edn. Wetland Research Centre Publication, Waterloo, Ontario

Poulin ML, Rochefort L, Desrochers A (1999) Conservation of bog plant species assemblages: assessing the role of natural remnants in mined sites. Appl Veg Sci 2:169–180

Price JS, Heathwaite AL, Baird AJ (2003) Hydrological processes in abandoned and restored peatlands: an overview of management approaches. Wetlands Ecol Manage 11:65–83

Price JS (1996) Hydrology and microclimate of a partly restored cutover bog, Québec. Hydrol Process 10:1263–1272

Price JS, Schlotzhauer SM (1999) Importance of shrinkage and compression in determining water storage changes in peat: the case of a mined peatland. Hydrol Process 13:2591–2601

R Core Team (2013) R: a language and environment for statistical computing. R Foundation for Statistical Computing, Vienna, Austria. http://www.R-project.org/

Stea RR, Conley H, Brown Y (1992) Surficial geology map of the Province of Nova Scotia. Nova Scotia Department of Natural Resources, Map ME 1992-3, scale 1:500,000, available online as DP ME 36, version 2, 2006. http://novascotia.ca/natr/meb/download/dp036.asp

Van Seters TE, Price JS (2001) The impact of peat harvesting and natural regeneration on the water balance of an abandoned cutover bog, Quebec. Hydrol Process 15:233–248

Waddington M, Warner K, Kennedy G (2002) Cutover Peatlands: a persistent source of atmospheric CO_2. Global Biogeochem Cycles 16:1–7

Wilson SE, Walker IR, Mott RJ, Smol JP (1993) Climatic and limnological changes associated with the Younger Dryas in Atlantic Canada. Clim Dyn 15:233–248

Webster T, Collins K, Crowell N, MacDonald C, McGuigan K (2015) Hydrological modelling of surface runoff on Brier Island, Nova Scotia. Technical Report, Applied Geomatics Research Group, Nova Scotia Community College, Middleton, Nova Scotia

Chapter 8
Differences in Shallow Groundwater Dynamics Between Disturbed and Undisturbed Peatlands on Brier Island, Nova Scotia: Implications for Persistence of *Geum peckii* in Big Meadow Bog

John C. Brazner, Makeala MacIntyre, Nicholas M. Hill, and John Drage

Abstract The goal of our project was to provide data that would help identify what hydrologic conditions were typical of relatively undisturbed and disturbed peatlands on Brier Island and what aspects of those conditions were most conducive to supporting healthy populations of *Geum peckii*. We instrumented 8 sites with shallow wells and automated data loggers spread widely across Brier Island in 2014 and 2015 to monitor shallow groundwater dynamics. Seven of the sites were reference peatlands—two horizontal fens, three basin bogs, one domed bog-fen complex. The other site was the Big Meadow Bog complex within which we incorporated data from 4 wells. One of the most important findings from our study was that there were clear differences in shallow water level dynamics among our four peatland classes. Mean

J. C. Brazner (✉)
Nova Scotia Department of Natural Resources—Wildlife Division, Kentville, NS B4N 4E5, Canada
e-mail: johnbrazner@gmail.com

M. MacIntyre
Antipa Minerals Ltd., Perth, WA 6005, Australia

J. C. Brazner
Fern Hill Institute of Plant Conservation, South Berwick, NS B0P 1E0, Canada

J. Drage
Halifax, NS, Canada

Present Address:
J. C. Brazner
Brazner Ecological Services, 1181 Russia Road, Black Rock, NS B0P 1V0, Canada

N. M. Hill
UNESCO Southwest Biosphere Association, 399 Wyman Road, Sand Beach, NS B5A 5G4, Canada
e-mail: nick@swnovabiosphere.ca

N. M. Hill et al. (eds.), *Wetland Restoration for Endangered Species Recovery*, Environmental Contamination Remediation and Management,
https://doi.org/10.1007/978-3-031-71344-6_8

water levels, monthly fluctuations and water table depth durations ranged along a spectrum from highly consistent among the three undisturbed fens to highly variable among the three disturbed bogs. Peatlands with intermediate hydrological patterns between the greater constancy and wetness of the undisturbed fens and the hypervariability and dryness of the disturbed bogs are largely those that supported substantial populations of *Geum peckii*. Identifying the hydrologic sweet spot for *Geum peckii* success was not the primary objective of this study, but our results suggest that mean water levels of −10 to −20 cm and overall water levels in the saturated root zone for about 75% of the time are two key aspects of the hydrologic niche space that generally lead to success for this species. However, it is important to note that the results of other studies included in this volume strongly suggest that while suitable hydrology is critical to the success of *Geum peckii*, other habitat factors (e.g., shrub density, microtopography) also play key roles.

Keywords Groundwater · Peatlands · Hydrology · Water level

8.1 Introduction

There are about 160,000 ha of peatland currently mapped in the Nova Scotia provincial wetland inventory (https://novascotia.ca/natr/wildlife/habitats/wetlands.asp). Even though this is likely a substantial underestimate given inaccuracies in the inventory (e.g., many bogs and fens are misclassified as swamps; Sheehy et al. 2023), peatlands compromise at least 3% of the provincial land base. In addition to being biodiversity hotspots for birds (e.g., Calme et al. 2002; Brazner and Mackinnon 2020), peatlands are known to store large amounts of carbon (e.g., Mitsch et al 2013), and as such, are one of the ecosystems that play a critical role in mitigating the effects of climate change in Nova Scotia (e.g., Temmink et al. 2022).

We know little about the history of peatland alteration and loss in Nova Scotia. There are bits and pieces of this history that have been documented (e.g., Andersen and Broughm 1988; Reid 2014), but as in much of the rest of Canada, data on historical wetland loss is generally lacking. With respect to current alteration of peatlands in Nova Scotia, there are 6 ongoing peat harvesting operations in the province (https://fletcher.novascotia.ca/arcgis/rest/services/databases/energy_databases/MapServer). These encompass 454 ha and evidence from aerial imagery (Frances MacKinnon, Nova Scotia Department of Natural Resources and Renewables, personal communication) and limited reporting (Andersen and Broughm 1988) suggests that a number of other peat harvesting operations have been shut down in recent decades. Compared to other provinces in the region (e.g., Poulin et al 2004; Graf et al. 2012), the practice does not seem to have been widespread in Nova Scotia. Substantial alteration of peatlands has occurred in New Brunswick and Quebec (LaVoie and Rochefort 1996; González and Rochefort 2014; Chimner et al. 2017), mainly through ditching, to facilitate the process of commercial peat extraction and to provide better conditions for agriculture.

Whether for peat extraction or to create farmland, peatlands are typically modified by creating a network of ditches which lower the water table and alter hydrology (e.g., Money et al. 2009). This kind of alteration of wetland hydrology is one of main forms of anthropogenic disturbance to wetland ecosystems that can dramatically affect their function and biodiversity (e.g., Richter et al. 1996; Hunt et al. 1999; Gopal 2009). In peatlands, alteration of hydrology alters peat structure, changes soil and water chemistry and general ecosystem dynamics (Price et al. 2003; Waddington et al. 2009; Holden et al. 2011; Landry and Rochefort 2012). Even small changes to peatland hydrology can result in significant biotic changes including changes to community composition, species richness, and overall productivity (e.g., Mitsch and Gosselink 2015; Gopal 2009; Acreman and McCartney 2009). As a result, understanding typical peatland hydrologic regime is an important first step in assessing the degree of impairment of any wetland and developing an effective approach for restoration if the site has been altered (e.g., van der Valk 2009).

The Big Meadow Bog (BMB) wetland complex on Brier Island, Nova Scotia (NS) is one of the peatlands in the province that has been modified using a network of ditches to lower the water table as part of an attempt to provide arable land for agriculture. The details associated with the alteration of BMB are described elsewhere (Hill et al. 2019; Chap. 1, this volume), so will not be repeated here. However, it is important to note that the alteration of hydrology at BMB has led to a cascade of events that have left the bog in poor condition. As ditching led to drying, a sequence of events described in detail by Hill et al. (2019), unfolded that included the colonization of a high density of herring gulls that now nest in BMB, the exposure of peat soils and the subsequent invasion of exotic grasses (e.g., *Holcus lanatus*), raspberry and blackberry (*Rubus* spp.) canes and Rosaceae shrubs, especially in the northern portion of BMB. There has also been a concomitant decline in the critically endangered population of Eastern Mountain Avens (*Geum peckii*). The status and distribution details related to *Geum peckii* have been described elsewhere (e.g., Environment and Climate Change Canada 2018), but it was first identified in BMB in 1949. It is now a globally rare species and BMB is the most important Canadian habitat. Before water levels were manipulated at BMB in 2017, the fen habitat along the margins of BMB were where the healthiest populations of *Geum peckii* occurred and these fens were considered to be the habitat most critical to the long-term persistence of the species (Hill et al. 2016). However, the specific hydrologic and habitat requirements for this species are not fully understood (Hill 2013; Environment and Climate Change Canada 2018), so the specific response of *Geum peckii* to the planned restoration of historic water levels in BMB was uncertain.

The goal of our project was to provide data that would help identify what hydrologic conditions were typical of relatively undisturbed and disturbed peatlands on Brier Island and what aspects of those conditions were most conducive to supporting healthy populations of *Geum peckii.* As it is, there was almost no research on baseline hydrologic conditions for peatlands anywhere in Nova Scotia at the time we were developing this project (but see, Brazner et al. 2015; Bell 2021) that could be used to develop meaningful indicators of typical undisturbed peatland hydrology, so our study was designed to characterize the hydrodynamics across the range of peatlands that occur on Brier Island in and adjacent to BMB (e.g., horizontal fens, basin bogs, domed bog). Our intent was to compare water level dynamics in these different peatland types as reference systems and use that to estimate the degree of hydrologic impairment of the ditched BMB complex. In addition, we also wanted to examine the relationship between the hydrodynamics in all of these peatlands and the known distribution of populations of the endangered *Geum peckii* on Brier Island to assess which hydrologic conditions may be optimal for *Geum peckii* persistence. Since the overall goal of the BMB restoration project is to reverse habitat degradation throughout the BMB complex and recover populations of this globally rare species (Hill et al. 2016), we hoped our results would be helpful for predicting possible changes in the distribution of *Geum peckii* within BMB over time as water levels were restored to historic levels.

8.2 Methods

8.2.1 Study Sites and Monitoring Wells

We instrumented 8 sites with shallow wells and automated data loggers spread widely across Brier Island (Fig. 8.1). Seven of the sites were reference peatlands—two horizontal fens, three basin bogs, one domed bog-fen complex, and the other site was the BMB complex within which we incorporated data from 4 wells (see Table 8.1 for more site details).

8.2.2 BMB Complex Monitoring Wells

We used four wells along the three longitudinal transects established by Kennedy and Drage (2015) in the BMB complex for this study. Monitoring Wells 3 and 9 were located in the central bog area and Monitoring Wells 14 and 23 were located on the peripheral lagg-fen where *Geum peckii* populations are found (Fig. 8.2). The wells were located at various distances from the central drainage ditch and along different transects, therefore having different degrees of hydrological impairment based on distance from the ditch (see Table 8.1 for more site detail). All of these sites were classified as disturbed due to ditching.

- Monitoring Well 3 (MW3) (44.253413, – 66.3578334) was located in the central bog along the first transect. The well was located 3.8 m west of the central drainage ditch.
- Monitoring Well 9 (MW9) (44.250055, – 66.3614195) was located in the central bog along the second transect. The well was 7.8 m west of the central drainage ditch. MW9 was also believed to be influenced by a ditch (Kennedy and Drage

Fig. 8.1 Locations of Big Meadow Bog complex and reference peatland monitoring wells

2015) excavated for a private airport runway located adjacent to BMB. The end of the runway was approximately 350 m northwest of the well.

- Monitoring Well 14 (MW14) (44.247687, – 66.3648813) was located near the treeline along the western fen margin/lagg in Transect 3. The well was 56 m from the central drainage ditch. This well was near a population of 100 *Geum peckii* plants (Toms and LaRue 2015).
- Monitoring Well 23 (MW23) (44.254624, – 66.3581186) was located in the western fen margin approximately 100 m northeast of the western end of Transect 1. The well was located in the lagg of the complex where there is a vigorous population of *Geum peckii* (747 plants; Toms and LaRue 2015).

8.2.3 Reference Peatlands Monitoring Wells

Reference peatlands were selected based on similarity to wetland classes present in the BMB complex, and also because they supported a range of *Geum peckii* population sizes. Reference peatlands consisted of two horizontal fens, three basin bogs, and a domed bog-fen complex (see Table 8.1 for more site detail).

Table 8.1 General site description of reference fens, reference bogs and the four monitoring wells in Big Meadow Bog complex based on field measurements and observations from November 3rd and 4th, 2016 (site code is the abbreviation used for each site in the text; *Geum peckii* counts reported by Toms and LaRue (2015); low shrub cover ≤ 2 m height and tall shrub cover ≥ 2 m height visually estimated; surface wetness was a subjective assessment of wetness underfoot at the time each site was visited)

Monitoring well	Site code	*Geum peckii* Count (# plants)	Dominant vegetation (% cover by taxa)	Mean shrub height (cm)	Sphagnum cover (%)	Ground-water pH	Conductivity (umhos/cm)	Surface wetness
Coastal Fen	CF	156	*Carex* spp. (80–90), cranberry (30), bog laurel, bayberry, cinnamon fern, goldenrod, and raspberry (1–5)	32.8	40–50	5.28	202	Wet
Inkberry Fen	IF	None	Low shrub cover (30), including common juniper (5–10), inkberry (10–15) and black spruce (15). Graminoid (55); scattered reindeer lichen, crowberry, cotton grass, and pitcher plants (1–5)	30.5	60	5.07	143	Wet
Lost Bog	LoB	1	Low ericaceous shrub includes huckleberry (30), crowberry (15), lambkill (5–10), *Carex* spp. (10–20), scattered common juniper, low black spruce, and bog laurel (1–5). Scattered herbaceous; golden rod, cotton grass, and pitcher plant (~1). Tall black spruce (>2 m) and white birch (5–10)	85.5	80	4.83	127	Wet
Brad's Bog	BB	1336	Low shrub cover (60), including common juniper (30) lambkill (5) low black spruce (5–10), scattered crowberry, dwarf raspberry, bog rosemary, bog laurel, leatherleaf (1–5). Tall black spruce (5–10)	–	30–40	4.92	157	Damp

(continued)

Table 8.1 (continued)

Monitoring well	Site code	*Geum peckii* Count (# plants)	Dominant vegetation (% cover by taxa)	Mean shrub height (cm)	Sphagnum cover (%)	Ground-water pH	Conductivity (umhos/cm)	Surface wetness
South Camp Bog	SB	805	Sedges (25), common juniper (10), black spruce (5–10), with scattered lambkill, leatherleaf, Labrador tea, speckled alder, and witherod (1–5)	37	50–60	5.24	130	Damp
Reference Bog	RB	None	Low shrub cover (95), including common juniper (5–10), crowberry (10–20), leather leaf (5–10), lambkill (10–20), black spruce, and Labrador tea (1–5). Reindeer moss, *Carex* spp., bunchberry, and pitcher plant (1–5)	31.5	10–20	3.82	132	Dry
Lighthouse Bog	LiB	1	Ericaceous shrub cover (90), *Carex* spp. (10), dwarf raspberry (10), with scattered crowberry, Labrador tea, Canada holly, cinnamon fern, and black spruce (1–5)	41	5–10	4.72	215	Dry
BMB-MW3	MW3	None	No herbaceous. Raspberry thicket (80–90), dense patch of bayberry (1–5), black spruce along ditch (1)	152.5	0	4.8	190	Dry
BMB-MW9	MW9	None	*Juncus* spp. (40), raspberry (5–10), blackberry (10–20) meadowsweet (5), scattered wood/ cinnamon fern, goldenrod, unidentified grass. Tall shrub on ditch edge (withered—10%)	105	0	3.6	182	Damp
BMB-MW14	MW14	100	Labrador tea (40), leatherleaf (5), lambkill (5–10), bog laurel, marsh fern, *Carex* spp. (1–5), tall shrub; witherod, and black spruce (1–5)	96.8	75	3.94	113	Wet

(continued)

Table 8.1 (continued)

Monitoring well	Site code	*Geum peckii* Count (# plants)	Dominant vegetation (% cover by taxa)	Mean shrub height (cm)	Sphagnum cover (%)	Ground-water pH	Conductivity (umhos/cm)	Surface wetness
BMB-MW23	MW23	747	Low shrub (80–90) lambkill, huckleberry, Labrador tea, dwarf raspberry. Low (<2 m) black spruce (5) tall (>2 m) black spruce (1–5), *Carex* spp. (1–5)	60.4	1–5	4.31	230	Dry

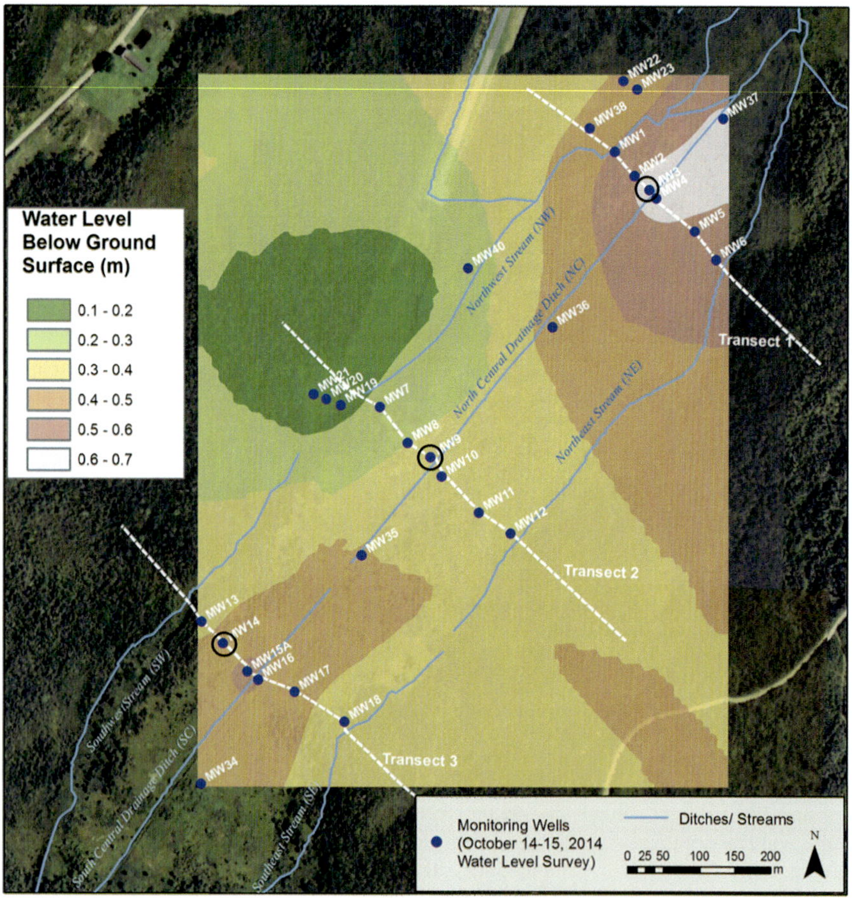

Fig. 8.2 Interpolated depth to water level below ground surface for October 2014 (Kennedy and Drage 2015). Monitoring wells used for this study are circled in black

- Coastal Fen (CF) (44.233418, – 66.368235) was classified as a horizontal fen estimated to be 2.9 ha. The site has medium abundance of *Geum peckii* (156 plants in 2013; Toms and LaRue 2015). This site was classified as undisturbed.
- Inkberry Fen (IF) (44.249759, – 66.379080) was classified as a horizontal fen estimated to be 4.6 ha. The site does not have any populations of *Geum peckii.* This site was classified as undisturbed.
- Lost Bog (LoB) (44.246996, – 66.373223) was classified as a basin bog and is estimated to be 11 ha. The site has one *Geum peckii* plant. This site was classified as undisturbed.
- Brad's Bog (BB) (44.248703, – 66.357186) was classified as a basin bog and was estimated to be 0.79 ha. The site has a high abundance of *Geum peckii* (1205 plants in 2013; Toms and LaRue 2015). This site was classified as undisturbed.
- South Camp Bog (SB) (44.239517, – 66.386956) was classified as a basin bog and is estimated to be 0.5 ha. The site has medium abundance of *Geum peckii* (805 plants in 2013; Toms and LaRue 2015). This site was classified as undisturbed.
- Lighthouse Bog (LiB) (44.246562, –66.388719) was classified as a basin bog and is estimated to be 3.4 ha. The site has a low abundance of *Geum peckii.* There is only one *Geum peckii* plant at this site (Toms and LaRue 2015). It was classified as disturbed due to extensive ditching along the adjacent access road.
- Reference Bog—Monitoring Well 29 (RB) (44.255396, – 66.3702686) is located northwest of BMB (Fig. 8.3). It is a domed bog and is part of a peatland complex that includes an adjacent fen (Hill et al. 2016). The complex was estimated to be 20.2 ha. There is no *Geum peckii* at this site and it was classified as undisturbed.

8.2.4 Well Installation, Set-Up, and Deployment

Automated water level recorders were deployed in the reference peatlands and in the BMB wetland complex in 2013 to collect shallow groundwater level measurements. The data used for this project was collected in 2014 and 2015. Two types of loggers were used. The loggers at monitoring wells 3, 9, 14, 23 in the BMB complex and at monitoring well 29 at RB were Solinst Model 3001 M5 water level loggers (Solinst Canada Ltd., Georgetown, Ontario, CA). The loggers from wells in BB, CF, IF, LoB, LiB, and SB were Hobo pressure transducer-based dataloggers (Hobo U20L-04, Onset Corp., Bourne, Massachusetts, USA). The loggers have similar accuracies ($\pm$ 3 to 5 mm). For both devices, wells were 1.5 m in length and installed in hand-augered boreholes (Brazner et al. 2015; Kennedy et al. 2015) and loggers were set to collect data hourly. Wells used for Solinst recording devices were constructed from 25 mm PVC pipes that were perforated starting 30 cm from the top of the well to the bottom with 5 mm diameter holes at 5 cm intervals (Kennedy et al. 2015). To prevent wells from filling with peat when installed the bottom was capped with a PVC pipe cap. Wells instrumented with Hobo recording devices were constructed from 38 mm ABS pipes with one hundred 5 mm diameter holes drilled in the lower 1 m of the pipe (Brazner et al. 2015). The submerged portion of each well was wrapped with

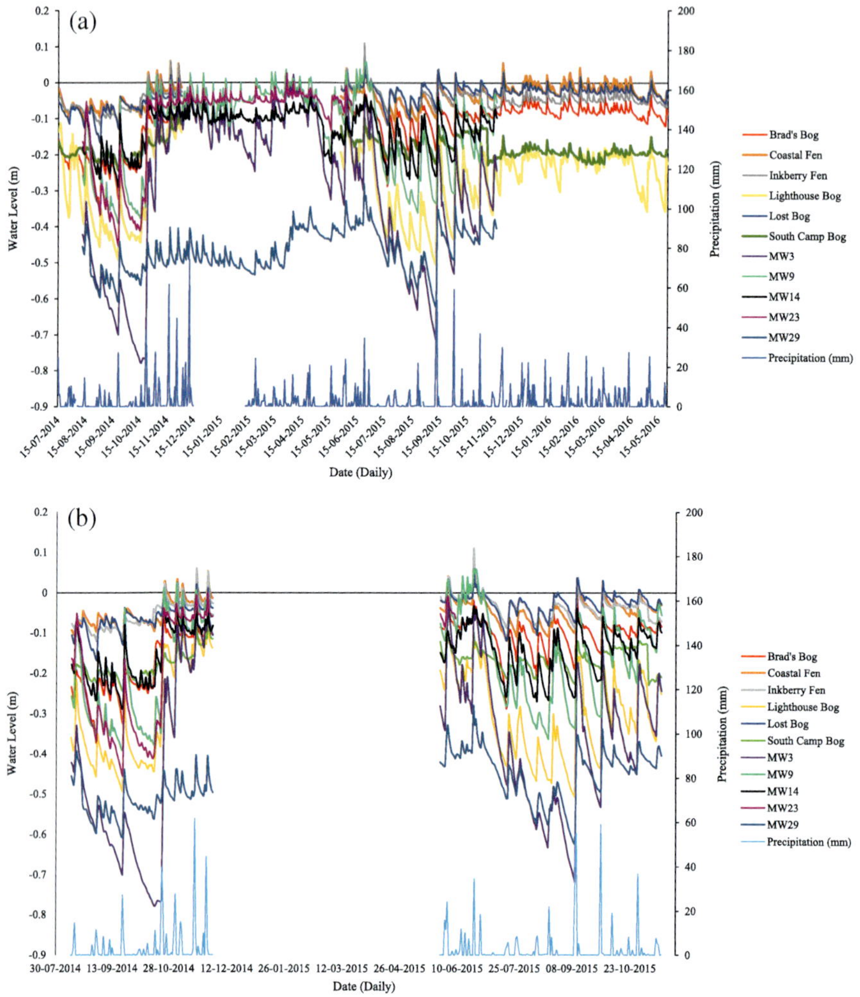

Fig. 8.3 **a** Hydrograph of all wells including all available data based on daily water level medians (m) with precipitation (mm; light blue lines) from July 15, 2014 to May 24, 2016. Includes data from December 1, 2014 to May 27, 2015 for the BMB wells and RB and data from July 15 to August 11, 2014 and November 16 to May 24, 2016 from the remaining reference wetlands. Water level is relative to ground surface (0 m). **b** Hydrograph of all wells based on daily water level medians (m) with precipitation (mm; light blue lines) from August 11 to December 1, 2014 and May 27 to November 16, 2015—the full period of analysis. It only includes data during which all monitoring wells were simultaneously recording. The data logger at MW23 was removed June 8, 2015. Water level is relative to ground surface (0 m)

a 2 mm mesh landscape fabric to prevent peat in-filling during installation and to reduce peat sediment in water quality samples. Water level was measured in relation to ground surface (see Brazner et al. 2015 for details). Data from all well loggers was corrected for differences in barometric pressure using an additional Hobo logger installed 1.5 m aboveground at Brad's Bog ("barologger").

Daily precipitation records from July 2014 to May 2016 were obtained from the Environment Canada climate station (ID 8200604) on the northern tip of Brier Island. Hourly precipitation data was collected by East Coast Aquatics Inc. Rain-logger station (Solinst Canada Ltd., Georgetown, Ontario, CA) from September 8th to December 3rd, 2015 on Brier Island.

8.2.5 General Habitat and Site Characterization

A general habitat characterization (Table 8.1) was completed in November 2016 within a 5 m radius around each well. A YSI 1030 sonde unit (Xylem, Yellow Springs, OH, USA) was used to collect a single groundwater pH and conductivity measurement from inside the monitoring well. Surface wetness was assessed subjectively as wet, damp, or dry based on examination of augered peat samples next to the monitoring well. Average shrub height was calculated from four measurements taken within 5 m of every well, one measure in each cardinal direction. Percent cover of low shrub, tall shrub, herbaceous, and sphagnum cover was visually estimated in the same area as the shrub height measurements. Low shrub cover was designated as percent cover of all woody species $\leq$ 2 m in height. Tall shrub cover was designated as percent cover of all woody species $\geq$ 2 m in height.

Data Analysis

We examined patterns in the hydrology of these wetlands to determine if there were differences in hydrodynamics associated with differences in human disturbance or wetland type at these sites. To assure comparability of the data collected within BMB and RB with those data from the other study sites, some data correction was required. Data from MW3, MW9, MW14, and RB were originally recorded as meters above sea level (mAsl). Elevation surveys were conducted by NSDNR in 2014 using a Leica Viva GS14 Global Navigation Satellite System to give wetland surface elevation in relation to sea level (see Kennedy et al. 2015 for more details). Values were converted to centimeters below ground level by subtracting the ground surface elevation specific to each well from the original mAsl values.

Once the water levels were processed from both types of loggers, water levels were compiled by the hour for all available dates. However, monitoring wells were not simultaneously recording at all times. Available data from the HOBO loggers was from July 15, 2014 00:00 AM until December 1, 2014 12:00 and from May 27, 2015 17:00 until May 24, 2016 13:00. The availability of data for the monitoring wells instrumented with Solinst loggers was from August 11, 2014 19:00 until November 16, 2015 17:00, except for MW23, which was removed on June 8,

2015 15:00 (Fig. 8.3a). Statistical comparisons were restricted to dates where there was data from all wells. The time series in which all wells were recording was from August 11, 2014 19:00 to December 1, 2014 12:00 and May 27, 2015 17:00 to November 16, 2015 17:00 and is referred to as the *period of analysis* throughout this document (Fig. 8.3b). Due to MW23's shorter recording period, it was left out of statistical comparisons where data was unavailable.

To facilitate certain statistical comparisons, sites were grouped into classes by disturbance and wetland type (e.g., undisturbed fens, undisturbed bogs, disturbed fens, and disturbed bogs). Sites were classified as undisturbed when there was no obvious sign of human drainage features. Undisturbed fens were designated as CF, IF, and LoB. LoB, despite its name, was classified as a fen after preliminary data analysis revealed the low fluctuation water level dynamics typical of groundwater influences, which are normally absent in bogs (Hunt et al. 1999; Tiner 2016). MW14 and MW23, the two monitoring wells located on the peripheral lagg-fen of the BMB complex were designated as disturbed fens due to nearby drainage features in BMB, while BB, SB, and RB (also referred to as MW29) were designated as undisturbed bogs. RB was excluded from the undisturbed bog group for some of the analyses that involved comparison of mean or median water levels due to the raised peat structure at this site and the inherent difference in water levels that results from this geomorphology relative to undomed bogs. MW3 and MW9, near the central ditch in the BMB complex, were classified as disturbed bogs, as was LiB due to extensive ditching along the adjacent access road.

For each well, we determined mean, standard deviation, interquartile range, median, maximum, and minimum across all water table depths. Range of water level fluctuations for the period of analysis was calculated as the difference between the maximum and minimum water table depths. Monthly fluctuations were calculated as the difference between the highest and lowest water level depths for the corresponding month in each well. Summer (June, July, and August) and Fall (September and October) mean fluctuations were also calculated for each wetland disturbance class. The amount of time water levels in each well were within various depth intervals (water table depth duration) was quantified from the original hourly dataset by summing the number of measurements in a particular depth stratum and converting that to a percentage of time for the period of analysis. Depth strata were designated as > 0 cm ("inundated"), in the shallow root zone (0 to − 20 cm—saturated"), between − 20 cm and − 40 cm ("dry"), and below 40 cm ("very dry"). Hydrologic behavior within the shallow root zone was of particular importance since it is thought to be a key factor in determining the success of *Geum peckii* populations (Environment and Climate Change Canada 2018). In addition, daily water level medians were calculated for each day that data was available for every well. A hydrograph of daily medians with daily precipitation from Environment and Climate Change Canada's climate station (ID 8200604) was used to visualize overall hydrodynamics, water level fluctuations and water table depth durations for the period of analysis. Differences in all water level characteristics were compared with one-way analysis of variance (ANOVA) using Minitab Statistical Software (Minitab, State College, PA,

USA). Tukey's HSD was used to test for significant ($p \leq 0.05$) pairwise differences among each wetland disturbance class.

8.3 Results

8.3.1 Water Table Characteristics—Across the Full Period of Analysis

Mean and median water table depths across the entire period of analysis were significantly shallower in undisturbed fens compared to all other sites and their range was also the smallest (Fig. 8.4, Table 8.2). Mean and median water depth at the disturbed fens, MW14 and MW23, was typically 10 cm lower than at the undisturbed fens, while maximum water levels were generally 10–15 cm lower and minimum depths 15–35 cm deeper. Although data availability was more limited for MW23, the range in values was more comparable to the range observed at undisturbed bogs (Table 8.2). RB had the deepest mean, median, maximum, and minimum depths among all sites due to the domed structure at this site and the fact that the location of the monitoring well was at the highest elevation point within the bog. Mean water levels at the undisturbed basin bogs, SB and BB, were significantly deeper than undisturbed fens, about the same as disturbed fens and generally shallower than in disturbed bogs (Fig. 8.4, Table 8.2). SB stood out among all bogs in that the overall range in water levels was the smallest (19.9 cm). It is also important to note that the range in values at BB (53.1 cm) was higher than at all other sites except MW3 near the ditch in BMB, that the maximum value at LiB, SB, and RB never rose to inundation and that the range for MW3, a disturbed bog (85.3 cm) was over 30 cm greater than any other well. We also analyzed the daily mean and median water levels for the summer months of 2015 to determine if differences among site types might be more apparent during the growing season, but the patterns were virtually identical to those observed for the entire period of analysis, so we have omitted them here.

8.3.2 Monthly Water Table Fluctuations

Examining differences across all months, mean monthly water table fluctuations were significantly higher at disturbed sites with disturbed bogs having the most dramatic fluctuations and undisturbed fens and SB having the smallest (Table 8.3). In fact, there was little to no difference between the overall, summer and fall means for the undisturbed fens—the largest monthly range was ~ 20 cm in June and September 2015. Although not statistically significant, MW14's mean monthly fluctuation was nearly double that observed at undisturbed fens (Table 8.3) and for the limited period that MW23 was recording, the mean monthly fluctuation was nearly triple the fluctuation

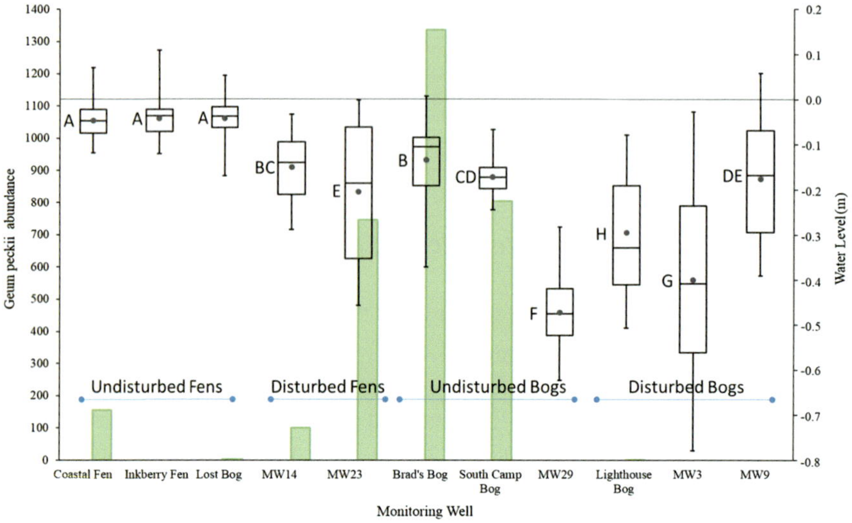

Fig. 8.4 Boxplot of daily median water level (m) relative to ground surface (0 m) for all monitoring wells during the period of analysis and *Geum peckii* abundance (# of individual plants as green bars) at each well location. Black points inside each box depict mean value for the corresponding wells (means that do not share a letter in their superscripts were significantly different, $p \leq 0.05$)

in the undisturbed fens. The pattern of lowest monthly fluctuations at undisturbed fens and highest at disturbed bogs held for both summer and fall means, with fluctuation at undisturbed bogs and disturbed fens being intermediate (Table 8.3). It is worth noting that differences in the mean summer and fall fluctuations was relatively small for each of the disturbance groups (11 cm at disturbed bogs was the maximum observed; Table 8.3), suggesting analysis of these data by season was probably unnecessary. However, it is interesting that the largest monthly fluctuation was during September for all sites except LoB. This result is likely due to the occurrence of the largest rain events at this time of year (Fig. 8.3b) and the transition from dry to wet seasons typical of this time of year in Nova Scotia.

8.3.3 *Water Table Depth Duration*

Examining the percentage of time water levels in different wells were in different inundation conditions revealed that undisturbed fens had a significantly higher proportion of inundated and saturated water levels. Disturbed bogs had the least number of observations in these categories and undisturbed bogs and disturbed fens being intermediate, especially with respect to time in saturated conditions (Table 8.4). Inundation at undisturbed fens occurred 7–10% of the time, while the remaining observations were saturated (> 90%) (Table 8.4). Wells in undisturbed bogs and

Table 8.2 Summary of main hydrologic characteristics based on water level data (in cm) collected hourly at eight peatlands on Brier Island during the full period of analysis (August 11th to December 1, 2014 and May 27th to November 16, 2015; N is the number of water level measurements for each well; IQR is the interquartile range; Range is the difference between the highest and lowest water level depths; *Geum peckii* count is the number of plants known to occur at each site at the time of the study)

Monitoring well	N	Median depth	IQR	Maximum depth	Minimum depth	Depth range	*Geum peckii* count
Undisturbed fens							
Coastal fen	6817	– 4.7	5.1	16.40	– 12.3	28.7	156
Inkberry fen	6817	– 3.5	4.9	14.30	– 12.3	26.6	0
Lost bog	6817	– 3.6	4.7	13.80	– 17.7	31.5	1
Disturbed fens							
MW14	6817	– 13.9	11.3	0.38	– 30.5	30.9	100
MW23	2951	– 21.8	29.3	3.85	– 46.2	50.0	747
Undisturbed bogs							
Brad's bog	6817	– 10.5	10.8	15.10	– 38.0	53.1	1336
South camp bog	6817	– 17.2	4.6	– 5.20	– 25.1	19.9	805
RB (MW29)	6817	– 47.5	10.7	– 19.56	– 62.8	43.2	0
Disturbed bogs							
Lighthouse bog	6817	– 33.1	22.2	– 3.50	– 56.0	52.5	1
MW3	6817	– 40.5	32.5	7.23	– 78.1	85.3	0
MW9	6817	– 16.4	22.9	9.88	– 40.3	50.1	0

disturbed fens were most commonly in a saturated condition (60–77%), but along with disturbed bogs, were dry between 20 and 40% of the time which was significantly lower than at undisturbed fens. Disturbed bogs were the only wetland disturbance group that was in a very dry condition for any significant period of time (~40%).

It is important to note that water table depth durations in MW9 were different from the other disturbed bogs due to the confounding affects of runoff from the airport runway ditch (Kennedy and Drage 2015). These effects were unique to this site and resulted in a greater percentage of water levels in the inundated and saturated depth stratum than the other disturbed bogs, but MW9 was also in a dry condition longer than any other site suggesting the effects of the central ditch were especially pronounced here (Table 8.4). Because of the anomaly that the additional drainage effect from the airport runway introduced, we decided to exclude MW9 from the analysis of depth duration differences among wetland disturbance groups, but highlight the patterns observed because they are informative.

Table 8.3 Monthly water level fluctuation for individual monitoring wells, mean monthly fluctuations for summer and fall, and overall mean by undisturbed and disturbed wetland group

Date (mm/yy)	Undisturbed fens			Disturbed fens		Undisturbed bogs			Disturbed bogs		
	CF	IF	LoB	MW14	MW23	BB	SB	RB	LiB	MW3	MW9
Aug-14	7.0	6.3	12.5	21.6	33.1	20.0	5.8	21.4	25.4	27.9	31.0
Sep-14	8.9	6.4	13.9	24.4	33.9	21.8	7.1	18.5	36.0	17.5	39.3
Oct-14	14.1	11.6	10.1	23.1	39.8	21.7	8.9	16.8	45.1	60.3	44.2
Nov-14	11.7	11.7	9.6	11.9	12.1	14.7	12.4	16.2	18.1	39.7	19.4
–	–	–	–	–	–	–	–	–	–	–	–
Jun-15	20.3	16.8	17.1	16.3	–	24.5	12.0	21.9	21.3	37.5	20.3
Jul-15	10.7	9.0	12.3	20.4	–	23.2	7.1	14.0	27.4	38.2	27.4
Aug-15	9.5	7.6	12.5	22.6	–	19.9	7.6	12.0	35.2	19.2	29.0
Sep-15	18.9	17.1	16.9	25.8	–	26.7	13.6	34.7	41.3	67.8	36.2
Oct-15	10.9	6.3	8.1	14.4	–	13.4	4.9	13.7	21.6	39.6	23.3
Mean	12.4^{C}	10.3^{C}	12.6^{C}	20.0^{BC}	29.7^{AB}	20.7^{BC}	8.8^{C}	18.8^{BC}	30.2^{AB}	38.6^{A}	30.0^{AB}
All month group mean	11.8^{A}			23.0^{B}		16.1^{AB}			32.9^{C}		
Summer group mean	11.8^{B}			20.2^{AB}		15.8^{B}			28.3^{A}		
Fall group mean	11.9^{B}			21.9^{B}		16.8^{B}			39.3^{A}		

Fluctuation was calculated as the difference between the highest and lowest monthly value in cm. Values are for months where all loggers were recording, except MW23 which was only recording from August to November 2014. The group mean for Disturbed Fens includes MW23 values, but summer and fall group means do not. June, July, and August are designated as summer months. September and October are designated as fall months. Means, within the same contextual comparison, that do not share a letter were significantly different ($p \leq 0.05$)

8.4 Discussion

One of the most important findings from our study was that there were clear differences in shallow water level dynamics among our four peatland classes. Mean water levels, monthly fluctuations, and water table depth durations ranged along a spectrum from highly consistent among the three undisturbed fens to highly variable among the three disturbed bogs. Disturbed fens and undisturbed bogs were typically intermediate for these parameters relative to the other two peatland classes. The peatlands with intermediate hydrological patterns between the greater constancy and wetness of the undisturbed fens and the hypervariability and dryness of the disturbed bogs are largely—with one exception noted below—those that supported substantial populations of *Geum peckii*. Typical water table depths of these intermediate systems

Table 8.4 Water table depth duration from hourly water levels during the period of analysis

Inundation category	Undisturbed fens			Disturbed fens		Undisturbed bogs		Disturbed bogs		
	CF	IF	LoB	MW14	MW23	BB	SB	LiB	MW3	MW9
% Inundated (> 0 cm)	9.66	8.22	7.82	0.03	1.05	0.76	0.00	0.00	0.34	6.83
Group mean	8.57^{A}			0.54^{B}		0.38^{B}		0.17^{B}		–
% Saturated (-20 cm $\geq x < 0$ cm)	90.35	91.78	93.64	72.47	48.18	75.67	78.56	27.38	19.28	49.43
Group mean	91.92^{A}			60.30^{B}		77.12^{AB}		23.33^{C}		–
% Dry (-40 cm $\geq x < -20$ cm)	0.00	0.00	0.00	27.50	42.86	23.57	21.44	43.41	29.77	43.73
Group mean	0.00^{A}			35.18^{B}		22.51^{B}		36.59^{B}		–
% Very dry $x > -40$ cm	0.00	0.00	0.00	0.00	7.90	0.00	0.00	28.93	50.61	0.02
Group mean	0.00^{A}			3.95^{A}		0.00^{A}		39.8^{B}		–

Depth duration is expressed as a percentage (%) of overall values. Data from the domed bog, RB, was not included in this comparison due to the inherent differences in depth duration between domed and basin bogs (see Methods for details). Due to the confounding effect of the runway ditch to MW9, it was left out of the Disturbed Bogs group mean for this analysis. Means, within the same inundation category, that do not share a letter in their superscripts were significantly different ($p \leq 0.05$)

ranged from 10 to 22 cm below ground surface (with the exception of the 47 cm mean depth recorded for RB), values that bracket the average within-fen water tables (14 and 17 cm, Chap. 4) that were recorded next to individual *Geum peckii* plants. The exception, the undisturbed RB (MW29), did not support *Geum peckii* because unlike the other bogs in this class, its center is conspicuously raised above a sedge fen margin along a long rocky ridge. Indeed, the Transplant Experiment (Chap. 4) demonstrated that Avens performed poorly at both ombrotrophic stations within RB (in comparison with fens and flat bogs, RB leaf widths were 33% narrower, $p < 0.01$; transplant mortality rate, 3 of 6 vs. 2 of 15, was greater, $p < 0.1$). Higher water level variation in bogs and deeper mean depths, particularly in those affected by drainage, compared to most fens is consistent with findings reported by Rydin (1986), Price et al. (2003) and Rydin and Jeglum (2006) which suggest the dependence of bogs on precipitation in large part leads to this pattern. Greater variation in the hydrodynamics of disturbed/drained bogs relative to undisturbed bogs has also been observed (e.g., Rothwell et al. 1996; Strack et al. 2008; Holden et al. 2011; Ketcheson and Price 2011) and attributed to factors such as the age and depth of the drainage ditch and degraded peat conditions, particularly in the catotelmic zone. So, it was not surprising that the greatest water level variation in the present study was recorded north of the ditch in Big Meadow Bog.

Another key result from our study was that mean water levels at undisturbed fens were at or very near the surface (< 5 cm below ground surface) most of the time and monthly fluctuations were minimal (<15 cm), while in disturbed bogs, mean water levels were considerably deeper (18–40 cm below ground surface), monthly fluctuations were nearly three times higher, and conditions were dry or very dry most of the time. *Geum peckii* was absent or poorly represented in both of these hydrological settings (at undisturbed fens with water tables near the surface, *Geum peckii* occurred at CF as 250 plants and was sparse at LoB; at disturbed bogs with water tables typically much further below ground, *Geum peckii* was absent except for a very sparse population at LiB). The paucity of *Geum peckii* at sites with water tables consistently near the surface is consistent with what was observed in another wet fen that was not monitored as part of our study (Chap. 4). Although considered a wet fen similar to IF and LoB, the average water table associated with *Geum peckii* occurrences at RF was further below ground than the mean water level condition, indicating a greater survival in microsites with a greater potential aerobic rooting zone. The absence or near absence of *Geum peckii* from IF and in the wet fen skirting RB (Chap. 4) as well as in adjacent nearby fens in the central part of the island (e.g., SB and LB), just a few hundred meters from healthy *Geum peckii* populations, argues that habitat limitation and not propagule limitation prevents the colonization of these undisturbed fens. In recent decades, ATV travel, at its peak in the 1980s, caused disturbance in most open peatlands on Brier Island and undoubtedly moved seeds (Hill et al. 2019), simultaneously creating the scarification needed for seedling establishment. On a longer time-scale, it appears (Chap. 3) that *Geum peckii* has had millennia to become established throughout the network of peatlands on the island. While the hydrology of RF and its hollow and hummock microtopography meets the requirements of *Geum peckii*, *Geum peckii* populations can be decimated by sudden changes in water table. Transplants placed directly in a bog pool in Runway Fen died within a month, and between 2017 and 2021, three quarters of an *Geum peckii* population in a low-lying fen were lost coincident with the ditch blocking restoration that raised the water table in the northeast of BMB by 20 cm (Chap. 7).

With respect to disturbed bogs, the low water table at these sites favors higher shrub colonization and growth (e.g., Bradford 2016), resulting in a higher rate of evapotranspiration (Morgan-Jones et al. 2005) and potentially leading to further lowering of the water table. In a study of hydrologic change in a disturbed bog in Quebec, Van Seters and Price (2002) observed increased water loss due to evapotranspiration and greater summer water deficit when there was a higher abundance of vascular plants. Similarly, Anderson et al. (2000) found lower mean water table depths in bogs when robust vegetation had established on drained peat. Dense shrub cover at MW3 in our study is likely also contributing to the greater amount of drawdown we observed during the growing season and creating competition for habitat that is limiting *Geum peckii* colonization (Chap. 4) Howie and van Meerveld (2011) stated that successfully restoring the hydrology in the center of a raised bog, at sites such as MW3, is dependent on high water levels in the lagg-fen. Since the runoff from the central bog in BMB complex is mostly redirected into the central drainage ditch or other peripheral ditches, the high water needed in the lagg-fen to maintain levels in the adjacent bog is no longer available and will need to be restored if the overall restoration of BMB is to be a success. The hydrologic stability that we observed in undisturbed fens outside of BMB is thought to be in part due to the inherent water holding capacity of the deep peat layer that is typical of certain fen forms (National

Wetland Working Group 1997; Rydin and Jeglum 2006; Money et al. 2009), but is likely also due to the more consistent supply of water inputs that fens receive from groundwater and runoff from the adjacent terrain (Ingram 1983; Gorham and Janssen 1992) as well as the lack of any drainage at these sites.

It is important to point out that there were exceptions to the general differences in the water level dynamics patterns noted above, particularly for some of the bog sites. For example, the dynamics at MW9, one of the three disturbed bogs, differed from the other two disturbed bogs in that mean levels were shallower and time in a saturated condition was much greater. It is likely that this was the result of redirected runoff originating from the airport runway that the other disturbed bog did not receive (Kennedy et al. 2015). Similarly, mean water levels in the undisturbed reference bog (MW29) were considerably deeper than in the other two undisturbed bogs, although this was not unexpected because the other two undisturbed bogs were basin bogs and RB was domed (e.g., Damman 1986; Crum and Planisek 1992; Bragg 2002; Gilvear and Bradley 2009), with the well placed at the highest elevation on the dome. In addition, we observed that variability associated with mean water levels and monthly fluctuations, as well as pH values at SB, one of the undisturbed basin bogs, were more comparable to what we observed at the undisturbed fens than any other wetland class. This suggests there may have been important surface and/or groundwater influences on the hydrodynamics at this site (Rydin and Jeglum 2006; Money et al. 2009). Based on the mix of hydrologic characters we measured, it is possible that SB is in transition from fen to bog, something that is not an unexpected trajectory for peatland succession under the right local conditions, such as when peat accumulates sufficiently to isolate the fen surface from most minerotrophic inputs as the result of changes in local hydrology (Ingram 1983; Gorham and Janssen 1992; Kolari et al. 2022).

Given the somewhat unique hydrodynamics at SB, it is interesting that it was one of the sites where there were healthy populations of *Geum peckii*. It was the only one of the three sites with an *Geum peckii* population above 700 that had such stable mean water levels (Fig. 8.4). The other two sites with high *Geum peckii* populations, BB and MW23, had mean levels between -13 and -21 cm, and more variable mean water levels that were in the shallow root zone (the saturated strata) for much of the growing season (~50–80%), only in the dry root zone for short periods (~20–40%) and rarely in the very dry zone if at all. However, it is important to note that the *Geum peckii* population located near MW23, although currently high relative to other sites we monitored, has experienced a 70% decline since 2012. It seems likely that this is at least in part due to the greater amplitude of water level fluctuations *Geum peckii* plants have been experiencing in this part of BMB for many years (Hill et al. 2016). Water levels at MW23 were below the shallow root zone for greater than 40% of this study. Based on observations at our other study sites, the highly variable and often low water levels at MW23 may not be sustainable for the *Geum peckii* populations and may favor increasing encroachment by shrubs, further reducing the suitability for *Geum peckii* unless the recent ditch blocking is successful in restoring higher water levels with more appropriate fluctuations. All this suggests that moderately variable mean water levels in the − 10 to − 20 cm range with regular but short duration fluctuations into the dry root zone are likely most conducive to the success of *Geum peckii*. However, moderately variable mean levels are apparently not a prerequisite given the success of *Geum peckii* at SB, where mean water levels were quite stable (standard deviation was only 3.2). Even so, mean water levels at SB were − 17 cm, in the shallow root zone 78% of the time and in the dry root zone 21% of the time, which appears to be the sweet spot for *Geum peckii*. Although it was unmeasured, it appears the microtopography at SB was considerably more varied (e.g., more pronounce hummocks) than at both IB and LB (two of the most consistently saturated undisturbed fens), which likely provided *Geum peckii* with colonization options at SB that are not available at these other sites (Chap. 4) that had no or very sparse *Geum peckii* populations.

That said, it was a bit puzzling that *Geum peckii* was as abundant as it was at CF, given the extremely stable water levels and the consistently high water table at this site where water levels were always in the saturated or inundated zone. However, this population of *Geum peckii* was relatively small (156 plants) and may be transitory. It is possible that it was temporarily favored by a recent fire at this site which reduced shrub cover and interspecific competition. If this is true, we would expect numbers to gradually decline at CF in the coming years as the site reaches a post-fire equilibrium. The other site with low to moderate *Geum peckii* numbers was MW14, which had water level dynamics that appear to be in the sweet spot for *Geum peckii*, but the site had dense sphagnum cover, extremely low groundwater pH (3.94) and tall mean shrub height (~1 m) which may not be as conducive to supporting higher *Geum peckii* numbers. Certainly, competition from shrubs can lead to reduced *Geum peckii* success (Chap. 4) and it seems likely that competition from sphagnum might be a deterrent as well, but these relationships are not yet well-established for *Geum peckii*. One other site where water level conditions appear to be at least close to the sweet spot for *Geum peckii* was MW9, where mean levels were − 17 cm, moderately variable and in the saturated zone for about 50% of the time. However, the vegetation was highly degraded at this site and gull activity relatively high (LaRue and Toms 2013), so it is not conducive to supporting *Geum peckii* and none has been detected there (Toms and LaRue 2015). The other two disturbed bogs, MW3 and LiB were clearly outside the hydrologic sweet spot for *Geum peckii* with highly variable and often very dry conditions, although LiB still supports one *Geum peckii* plant. MW3 was the most degraded of all the sites we monitored due to its close proximity to the central ditch, high gull activity, and poor vegetation quality in that area (Hill et al. 2019). We expect the *Geum peckii* at LiB will disappear in the very near future unless the drying effects of ditching at this site are remediated.

Identifying the hydrologic sweet spot for *Geum peckii* success was not the primary objective of this study (see MacIntyre 2017), but our results related to this may have important implications for evaluating the success of the overall restoration at BMB. From the outset of discussions about returning historic water levels as part of the restoration of BMB, those involved in the restoration planning (Hill et al. 2016) were aware that changes to water levels in BMB would almost certainly lead to a redistribution of *Geum peckii* populations within BMB. The ideal hydrologic niche space for *Geum peckii* was expected to be redistributed spatially across the

BMB complex even though the characteristics of that idealized niche space were unknown at the time. Botanists have been investigating the interrelationships between hydrology and vegetation in peatlands for many years (e.g., Gignac 1992; Bragazza and Gerdol 1999; Dimick et al. 2010; Caldwell et al. 2011; Johansen et al. 2018), but *Geum peckii* populations had not been examined in this fashion at the time we initiated this study. The results of our study have begun to elucidate what some aspects of the hydrologic niche for *Geum peckii* might be (e.g., mean water levels − 10 to − 20 cm, levels in the saturated root zone for about 75% of the time and in the dry root about 25% of the time), and as such, should be useful for predicting where *Geum peckii* is likely to do best in a restored BMB under the new water level regime. However, it is important to note that the results of Hill et al. (Chap. 4) strongly suggest that while suitable hydrology is critical to the success of *Geum peckii*, other habitat factors, such as shrub density, microtopography, and the presence of aggressive and/or invasive vegetation also play key roles in the success of this species.

Acknowledgements We acknowledge that we conducted this study in Mi'kma'ki, the ancestral and unceded territory of the Mi'kmaq people, that we are all treaty people and bound by mutual obligations under the Peace and Friendship Treaties of 1760-61. We thank Randy Milton for ongoing advice and assistance during the development and completion of the fieldwork for this project and Wally de Vries, owner of the R.E Robicheau General Store for providing accommodations and logistical support on Brier Island. Thanks also Tony Bowron for assistance as Co-Advisor to Makeala MacIntyre's Honour's Thesis project at St Mary's University, to Andrew Sharpe and Mike Parker from East Coast Aquatics Inc. for access to data from their rain gauge station and to NSDNRR Librarian, Tracy Lenfesty help locating supporting references and Frances MacKinnon for GIS support. This project has been funded wholly by the NSDNRR, but many of the water level loggers we used for this study were originally purchased under a National Wetland Conservation Fund grant provided by Environment and Climate Change Canada in 2014 and we are grateful to have had access to this equipment. This manuscript has been reviewed by the Wildlife Division of NSDNRR and approved for publication. Approval does not signify that the contents reflect the views of the Department, nor does mention of trade names or commercial products constitute endorsement or recommendation for use.

References

Acreman MC, McCartney MP (2009) Hydrological impacts in and around wetlands. The Wetlands Handbook, pp 643–666

Andersen AR, Broughm WA (1988) Evaluation of Nova Scotia peatland resources. Bulletin 6, Nova Scotia Department of Mines and Energy

Anderson AR, Ray D, Pyatt DG (2000) Physical and hydrological impacts of blanket bog afforestation at Bada′ Cheo, Caithness: the first 5 years. Forestry 73(5):467–478

Bell N (2021) Establishing baseline hydrologic conditions of forested wetlands in Nova Scotia, Canada. Master's Thesis, Dalhousie University, Halifax, NS, Canada

Bradford A (2016) Averting degradation of southern Ontario wetlands due to hydrologic alterations associated with development. Canadian Water Resour J 41(4):543–553

Bragazza L, Gerdol R (1999) Hydrology, groundwater chemistry and peat chemistry in relation to habitat conditions in a mire on the Southeastern Alps of Italy. Plant Ecol 144:243–256

Bragg OM (2002) Hydrology of peat-forming wetlands in Scotland. Sci Total Environ 294(1–3):111–129
Brazner J, Kennedy G, Drage J (2015) Understanding hydrology and water quality to increase restoration effectiveness—Phase I: Determining methodological best practices. National Wetland Conservation Fund Final report. NCP-06-EC
Brazner J, MacKinnon F (2020) Relative conservation value of Nova Scotia's forests: forested wetlands as avian diversity hotspots. Can J for Res 50(12):1307–1322
Caldwell PV, Vepraskas MJ, Gregory JD, Skaggs RW, Huffman RL (2011) Linking plant ecology and long-term hydrology to improve wetland restoration success. Trans ASABE 54(6):2129–2137
Calme S, Desrochers A, Savard JPL (2002) Regional significance of peatlands for avifaunal diversity in southern Québec. Biol Conserv 107:273–281
Chimner RA, Cooper DJ, Wurster FC, Rochefort L (2017) An overview of peatland restoration in North America: where are we after 25 years? Restor Ecol 25(2):283–292
Crum H, Planisek S (1992) A focus on peatlands and peat mosses. University of Michigan Press
Damman AW (1986) Hydrology, development, and biogeochemistry of ombrogenous peat bogs with special reference to nutrient relocation in a western Newfoundland bog. Can J Bot 64(2):384–394
Dimick BP, Stucky JM, Wall W, Vepraskas MJ, Wentworth TR, Arellano C (2010) Plant-soil-hydrology relationships in three Carolina bays in Bladen County. North Carolina. Castanea 75(4):407–420
Environment and Climate Change Canada (2018) Action Plan for the Eastern Mountain Avens (*Geum peckii*) in Canada. Species at Risk Act Action Plan Series. Environment and Climate Change Canada, Ottawa
Gignac LD (1992) Niche structure, resource partitioning, and species interactions of mire bryophytes relative to climatic and ecological gradients in western Canada. Bryologist, pp 406–418
Gilvear DJ, Bradley C (2009) Hydrological dynamics II: groundwater and hydrological connectivity. The Wetlands Handbook, pp 169–212
González E, Rochefort L (2014) Drivers of success in 53 cutover bogs restored by a moss layer transfer technique. Ecol Eng 68:279–290
Gopal B (2009) Biodiversity in wetlands. The Wetlands Handbook, pp 65–95
Gorham E, Janssen JA (1992) Concepts of fen and bog re-examined in relation to bryophyte cover and the acidity of surface water. Acta Soc Bot Pol 61:7–20
Graf M, Bérubé V, Rochefort L (2012) Restoration of peatlands after peat extraction. Restoration and reclamation of boreal ecosystems. Cambridge University Press, Cambridge, pp 259–280
Hill NM, Chinner K, Toms B, Garbary DJ (2019) Recovery succession of native plant communities at Big Meadow Bog after disturbance and eutrophication from herring gulls. Wetlands 39(3):533–548
Hill N, Kennedy G, Smith C, Horne D, Brazner J, Drage J, Boates S, Eaton S (2016) A restoration plan for Big Meadow Bog. Application to Nova Scotia Environment, Parts I and II
Hill N (2013) Conservation management report on Eastern Mountain Avens on Brier Island. Prepared for the Canadian Wildlife Service, Contract K4B20–12–0267
Holden J, Wallage ZE, Lane SN, McDonald AT (2011) Water table dynamics in undisturbed, drained and restored blanket peat. J Hydrol 402(1–2):103–114
Howie SA, van Meerveld ITV (2011) The essential role of the lagg in raised bog function and restoration: a review. Wetlands 31:613–622
Hunt RJ, Walker JF, Krabbenhoft DP (1999) Characterizing hydrology and the importance of ground-water discharge in natural and constructed wetlands. Wetlands 19(2):458–472
Ingram HAP (1983) Hydrology. In: Gore AJP (ed) Ecosystems of the world, vol 4B. Mires, swamp, bog, fen and moor, Regional studies, Elsevier, Oxford, pp 67–158
Johansen OM, Andersen DK, Ejrnæs R, Pedersen ML (2018) Relations between vegetation and water level in groundwater dependent terrestrial ecosystems (GWDTEs). Limnologica 68:130–141

Kennedy GW, Drage J (2015) Preliminary results of baseline hydrological monitoring at Big Meadow Bog, Brier Island, Digby County, Nova Scotia; Nova Scotia Department of Natural Resources. Open File Report 2015–001:17
Kennedy G, Drage J, Nixon C (2015) 2014 Update on baseline hydrological monitoring at Big Meadow Bog, Brier Island, Nova Scotia. Open file report, Nova Scotia Natural Resources, Geoscience and mines branch. Open File Report ME 2015-005
Ketcheson S, Price J (2011) The impact of peatland restoration on the site hydrology of an abandoned block-cut bog. Wetlands 31:1263–1274
Kolari TH, Sallinen A, Wolff F, Kumpula T, Tolonen K, Tahvanainen T (2022) Ongoing fen–bog transition in a boreal aapa mire inferred from repeated field sampling, aerial images, and Landsat data. Ecosystems 25(5):1166–1188
LaRue D, Toms B (2013) Building for the future: establishing a foundation for the recovery of Eastern Mountain Avens. Final Report to Environment Canada and the Eastern Mountain Avens Recovery Team
Lavoie C, Rochefort L (1996) The natural revegetation of a harvested peatland in southern Québec: a spatial and dendroecological analysis. Ecoscience 3(1):101–111
MacIntyre, M (2017) Comparing shallow groundwater hydrodynamics in reference peatlands and the ditched Big Meadow Bog complex on Brier Island, Nova Scotia. Honour's Thesis, St. Mary's University, Halifax, Nova Scotia, Canada.
Mitsch WJ, Gosselink JG (2015) Wetlands, 5th edn. John Wiley & Sons Inc., Hoboken, NJ
Mitsch WJ, Bernal B, Nahlik AM, Mander Ü, Zhang L, Anderson CJ, Jørgensen SE, Brix H (2013) Wetlands, carbon, and climate change. Landsc Ecol 28:583–597
Money RP, Wheeler BD, Baird AJ, Heathwaite AL (2009) Replumbing Wetlands–managing water for the restoration of bogs and fens. The Wetlands Handbook, pp 755–779
Morgan-Jones W, Poole JS, Goodall R (2005) Characterization of hydrological protection zones at the margins of designated lowland raised peat bog sites. Joint Nature Conservation Committee, Peterborough. JNCC Report no. 365
National Wetland Working Group (1997) The Canadian wetland classification system. Wetlands Research Centre, University of Waterloo, Waterloo, ON
Poulin M, Rochefort L, Pellerin S, Thibault J (2004) Threats and protection for peatlands in Eastern Canada. Géocarrefour 79(4):331–344
Price JS, Heathwaite AL, Baird AJ (2003) Hydrological processes in abandoned and restored peatlands: an overview of management approaches. Wetlands Ecol Manag 11:65–83
Reid M (2014) The advantages of incorporating historical geographic information systems (H-GIS) into modern coastal management planning. J Map Geogr Libr 10(2):157–172
Richter BD, Baumgartner JV, Powell J, Braun DP (1996) A method for assessing hydrologic alteration within ecosystems. Conserv Biol 10(4):1163–1174
Rothwell RL, Silins U, Hillman GR (1996) The effects of drainage on substrate water content at several forested Alberta peatlands. Can J for Res 26(1):53–62
Rydin H (1986) Competition and niche separation in Sphagnum. Can J Bot 64(8):1817–1824
Rydin H, Jeglum J (2006) The biology of peatlands. Oxford University Press, Oxford, p 343
Sheehy C, Brazner JC, Basquill S, MacKinnon F, Crewe T, Colyn R (2023) On-the-ground verification of forested wetland classes and locations attained using the Predictive Ecosystem Mapping (PEM) model—Year 1. Nova Scotia Department of Natural Resources and Renewables, Biodiversity Tech Report 2023–002
Strack M, Waddington JM, Bourbonniere RA, Buckton EL, Shaw K, Whittington P, Price JS (2008) Effect of water table drawdown on peatland dissolved organic carbon export and dynamics. Hydrol Process Int J 22(17):3373–3385
Temmink RJ, Lamers LP, Angelini C, Bouma TJ, Fritz C, van de Koppel J, Lexmond R, Rietkerk M, Silliman BR, Joosten H, van der Heide T (2022) Recovering wetland biogeomorphic feedbacks to restore the world's biotic carbon hotspots. Science 376(6593):1479
Tiner RW (2016) Wetland indicators: a guide to wetland identification, delineation, classification and mapping. Lewis Publishers, Boca Raton, FL

Toms R, LaRue D (2015) Counts of *Geum peckii* on Brier Island within permanent monitoring plots, 2014. Morsey Tobeatic Research Institute, Kempt, NS

van der Valk AG (2009) Restoration of wetland environments: lessons and successes. The Wetlands Handbook, pp 729–754

Van Seters TE, Price JS (2002) Towards a conceptual model of hydrological change on an abandoned cutover bog, Quebec. Hydrol Process 16(10):1965–1981

Waddington JM, Quinton WL, Price JS, Lafleur PM (2009) Advances in Canadian peatland hydrology, 2003–2007. Can Water Resour J 34(2):139–148

Part II
Post-Restoration Changes, Intervention, and Management

Chapter 9
Greenhouse Gas Emissions from Big Meadow Bog: Influence of Gull-Derived Phosphate as a Biogeochemical Driver

Nicholas M. Hill, David L. Burton, Drucie Janes, Emil H. Jurga, and David J. Garbary

Abstract Greenhouse gas (GHG) emissions (CO_2, CH_4, N_2O) were measured in the two years prior to, and the year of, the hydrological restoration of the ditched Big Meadow Bog (BMB). The total greenhouse gas emission from the south of BMB in recovery succession after abandonment of the area by nesting Herring Gulls 20 years earlier ('BMB Recovering') was twice that of the pristine raised 'Reference Bog'. The emissions from the north of BMB, the zone of current gull nesting ('BMB Gull'), was twice that of the 'BMB Recovering' site. The coefficient of variation (CV) for CO_2 evolution measurements at all sites was comparatively low (average = 36%) but extreme for N_2O and CH_4 (average in each case = 139%); regardless, for each measurement year, the N_2O values (17–92 CO_2 equivalent units) at BMB Gull were significantly greater than either the Reference Bog or BMB Recovering sites. While CO_2 evolution accounted for nearly all (88–100%) of total GHG emissions prior to blocking the ditches (hydrological restoration) N_2O levels were greatest (92 CO_2 eq.) in the restoration year when the water table was elevated. Phosphate levels in peat at the two BMB sites—the legacy of gull guano inputs—accounted for 73% of

N. M. Hill (✉)
UNESCO Southwest Biosphere Association, 399 Wyman Road, Sand Beach, NS B5A 5G4, Canada
e-mail: nick@swnovabiosphere.ca

D. L. Burton · D. Janes
Department of Plant, Food, and Environmental Sciences, Dalhousie University, Agriculture Campus, Halifax, NS B2N 5E3, Canada
e-mail: dburton@dal.ca

D. Janes
e-mail: drucie.janes@dal.ca

E. H. Jurga · D. J. Garbary
Department of Biology, St. Francis Xavier University, Antigonish, NS B2G 2W5, Canada
e-mail: ejurga@proton.me

D. J. Garbary
e-mail: dgarbary@gmail.com

N. M. Hill et al. (eds.), *Wetland Restoration for Endangered Species Recovery*, Environmental Contamination Remediation and Management,
https://doi.org/10.1007/978-3-031-71344-6_9

CO_2 evolution variation in its measurement year (2016). Three types of nitrogen in peat (NO_3, NH_4 and total soluble N) accounted for less variation ($r^2 = 42$–52%) in N_2O than phosphate ($r^2 = 66\%$). Overall, greenhouse gas emissions have been exacerbated by the changing hydrology and eutrophication of BMB; however, the long-term management goal of reducing gull numbers could return BMB to a carbon sink.

Keywords Green house gas · Phosphate · Nitrate · Decomposition · Bog

9.1 Introduction

Over the past 2500 years, enough carbon dioxide was fixed and retained in Big Meadow Bog (BMB) to deposit a layer of peat more than a meter thick over layers of marine sands and muck laid down in the preceding 7000 years (Spooner et al. 2017). The peat was produced in three successive wetland stages—swamp, fen and then bog—where the production of organic matter exceeded its decomposition (Chap. 7). Decomposition of the organic materials was restricted in the waterlogged environments of the swamp and fen by anaerobiosis and by a shortage of mineral nutrients. Over the past 750 years in the center of Big Meadow, there was a shift in the moss community to one dominated by *Sphagnum* section *Acutifolia* (Spooner et al. 2017). As this *Sphagnum* community elevated the surface of the bog about half a meter above the water table, precipitation became the sole source of water resulting in a plant and a decomposer community limited by desiccation (height above water table), acidity and shortage of mineral nutrients (Rydin et al. 2006; Vitt and Slack 1984).

The global contribution of peatlands to carbon storage is ten-fold greater than their area (Gorman 1991; Leifeld and Menechetti 2018; Xu et al. 2018) and peatlands contain 44% of all soil carbon (IUCN 2024). They are thus a critical part of climate change mitigation strategies. Despite this contribution, however, these valuable ecosystems that began to accumulate carbon as peat over the millennia following deglaciation (Vitt et al. 2000) are now being undermined by drainage and eutrophication which increase decomposition and the mineralization of reduced carbon. The IUCN (2024) calculates that damaged peatlands are major greenhouse gas contributors and are responsible for 5% of the total global anthropogenic emissions. Big Meadow Bog is a 65-ha peatland with a central raised bog that has been drained since 1958 and substantially fertilized by Herring Gulls from 1978 to the present (Geizer et al. 2021; Hill et al. 2019; Kickbush et al. 2018). The present study contrasts the greenhouse gas emissions from the intact raised Reference Bog with the ditched Big Meadow Bog in its two states: Recovering from gull disturbance, BMB Recovery, and actively disturbed and enriched by gulls, BMB Gull.

9.2 Methods and Materials

9.2.1 *Field Sites*

Four greenhouse gas measuring stations were installed along two Big Meadow Bog transects: BMB Gull and BMB Recovering, and in a low shrub, elevated area of the Reference Bog. The Reference Bog stations were situated 30 m apart in a typical ombrotrophic plant community (*Rubus chamaemorus, Empetrum nigrum, Eriophorum spissum* and *Sphagnum fuscum,* and *S. flavicomans*). The Big Meadow Bog stations were spaced at 30-m intervals along transects perpendicular to the long axis of the bog. The northern transect, BMB Gull, is a zone of current Herring gull nesting, whereas the southern transect, BMB Recovering, was last disturbed by gulls 25 years prior to sampling. There are strong vegetation differences between the BMB transects. Whereas the currently gull-disturbed area is dominated by weedy vegetation (raspberry; *Rubus idaeaus*, and the exotic, Yorkshire fog, *Holcus lanatus*), the south transect, a 25-year post-gull disturbance, undergoing secondary succession, is vegetated by many species typical of swamps (e.g., *Picea mariana, Photinia* spp., *Viburnum nudum, Rhododendron groenlandicum, Maianthemum trifolium* and *Sphagnum magellanicum*). The recovery succession of the vegetation of Big Meadow Bog has been analyzed with respect to the target restoration state of the pristine Reference Bog (Hill et al. 2019). The current study was designed to contrast the differences in the greenhouse gas evolution between the ditched peatland (Big Meadow) and the intact ecosystem (Reference Bog), in addition to the extreme levels of gull-induced eutrophication in BMB.

9.2.2 *Greenhouse Gas Collection*

At each site, four plastic collars, 20 cm in diameter and 40 cm in height, were inserted into the peat. The moss and above-ground vegetation at each site were clipped away and the collar was pushed and turned into the peat to a depth of 15 cm. Collars were installed a minimum of 24 h before initial GHG flux measurements were taken and remained permanently installed at the sites throughout the season. Soil gas fluxes were measured using the static chamber method as described in Burton et al. (2008, 2013). On measurement days the chamber top was placed on each soil collar and a 20 ml disposable nylon syringe (Becton–Dickinson 14-823-2B) fitted with 25 gauge, 5/8″ length Luer-Lok tip needles (Becton–Dickinson 14-826AA) was used to obtain 20 ml air samples from the headspace of each chamber at 0, 10, 20, and 30 min after deployment. Samples were subsequently injected and stored in 12 ml pre-evacuated exetainers (Labco Limited 739W) fitted with butyl rubber septa. Air samples were stored in a cool dark location until they were shipped to the Greenhouse Gas Lab (Dalhousie University, Faculty of Agriculture) in Truro, Nova Scotia, for analysis.

9.2.3 Gas Analysis

Gas samples were analyzed for N_2O, CO_2, and CH_4 using a Varian Star 3800 Gas Chromatograph (Varian, Walnut Creek, CA) fitted with an electron capture detector (N_2O), thermal conductivity detector (CO_2), and a flame ionization detector (CH_4). Samples were transferred from the exetainers to the GC using a Combi-PAL Autosampler (CTC Analytics, Zwingen, Switzerland). The electron capture detector was operated at 350°C, 90% Ar, 10% CH_4 carrier gas at 20 ml min^{-1}, Haysep N 80/100 pre-column (0.32 cm diameter × 50 cm length) and Porapak QS 80/100 mesh analytical columns (0.32 cm diameter × 200 cm length) in a column oven operated at 80°C. The pre-column was used in combination with a four-port valve to remove water from samples. The thermal conductivity and flame ionization detectors were operated in series at 130°C and 200°C respectively, pre-purified helium (He) carrier gas at 20 psi, Haysep N 80/100 mesh (0.32 cm diameter × 50 cm length) pre-column followed by a Porapak QS 80/100 mesh (0.32 cm diameter × 200 cm length) analytical column maintained at 80°C.

Individual gas fluxes (F_c) for N_2O, CO_2, and CH_4 (ng N_2O m^{-2} s^{-1}, μg CO_2 m^{-2} s^{-1} and ng CH_4 m^{-2} s^{-1}) were calculated from changes in each gas concentration within the static chamber according to Hutchinson and Livingston (1993):

$$F_c = (dC/dt)((V_c{*}M_{mol})/(A{*}V_{mol}))$$

where dC/dt is rate of change in N_2O, CO_2, or CH_4 concentration (mol mol^{-1} h^{-1}), A is the surface area (m^2) of the collar, V_c is the total volume (in liters) of the enclosure (chamber volume plus collar volume), M_{mol} being the molar mass of N_2O, CO_2, or CH_4 (g mol^{-1}), and V_{mol} being the volume of one mole of N_2O, CO_2, or CH_4 (liters mol^{-1}) inside the chamber and corrected for temperature using the ideal gas law. The rate of change in concentration in the headspace (dC/dt) was calculated using a simple linear regression of the gas concentrations versus the time of deployment. Positive F_c values indicated the soil was emitting the gas, while negative F_c values indicated the soil was acting as a sink for the gas, CH_4 and N_2O sinks represent gas consumption by methanotrophs and denitrifiers, respectively. Cumulative growing season gas fluxes were calculated for each individual chamber by linear interpolation between sampling dates (Burton et al. 2008). Cumulative chamber emission values were averaged across chambers within a treatment to determine mean cumulative emission values for each treatment and sub-treatment and cumulative N_2O emissions expressed as mg N_2O m^{-2}, cumulative CO_2 as g CO_2 m^{-2}, and cumulative CH_4 as mg CH_4 m^{-2} over the monitoring period. N_2O and CH_4 were converted to CO_2 equivalents (CO_2e) using conversion factors of 298 and 25 to determine the global warming potential (GWP) of greenhouse gas emissions at the soil–atmosphere interface.

9.2.4 *Extractable* N and P from Soil Cores

In 2016, extractable N (NH_4, N_2O, total N) and P were determined for each GHG chamber ($n = 4$) at each site using a 4″ (ca. 10 cm) soil corer as follows: (1) each core was extracted separately using the Mehlich 3 (M3) extraction protocol (House 2005; Ring 2001; Simard and Tran 1993), and (2) N and P composition in each extract was determined at the Analytical Lab, Nova Scotia Department of Agriculture (https://novascotia.ca/agri/programs-and-services/lab-services/analytical-lab/).

9.2.5 *Statistical Analysis*

GHG flux differences among the three wetlands were determined using a 2-way ANOVA with sites and years as the variable followed by post-hoc Tukey's test to determine significant differences ($p < 0.05$). Linear regression of peat extractable *P* and CO_2 evolution was carried out in GraphPad (https://www.graphpad.com/quickcalcs/linear1/).

9.3 Results

9.3.1 *CO_2, CH_4, and N_2O Fluxes*

The evolution of three greenhouse gases, CO_2, CH_4, and N_2O, was assessed over three years, 2015–2017, at the ditched Big Meadow Bog and at the intact Reference Bog. Carbon dioxide evolution in all study years showed the same rank order: CO_2 evolution was greatest at BMB Gull, intermediate from BMB Recovering, and lowest from the pristine Reference Bog (Fig. 9.1). In 2015 and 2016, there was a significant difference in the CO_2 evolution from the Reference Bog and the BMB Gull but not between either of these sites and the BMB Recovering. In contrast, in 2017, after two additional gas measurement collars were added to each site (n = 6), there were significant differences in the CO_2 evolution between each site.

The coefficients of variation (CVs) for CO_2 averaged 35% for mean values between 2015 and 2017, the average CV for those dates for both CH_4 and N_2O was four times greater (139% for each gas's mean data). While the rank order of the average methane evolution changed between years, in no year were there significant differences (Fig. 9.2); however, the nitrous oxide measurements consistently showed near zero to negative values for both the Reference Bog and the BMB Recovering but substantially positive values (17–92 g CO_2 e.m^{-2}) for BMB Gull (Fig. 9.3); in each year, there were no significant differences between the two former sites yet significant differences between BMB Gull and Reference Bog and BMB Recovering. There was a significant difference in the total greenhouse gas emission (expressed

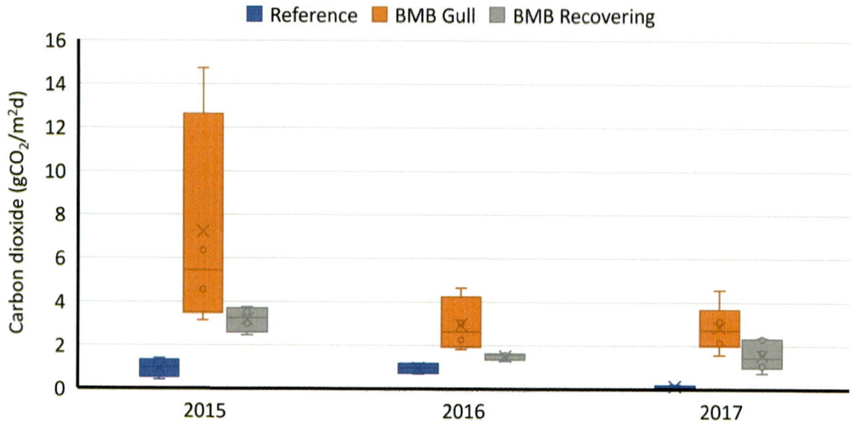

Fig. 9.1 Carbon dioxide evolution from Big Meadow Bog (BMB Gull and BMB Recovering) and from Reference Bog measured biweekly from June through August in 2015, 2016 and 2017. Box plot notation: X, mean; horizontal line, median; circles, interquartile range; whiskers, extreme values

as CO_2 equivalents) in 2015 between the BMB Gull area and the Reference Bog and no significance between the intermediate value of the BMB Recovering area and the other two stations. In the two following years, both the Reference Bog and the BMB Recovering area were significantly lower in their GHG emissions than the BMB Gull area but not from one another (Fig. 9.4).

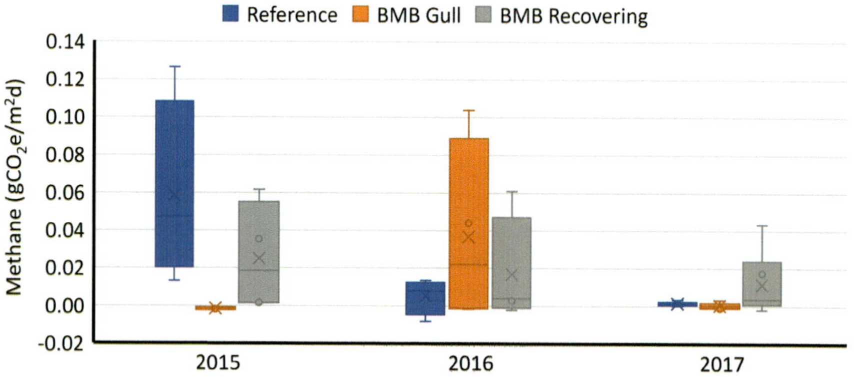

Fig. 9.2 Methane evolution (as CO_2 equivalents) from Big Meadow Bog (BMB Gulland BMB Recovering) and from Reference Bog measured from June through August in 2015, 2016, and 2017. Box plot notation: X, mean; horizontal line, median; circles, interquartile range; whiskers, extreme values

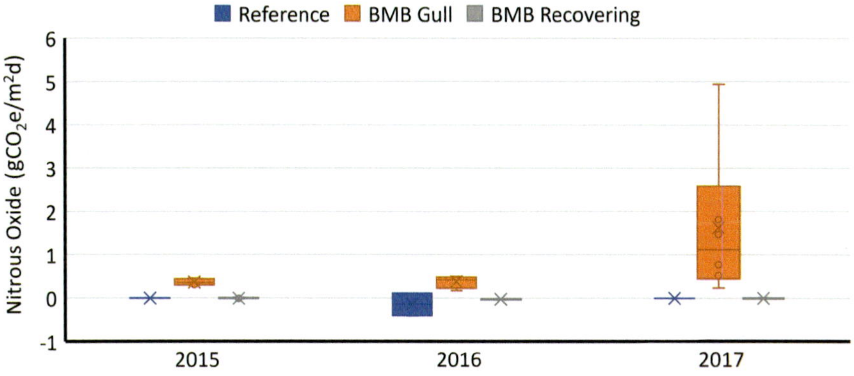

Fig. 9.3 Nitrous oxide evolution (as CO_2 equivalents) from Big Meadow Bog (BMB Gull and BMB Recovering) and from Reference Bog measured from June through August in 2015, 2016 and 2017. Box plot notation: X, mean; horizontal line, median; circles, interquartile range; whiskers, extreme values

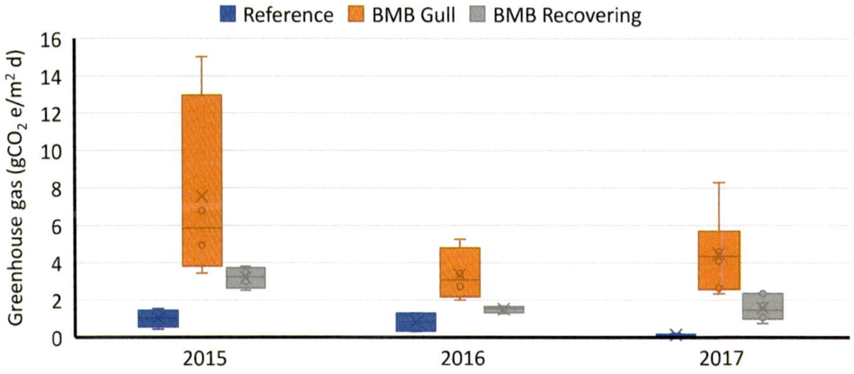

Fig. 9.4 Greenhouse gas production (as CO_2 equivalents) from Big Meadow Bog (BMB Gull and BMB Recovering) and from Reference Bog measured from June through August in 2015, 2016, and 2017. Box plot notation: X, mean; horizontal line, median; circles, interquartile range; whiskers, extreme values

9.3.2 Extractable N and P and CO_2 Evolution

Extractable soil N (NO_3-N, NH_4, total soluble N) and P from 2016 are given in Table 9.1. In all cases means from BMB Gull and Reference Bog were significantly different with BMB Recovering having intermediate values (Student's *t*-test, $p < 0.05$). Percent variation among sites was in the order of 100%, 600%, 300%, and 800% for NO_3, NH_4, total extractable N, and P, respectively. In all cases, BMB Gull had higher soil nutrient composition than the Reference Bog, with BMB Recovery having intermediate values.

Table 9.1 Peat nutrient composition for KCl extracted NO_3-N, NH_4-N and total soluble N, and maleic acid extraction of total P; values are mean ± s (n = 4; note that three of four values of NO3-N at Reference Bog were not distinguishable from blank values)

	Wetland site		
	BMB Gull	BMB recovery	Reference Bog
Mineral form	(mg/kg peat)		
NO_3-N	42.1 ± 15.3	3.51 ± 7.01	(1.44 ± 2.88)
NH_4-N	126 ± 73.5	68.9 ± 11.2	42.2 ± 14.6
Total soluble N	433 ± 159	266 ± 30.0	163 ± 78.2
Total phosphorus	197 ± 32.9	80.6 ± 10.7	26.2 ± 3.55

Much of the variation ($r^2 = 0.70$, $p < 0.001$, $n = 12$) in carbon dioxide evolution was accounted for by the total extractable phosphorus in the peat at each station of the three sites (Fig. 9.5). Strong relationships were also observed for soluble nitrogen (KCl extracted) as NH_4-N or total soluble nitrogen (r^2 = 0.53 and 0.57 for linear relationships between CO_2 evolution and NH_4-N and total soluble N respectively, $p < 0.01$, $n = 12$, in each case). Variation in nitrous oxide evolution was also closely related to soil phosphorus ($r^2 = 0.66$, $p < 0.005$, $F = 19.3$, $n = 12$). Soil nitrogen variables accounted for less variation in N_2O emission (r^2 and p values = 0.44/0.05, 0.40/0.05 for linear regressions where NH_4-N and total soluble N were respective, respective independent variables in linear regression equations with N_2O; note that in 6 of 12 cases, NO_3-N values were less than blank values).

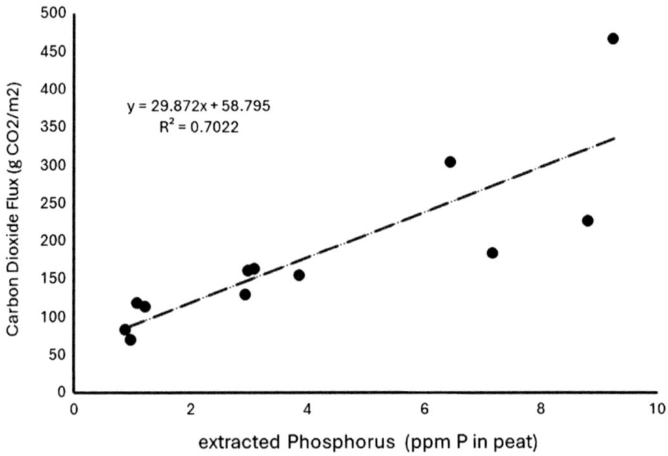

Fig. 9.5 The relationship between extracted phosphorous from peat (surface to 75 cm depth) and carbon dioxide evolution ($F = 23.5$, $p < 0.001$, $n = 12$)

9.4 Discussion

The recovering site in Big Meadow Bog (i.e., BMB Recovering) had double (average of the means from 2015 and 2016) the greenhouse gas emissions of the Reference Bog prior to the hydrological restoration of BMB (Fig. 9.4). This is consistent with studies that show an increase in as greenhouse gas evolution in ditched and disturbed peatlands (Tey et al. 2011; Krause et al. 2021; Wilson et al. 2016). We are reporting on the three greenhouse gases (CO_2, CH_4, and N_2O). However, prior to the hydrological restoration of Big Meadow Bog in 2017 when ditch blocking elevated the water table (Chap. 10), CO_2 alone accounted for 88–100% of total greenhouse gas emissions. In 2017, the water table was elevated throughout BMB, but particularly so in the northern end of the bog (Chap. 7), the area of greatest Herring Gull numbers and BMB Gull gas monitoring sites. The greater contribution in 2017 of N_2O (32%) to total greenhouse gas flux in the Ditched Gull site is likely the result of increased denitrification under waterlogged conditions of a peatland enriched in nitrogen (Knowles 1982; Takeda et al. 2023), in our case a peatland with decades of gull guano.

In addition to nitrogen inputs, the phosphate inputs via Herring Gull wastes to Big Meadow Bog stand out (ca. 6000 mg P m^{-2} y^{-1}; Geizer et al. 2021). Phosphate levels are more than two orders of magnitude greater than the maximum phosphorus deposition rates (25 mg m^{-2} y^{-1} in Schillereff et al. 2021) considered in an analysis of phosphate deposition into peatlands in industrialized regions of the northern hemisphere. Phosphorus has been identified as a limiting factor of carbon mineralization in peatlands (Hill et al. 2014). This is particularly the case at lower concentrations of P where there is a tight relationship between P and CO_2 flux, but this relationship becomes less so at higher concentration of P (Fig. 9.5). It is possible that at higher concentrations where P is in excess, N becomes the limiting factor in decomposition. Overall, P may be the driver of microbial activity, decomposition and N_2 fixation. High P levels lead to increased CO_2 evolution, decreased C burial and increased P burial, and increased dissolved C, N and P (Schillereff et al. 2021). In this study, total extractable phosphorus was positively correlated with CO_2 evolution ($r^2 = 0.73$) as well as with N_2O ($r^2 = 0.66$). That phosphorus accounted for more of the variation in N_2O flux than did all forms of nitrogen tested as independent variables (viz. NO_3, NH_4, or total N: respective r^2 values = 0.52, 0.43, 0.42), may suggest that the increased decomposition brought on by P, creates the anaerobiosis that permits denitrification (Knowles 1982). Mineral nutrients that are the limiting factors of the microbial decomposition of organic matter also will control the redox state and the partial pressure of oxygen (Fiedler et al. 2007; Takai et al. 1956). From the central role of P as regulator of decomposition, partial pressure of oxygen and the redox state of biochemical reactions, a list can be compiled of the negative biochemical impacts thought to be associated with the number of nesting Herring Gulls:

- Increased CO_2 evolution (increase in GHG).
- Decreased burial of organic C (decrease in C storage).
- Increased N_2O production (increase in GHG).
- Increased reduction of Hg to CH_3Hg^+ (toxin to food web in Chap. 13).

These GHG data clearly present another benefit coming from the restoration of Big Meadow Bog.

Our data show how different the GHG profile of the ditched bog was from a typical intact raised bog, the Reference Bog (Fig. 9.4). Even after a 25-year recovery from gull disturbance and enrichment where the vegetation is no longer weedy (Hill et al. 2019), the total GHG emissions are double that of the intact system and much more variable (Fig. 9.5). The data also suggest that the phosphate inputs in an active nesting gull colony act to double the amount of GHG emissions measured for the unfertilized ditched state (BMB Recovery). Taken together with the greatest N_2O emission in the year of restoration (2017), this study makes clear that the restoration of the ecological services of BMB will not be possible until the population of nesting Herring Gulls is greatly reduced.

The disrepair of peatlands the world over has altogether shifted these ecosystems from being a global sink for GHG to a source (Leifeld et al. 2019). The remedy for drained peatlands is not difficult conceptually and requires the return to the original pre-disturbance hydrology. Peatlands are resilient and their restoration can be brought about by returning the ecosystem to its original water table state (Loisel and Gallego-Sala 2022). In the case of BMB, restoration following the blocking of ditches may require several decades to establish the original water table caused by the ditching. Given the opportunity that drained peatlands offer climate change mitigation, prompt rewetting of peats is key to rapidly reducing CO_2 emissions and reinstating the onset of C sequestration (Günther et al. 2020; Tanneberger et al. 2021; Loisel and Gallego-Sala 2022). Restoration of water tables can restore conditions to reestablish specialist peatland organisms. Recolonization of *Sphagnum* depends upon appropriate water table conditions but once reestablished (Chap. 10), this keystone genus will direct restoration of GHG functions, structure, and biodiversity (Rochefort 2000; Loisel and Gallego-Sala 2022). There are now insights into the diversity of the microbiome of *Sphagnum* (Kostka et al. 2016; Wickaksono et al. 2021) and as *Sphagnum* recolonizes Big Meadow Bog, particularly in the depression associated with the blocked ditches, this invisible micro-diversity will return. There are glimmers that suggest a functional importance of the microbiome of the Eastern Mountain Avens (Chaps. 14 and 15), a plant intimately associated with *Sphagnum* at Big Meadow Bog. Notwithstanding the return of the historic water table in BMB, the effects of climate change (Chap. 5) and ongoing eutrophication caused by gulls in this wetland (Geizer et al. 2021) may preclude a return to healthy GHG fluxes. We have established that GHG emissions are linked to phosphorus levels as in Chap. 12. Mallory et al. have linked these phosphorus levels to gulls. If the phosphorus levels in Big Meadow Bog, historically a community peatland growing cloudberry and cranberries, were originally similar to the Reference Bog, then phosphorus has been enriched by almost an order of magnitude. With so much effort, community engagement and funding, all mustered to restore this peatland and recover the endangered Avens, a laissez-faire approach to gull management is contrary to biodiversity, community, and global biogeochemistry.

References

Burton DL, Ziebarth BJ, Gillam KM, MacLeod JA (2008) Effect of split application of fertilizer nitrogen on N_2O emissions from potatoes. Can J Soil Sci 88:229–239

Burton MR, Sawyer GM, Granieri D (2013) Deep carbon emissions from volcanoes. Rev Miner Geochem 75(1):323–354. https://doi.org/10.2138/rmg.2013.75.11

Fiedler S, Vepraskas MJ, Richardson JL (2007) Soil redox potential: importance, field measurements, and observations. Adv Agron 94:1–54. https://doi.org/10.1016/s0065-2113(06)94001-2

Geizer HD, Klapstein SJ, Mallory ML, O'Driscoll NJ (2021) Total mercury, methyl mercury, phosphate, and sulfate inputs into a bog ecosystem from herring gull (*Larus smithsoniansus*) guano. Ecotoxicol Environ Safety 226:112845. https://doi.org/10.1016/j.ecoenv.2021.112845

Gorman E (1991) Northern peatlands: role in the carbon cycle and probable responses to climatic warming. Ecol Appl 1(2):182–195. https://doi.org/10.2307/1941811

Günther A, Barthelmes A, Huth V, Joosten H, Jurasinki G, Koebsch F, Couwenberg J (2020) Prompt rewetting of drained peatlands reduces climate warming despite methane emissions. Nat Commun 11:1644. https://doi.org/10.1038/s41467-020-15499-z

Hill BH, Elonen CM, Jicha TM, Kolka RK, Lehto LLP, Sebestyen SD, Seifert-Monson LR (2014) Ecoenzymatic stoichiometry and microbial processing of organic matter in northern bogs and fens reveals a common P-limitation between peatland types. Biogeochemistry 120:203–224

Hill NM, Chinner K, Toms B, Garbary DJ (2019) Recovery succession of native plant communities at Big Meadow Bog after disturbance and eutrophication from Herring gulls. Wetlands 39:533–548. https://doi.org/10.1007/s13157-018-1099-1

House SK (2005) The effects of repeated fertilizer application of phosphorus cycling in wild blueberry stands. MSc thesis, Nova Scotia Agricultural College and Dalhousie University

Hutchinson GL, Livingston GP (1993) Use of chamber systems to measure trace gas fluxes. In: LA Harper, AR Moiser, JM Duxburg, DE Rolston (eds) Agricultural ecosystem effects on trace gases and global climate change. Am Soc Agron Spec Pub 55:63–78

IUCN (2024) Issues brief: Peatlands and climate change. https://www.iucn.org/resources/issues-brief/peatlands-and-climate-change

Kickbush JC, Mallory ML, Murimboh JD, Rand J, Klapstein S.J., Loder A.L., Hill NM, O'Driscoll NJ (2018) The influence of avian biovectors on mercury speciation in a bog ecosystem. Sci Tot Environ 637–638:264–273. https://doi.org/10.1016/j.scitotenv.2018.04.224

Knowles R (1982) Denitrification. Microbiol Rev 46(1):43–70. https://doi.org/10.1128/mr.46.1.43-70.1982

Kostka JE, Weston DJ, Glass JB, Lilleskov EA, Shaw AJ, Turetsky MR (2016) The *Sphagnum* microbiome: new insights from an ancient plant lineage. New Phytol 211(1):57–64. https://doi.org/10.1111/nph.13993

Krause L, McCullough KJ, Kane ES, Kolka RK, Chimner RA, Lilleskov EA (2021) Impacts of historical ditching on peat volume and carbon in northern Minnesota USA peatlands. J Environ Manage 296:113090. https://doi.org/10.1016/j.jenvman.2021.113090

Leifeld J, Menichetti L (2018) The underappreciated potential of peatlands in global climate change mitigation strategies. Nat Commun 9:1071. https://doi.org/10.1038/s41467.018-0340616

Leifeld J, Wust-Galley C, Page S (2019) Intact and managed peatland soils as a source and sink of GHGs from 1850 to 2100. Nat Clim Change 9(1):945–947. https://doi.org/10.1038/s41558-019-0615-5

Loisel J, Gallego-Sala A (2022) Ecological resilience of restored peatlands to climate change. Comm. Earth Environ 3:208. https://doi.org/10.1038/s43247-022-00547-x

Ring RA (2001) A comparison of five extraction methods for determining available soil phosphorus in Nova Scotia blueberry soils. MSc thesis. Nova Scotia Agricultural College and Dalhousie University

Rochefort L (2000) *Sphagnum*—a keystone genus in habitat restoration. The Bryol. 103(3):503–508. https://doi.org/10.1639/0007-2745(2000)103[0503:SAKGIH]2.0.CO;2

Rydin H, Gunnarsson U, Sundberg S (2006) The role of *Sphagnum* in peatland development and persistence. In: Weider RK, Vitt DH (eds) Boreal peatland ecosystems. Ecosystem studies, vol 188. Springer, Berlin, Heidelberg. https://doi.org/10.1007/978-3-540-31913-9_4
Schillereff DN, Chiverrell RC, Sjöström JK, Kylander ME, Boyle JF, Davies JAC, Toberman H, Tipping E (2021) Phosphorus supply affects long-term carbon accumulation in mid-latitude ombrotrophic peatlands. Comm Earth Environ 2:241. https://doi.org/10.1038/s43247-021-003 16-2
Simard RR, Tran TS (1993) Mehlich III extractable elements. In: Carter MR (ed) Soil sampling and methods of analysis. Lewis, Ann Arbor, MI, pp 449–511
Spooner I, Principato S, Hill N, White H, Dunnington D, Neily T, Stolze (2017) Late Holocene records of changing moisture regime from wetlands in southwestern Nova Scotia, Canada: implications for wetland conservation and restoration. Northeast Nat 24: 331–348
Tanneberger F, Appulo L, Ewert S, Lakner S, Brolchain NO, Peters J, Wichtmann W (2021) The power of nature-based solutions: how peatlands can help us to achieve key EU sustainability objectives. Adv Sustain Syst. https://doi.org/10.1002/adsu.202000146
Takai Y, Koyama T, Kamura T (1956) Microbial metabolism in reduction process of paddy soils (Part 1). Soil Sci Plant Nutr 2(1):63–66. https://doi.org/10.1080/00380768.1956.10431859
Takeda N, Friedl J, Kirkby R, Rowlings D, Scheer C, De Rosa D, Grace P (2023) Denitrification losses in response to N fertilizer rates—integrating high temporal resolution N_2O, in situ $^{15}N_2O$ and $^{15}N_2$ measurements and fertilizer ^{15}N recoveries in intensive sugarcane systems. J Geophys Res: Biogeosci 128:e2023JG007391. https://doi.org/10.1029/2023JG007391
Tey YA, Silver WL, Sonnentag O, Detto M, Kelly M (2011) Large greenhouse gas emissions from a temperate peatland pasture. Ecosystems 14:311–325. https://doi.org/10.1007/s10021-011-9411-4
Vitt DH, Halsey LA, Bauer IE, Campbell C (2000) Spatial and temporal trends in carbon storage of peatlands of continental western Canada through the Holocene. Can J Earth Sci 37(5):683–693. https://doi.org/10.1139/e99-097
Vitt DH, Slack NG (1984) Niche diversification of *Sphagnum* relative to environmental factors in northern Minnesota peatlands. Can J Bot 62:1409–1430
Wickaksono WA, Cernava T, Berg C, Berg G (2021) Bog ecosystems as a playground for plant–microbe coevolution: bryophytes and vascular plants harbour functionally adapted bacteria. Microbiome 9:article 170. https://doi.org/10.1186/s40168-021-01117-7
Wilson D, Farrell CA, Fallon D, Moser G, Muller C, Renou-Wilson F (2016) Multiyear greenhouse gas balances at a rewetted temperate peatland. Global Change Biol 22(12):4080–4095. https://doi.org/10.1111/gcb.13325
Xu J, Morris PJ, Lui J, Holden J (2018) PEATMAP: refining estimates of global peatland distribution based on a meta-analysis. Catena 160:134–140. https://doi.org/10.1016/j.catena.2017.09.010

Chapter 10
Restoration of Wetland Hydrology at Big Meadow Bog

Andy Sharpe and Michael Parker

Abstract In 2015 East Coast Aquatics (ECA) Inc developed, on behalf of the Eastern Mountain Avens Recovery Team, a hydrological restoration design for Big Meadow Bog, Brier Island, Nova Scotia. The restoration design was based on a review of available literature, extensive field surveys of the site, visits to comparable restoration sites in the United Kingdom and empirical data gathered from a ditch block trial conducted at Big Meadow Bog. ECA subsequently implemented the restoration design between 2016 and 2018, with a total of 123 ditch blocks being constructed, 3700 m of ditch being re-profiled as well as shrub and trees being cut across 34 ha of the bog. In 2023 ECA conducted a self-directed qualitative follow-up review to assess the effectiveness of hydrology restoration.

10.1 Introduction

10.1.1 Design Goals and Vision

East Coast Aquatics Inc (ECA) began working with the Eastern Mountain Avens Recovery Team in 2015 to jointly shape the restoration of Big Meadow Bog. ECA's role was to design, and subsequently construct, an adaptive hydrological restoration approach for Big Meadow Bog. This work was undertaken within the context of the planned recovery of the globally endangered Eastern Mountain Avens *Geum peckii*. The recovery plan for this small herbaceous plant had the following vision and goals.

A. Sharpe (✉)
Kejimkujik National Park and National Historic Site, Maitland Bridge, NS, Canada
e-mail: Andrew.sharpe@pc.gc.ca

M. Parker
East Coast Aquatics Inc., Bridgetown, NS, Canada
e-mail: mike@eastcoastaquatics.ca

N. M. Hill et al. (eds.), *Wetland Restoration for Endangered Species Recovery*, Environmental Contamination Remediation and Management,
https://doi.org/10.1007/978-3-031-71344-6_10

10.1.1.1 Project Vision

The shared vision of the project partners is that …. ***the Big Meadow Bog wetland ecosystem is restored, leading to the recovery of sustainable populations of Geum peckii and associated historic and rare plant communities….***
(Hill et al. 2015)

10.1.1.2 Project Goals

- By 2050, peatland vegetation community[1] is restored to a minimum of 75% of the pre-ditching extent of the Big Meadow Bog.
- By 2050, increase the Avens population numbers and total individuals within the Big Meadow from 2014 levels.[2] (Hill et al. 2015)

The hydrologic restoration design of Big Meadow Bog targeted the restoration of major ecosystem processes in support of these vision and goals through restoring ombrotrophy in the central raised bog and lagg hydrology in the zone between the central bog and the perimeter swamp. The lagg is the primary habitat for Eastern Mountain Avens, and a key to its long-term survival. It was intended that the hydrologic restoration would reduce the nutrient load of the raised bog by creating conditions undesirable for Herring gull nesting. This in turn would allow the bog to return to an ombrotrophic state, and generally raise the shallow water table in the bog and lagg through increased water retention. Water discharge from the bog would be shifted from predominantly flashy ditch flow to more gradual subsurface flows and shallow lagg channel surface flow.

Based on the *Geum peckii* context for hydrological restoration, the Recovery Team brought a number of specific restoration objectives to ECA that were based on scientific study of the project site to date and believed to mitigate particular identified threats to *G. peckii* recovery (Environment Canada 2011).

10.1.1.3 Restoration Objectives

- Restore summer low flow period shallow groundwater elevation to within 20 cm of the ground surface in lagg areas.
- Re-create natural ground saturation conditions that are likely unsuitable/ undesirable for ongoing nesting of Herring Gulls.
- Re-create a natural rate of discharge for the project site.
- Re-establish bog/fen hydrology through the use of measures that require little or no future maintenance, do not create liability situations for the project partners

[1] *Low biomass, sphagnum based community supporting dwarf ericads and fine sedges.*

[2] *In 2014, the Big Meadow Bog supported 6 populations and 2550 individual Geum peckii.*

or landowners, can be adaptively managed if necessary, and that blend with the natural landscape of the site.

In order to design the hydrologic restoration of Big Meadow Bog within the context and constraints of the Eastern Mountain Avens recovery plan, ECA carried out five distinct steps in the restoration design process.

1. Completion of both remote and onsite evaluation of past direct and indirect impacts to wetland hydrology.
2. Completion of temporary trial blockage and evaluation of the hydrological response.
3. Development of a hydrologic restoration design for Big Meadow Bog.
4. Implementation of restoration design.
5. Follow-up and effectiveness monitoring.

Brier Island, and Big Meadow Bog (BMB), as an internationally significant conservation site, have been the focus of considerable study and restoration efforts over the past decade. This chapter specifically addresses the design and restoration of hydrology at BMB, focusing on the impacts resulting from the 1950s ditching of the site. Over this period adjacent and complimentary restoration actions have occurred at Jimmy's Pond, the runway drainage features, the Big Meadow 1 (BM1) complex and on adjacent Nature Conservancy of Canada (NCC) properties, but are not addressed here due to space constraints.

10.1.2 Historic Alterations to BMB

An initial critical action taken by ECA was to review and expand the understanding of the various anthropogenic impacts on BMB over the preceding decades. This included a detailed review of both historic and current aerial photography of the site, oral interviews, and review of historic documents.

A number of authors have outlined various impacts to Big Meadow Bog (Kennedy and Drage 2015, Webster et al. 2015, Hill et al. 2015, Goodale 2015, LaRue and Toms 2013, Environment Canada 2011). The significance of individual impacts to the alteration of Big Meadow Bog was in some cases apparent and in others unclear. Within the context of the needs of the Eastern Mountain Avens, the Federal Action Plan for this species' recovery indicated the high priority threats at the Brier Island *Geum peckii* sites included drainage ditch creation, gull nesting, tree, and shrub encroachment (Environment Canada 2011). Not all of these threats (outlined below) impact hydrology, but all are ecologically tied to the site hydrology. These impacts can be view as either direct anthropogenic or indirect ecological impacts. The historic impacts to BMB are more fully addressed in Chap. 2.

Direct anthropogenic impacts	Indirect ecological/anthropogenic impacts
Ditching	Gulls
Foot and ATV traffic	Vegetation
Ploughing	Herbivory
Impoundment	Invasive species
Climate change	Runway construction

10.2 Development of Restoration Design

10.2.1 Literature Review

10.2.1.1 Feasibility of Restoring Drained Peatlands

As the multiple ecological functions of peatlands become increasingly recognized, so too has the significant proportion of wetlands altered through ditching and draining. There exists a growing body of literature from Canada (Gonzalez et al. 2013; Landry and Marcoux 2010; Landry and Rochefort 2012; McCarter and Price 2013), United Kingdom (Armstrong et al. 2009; Evans et al. 2005), Switzerland (Grosvernier and Staubli 2009), and Belarus (Kozulin et al. 2010) documenting successes in peatland restoration that provide guidelines for implementation. These sources and others were reviewed to provide best practice for the planned restoration of Big Meadow Bog at Brier Island.

10.2.1.2 Techniques Used to Restore Bogs

The rehabilitation of peatlands undertaken to date have largely focused on the restoration of hydrology as an initial and core element as part of a larger strategy. The blocking of ditches within drained peatlands has allowed the raising and stabilization of water levels, the reduction of water losses and progressive return of ecohydraulic functions characteristic of peatlands (Landry and Rochefort 2012).

Three broad categories of ditch blocking techniques have been used:

- Complete backfilling of drainage channels, which can be effective but cost-prohibitive with extensive ditch networks and can also result in extensive physical disturbance to site.
- Construction of small, low-head dams which have been effective at raising water levels, where the frequency and interval between dams is dependent on the site slope.

- Installation of flow regulation devices that allow control over the height of water levels, providing for gradual changes and modifications (Landry and Rochefort 2012).

Multiple materials have been successfully used to block drains and initiate the restoration of bog hydrology. Armstrong et al. (2009) documented the use of multiple dam materials at 32 restoration sites in the United Kingdom, with plywood and plastic pile dams being successful in more than 80% of the sites examined. Evans et al. (2005) similarly examined the use of stone, wood, wool, plastic pilings, heather brash, heather bales, and coir logs to block ditches in drained peatlands within the Peak District National Park, England. Wooden fencing, plastic piling, and stone walls were all found to be effective in the construction of small low-head dams. Kozulin et al. (2010), and Landry and Rochefort (2012) have indicated that ditch blocks with a wooden core can be made more water tight by packing locally excavated catotelmic peat on the up and downstream faces. Shepherd et al. (2013), in a systematic review of available evidence, concluded that "using peat dams to block grips (ditches) provides comparable or better success rates to retaining water to more expensive solutions".

10.2.2 Background 2015 Field Surveys by ECA

ECA completed multiple field visits to Brier Island and BMB over the period of August to October 2015 to gain a better understanding of the physical and biological characteristics of the site. Of particular interest for the restoration design were the size and substrate of the excavated ditches, the location and connectivity of the ditches, the presence and dimensions of natural watercourse features, and the thickness of peat and underlying material along the various ditchlines.

ECA recorded more than 30 observations of the excavated ditch dimensions, with the central and east ditches, in general, being wider and deeper than the south west ditch. Contemporary ditch dimensions were compared against the original (1958) ditches, to better understand change over time. Everett Powell, the equipment operator who excavated the original ditches, recalled that the east, west and central ditches were each 6′ (1.83 m) wide and 6′ (1.83 m) deep upon construction (LaRue and Toms 2013). While portions of the east ditch retained a rectangular profile, most of the ditches were found to have a steep "V" profile, most likely from the slumping of unstable soft peat on the sides and, and to a limited degree, downcutting (erosion).

Based on a review of mapping, historical air photos and field surveys, ECA has documented the location of numerous pre-alteration watercourse features at BMB. The dimensions of these watercourses were documented at 28 locations across the BMB site. These investigations led to several important conclusions, with respect to the hydrology restoration design:

- Prior to the 1958/59 ditching, an extensive network of water features formed the north and south drainages through the lagg at BMB. These features included meandering low-gradient watercourses, ponds, and wet depressions.

- The 1958/59 ditching activity directly intersected these natural watercourses at numerous locations, with the obvious goal of providing a more direct and efficient means of moving water off the site.
- Many of the pre 1957 water features were found on the landscape at BMB prior to restoration, as they were not destroyed by ditching, only intersected. For example, the meandering remnant eastern watercourse, located just east of the east ditch, was examined and found to have bankfull widths and depths consistent with the natural catchment area for the watercourse. Similarly, topographical depressions, which once formed the network of interconnected ponds in the south east / central east area of BMB and noted in the island's oral history, were also observed.

Kennedy and Drage (2015) reported the results of a peat depth survey conducted in 2014 along three transects with maximum peat depths of 4.5–5.0 m near the center of BMB tapering to < 1.0 m in depth at the north and south ends. ECA, as part of the autumn 2015 field surveys, probed peat depths at the north and south ends of BMB adjacent to and within the east, central, and west ditches and their spoils piles. ECA sought to gain a better understanding of local conditions which the 1958/59 ditch excavators faced, and to provide input to the restoration design on the nature of the alterations. The south draining ditches, within 50–150 m upslope of Big Pond, had channel bottoms composed of abundant gravel, cobble and stone, with stone also encountered in the adjacent spoil piles (Fig. 10.1). Stone within the spoils indicated that some depth of stoney substrate was excavated during ditching.

Fig. 10.1 Cobble and rock substrate encountered in southeast ditch, north of ATV trail crossing. Photo credit: Andy Sharpe

Fig. 10.2 North-draining, meandering low-gradient watercourse downslope of point of confluence of ditches. Photo credit: Andy Sharpe

Excavation of the south draining ditches would have been challenging with the equipment used in 1957 given the decreasing peat depth over rock as the ditches approached Big Pond. The rock and fractured bedrock encountered very likely limited the depth to which the ditches could be dug and represented a hard control for the ditch network, limiting the slope and hence overall drainage efficiency.

The north-draining ditches were examined over approximately a 200 m reach upstream of their point of confluence in the north central area of Big Meadow Bog. In general, the north-draining ditches were found to be less rocky than the south draining ditches over the area surveyed. Peat depths to the underlying gravel were in the range of 0.2–1.3 m. Little or no rock was encountered in the ditch spoil piles. Downstream of the confluence with the northwest channel, the meandering low-gradient watercourse had no adjacent spoil piles and appeared unaltered (Fig. 10.2).

10.2.3 Bog Restoration Experience in UK

ECA identified early during the literature review for this project that a considerable body of research, knowledge and practice existed in the United Kingdom within the field of peatland restoration. So that the BMB restoration could benefit from this experience, ECA staff contacted the Yorkshire Peat Partnership and arranged a site visit to a recently restored peatland, as part of a family holiday to England.

The Yorkshire Peat Partnership (YPP) works with contractors, landowners and farmers, and seeks to substantially increase the amount of peatlands restoration activity in the Yorkshire uplands through a combination of restoration, management, and monitoring. The partnership is composed of the Yorkshire Wildlife Trust, Natural England, Yorkshire Dales National Park Authority and North York Moors National Park Authority and the Environment Agency. Active since 2009, YPP had blocked 1110 km of drainage channels, re-vegetated 50 ha of exposed peat, monitored 7 restoration sites, and surveyed over 70,000 ha of peatlands at the time of inquiry (Yorkshire Dales National Park 2015).

An ECA staff member met with YPP Peatland Restoration Officer Matt Cross near Pateley Bridge, North Yorkshire on October 7, 2015, to tour the Nidderdale restoration site. Restoration activities composed of blocking man-made agricultural ditches, re-profiling ditch edges, and re-vegetation of eroded gullies had been completed at the site approximately 12 months earlier (Figs. 10.3, 10.4 and 10.5).

Fig. 10.3 View along ditch, showing three ditch blocks at Nidderdale restoration site, Yorkshire Dales. Photo credit: Andy Sharpe

Fig. 10.4 Matt Cross, Peatland Restoration Officer, standing on a vegetated ditch block for scale, with additional ditch blocks at the rear of the photo. Photo credit: Andy Sharpe

YPP Peatland Restoration Officer Matt Cross is a member of a United Kingdom-wide consortium of agencies which study and restore peatlands. The group provides a forum for the exchange of research findings and restoration best practice. YPP has, over several years, evaluated the use of a variety of materials for the construction of ditch blocks within degraded peatlands. YPP's restoration prescriptions had undergone a number of revisions based on these evaluations, resulting in the recommendation to use either locally sourced peat or stone in the construction of ditch blocks.

In summary, the Yorkshire Peat Partnership approach to bog hydrology restoration was directly applicable to the design considerations for Big Meadow Bog. BMB had drainage path slopes that were considerably less than that for which successful ditch block methodologies had been employed. The use of local peat met the cost and materials transportation constraints that existed for the Big Meadow Bog project. The approach had proven longevity and naturalized the restored landscape in a manner that man-made structures and structures with imported materials could not. The methods would create a number of small shallow open water areas, similar to those that existed historically at the site, which would provide a diversity of microhabitats

Fig. 10.5 Components of Yorkshire Peat Partnership Ditch Block, showing ditch alignment, ditch block, and adjacent borrow pit with a broad and shallow overflow area. Photo credit: Andy Sharpe

across the bog. Such complexity has been lost at Big Meadow as evidenced by review of historic air photos.

On a separate family visit to the United Kingdom, an ECA staff member contacted Andrew McBride, Peatland Action Manager with Scottish Natural Heritage, arranging a site visit to Blawhorn Moss National Nature Reserve on August 11, 2017 (Blawhorn Moss 2024).

Over the past 25 years Blawhorn Moss has been an important Scottish site for the trialing and review of peatland restoration techniques. Over this period, a variety of ditch blocking approaching had been constructed and evaluated, including:

- Corrugated steel sheet dams.
- Plastic sheet pile (both “S” and “Z” profile) dams.
- Wooden timber dams.
- Large-scale steel sheet pile dams.
- Placement of heather bales (small round) in ditches.
- Placement of cut regeneration wood waste (both stems and branches) in ditches.
- Construction of bunds across the slope.
- Peat dams (ditch blocks).

Andrew McBride (Fig. 10.6) kindly provided a tour of the site, identifying the pros, cons and unique features of each of the restoration measures. Being able to see a variety of ditch blocking approaches together, and to examine their functionality over a time span of years to decades, was enormously beneficial. The visit to Blawhorn

Fig. 10.6 Andrew McBride at Blawhorn Moss National Nature Reserve, Scotland with a dam constructed of sheet metal piling. Photo credit: Andy Sharpe

Moss consolidated the conclusion to use peat ditch blocks at BMB, based on their effectiveness, constructability, and longevity.

10.2.4 Constraints and Uncertainties

10.2.4.1 Guiding Principles for Design

Based on a review of available literature, consultations with the project team and past experience, ECA identified a number of guiding principles for the hydrology restoration design of Big Meadow Bog. These included:

- Low cost of implementation.
- Techniques shown to have been successful in previous peatland restoration projects.
- Limited future liability for project team and property owners.
- Limited future maintenance obligation for project team and property owners.
- Guided by the empirical data from Big Meadow Bog hydrology restoration pilot (completed Autumn 2015).

- Adaptability, as additional data becomes available and in response to a phased implementation of restoration.

10.2.4.2 Constraints for Design

A number of design constraints were considered in the development of the restoration design, some of which were unique to the Big Meadow Bog site. These included:

- Only a limited number of heavy equipment access points to site were available. Each of the available access points in turn had unique constraints, such as requiring prior consent from landowner, development of the access, and/or limitations on the timing of access.
- Given its protected status, heavy equipment movements at the site had to avoid areas with high *Geum* concentrations.
- Given the above two constraints, as well as the unconsolidated ground surface of the wetland, there were signification limitations on the ability to transport restoration materials out onto bog. The movement of conventional construction materials such as stone, wood, soil for use in restoration was thus severely curtailed.
- The presence of nesting Herring gulls, a bird species protected under the Migratory Birds Convention Act. Subject to subsection 5(9) of the act, "no person shall disturb, destroy or take a nest, egg, nest shelter, eider duck shelter or duck box of a migratory bird". The first Herring Gull *Larus agentatus* eggs were laid on Brier Island about the third week of May with fledging occurring by July 01 annually (Mills and Laviolette 2011). This provided an exclusion window for construction work.
- Limitations in working on the rewetted bog. As a result of the autumn 2015 hydrology restoration trial, ECA had established that it was possible for heavy equipment to traverse the bog with care. Following the commencement of restoration and raised water levels, it was anticipated that the bog surface would become increasingly unstable for heavy equipment. It would therefore be difficult for heavy equipment to access every structure to make significant alterations once the surrounding areas had been flooded. Similarly, given the limited number of access points, it would be difficult for equipment to traverse a previously restored area to access an un-restored area. This constraint was therefore to be considered in the sequencing of restoration activities.

10.2.4.3 Rehabilitation Uncertainties and Unknowns

Three broad headings of uncertainty were identified by ECA prior to implementation of the hydrology restoration of BMB.

- Nutrient response following rewetting.

  - *Will there be a pulse of nitrogen and phosphorus mobilized in surface waters? For how long will elevated nitrogen and phosphorus levels persist following the reduction of Herring Gull nesting at the bog?*

- Behavior/nesting changes of Herring Gulls following rewetting.
  - *Will Gulls persist in nesting in wetted areas? Will there be increased nest failure? Will Gulls relocate to adjacent drier areas? Will Gulls become a greater nuisance if they are displaced and relocate closer to residential areas?*

- Vegetation community response, in particular *Geum*, following rewetting, given variables of elevated nutrients, presence of competing species, physical disturbance of habitat due to gulls, etc.
  - *Will existing exotic plant species be aided or hindered by rewetting? Will Geum be aided or hindered by rewetting? Will typical bog/lagg/ericaceous species re-establish following re-wetting? Will Sphagnum species be able to re-establish in the central bog?*

Dr. Jonathon Price, professor emeritus (University of Waterloo) and bog specialist, in a visit to the site in May 2014, noted that numerical modeling may be of limited use in predicting quantitative changes to the system following restoration, in part due to the complexity of peatland systems. In 2015 ECA concluded that while additional research may address one or two of these questions, it is very unlikely that all questions and uncertainties could be addressed prior to restoration, particularly quantitatively in the short to medium term. The interactions of these uncertainties were highly complex and very likely unique to BMB. It was considered unrealistic to delay restoration until all uncertainties were addressed. Therefore, a combined approach of adaptive management, phased implementation, and ongoing monitoring was recommended as the means for managing uncertainties.

10.2.5 Northeast Ditch Trial

East Coast Aquatics, early during the development of the restoration design identified the absence of empirical data on the practicalities of blocking ditches and the likely response of shallow groundwater to the installation of ditch blocks. This recognition, and the need to achieve specific targets for *G. peckii* recovery, led ECA to propose an experimental pre-design trial of water regulation within several ditches at BMB. The hydrology restoration trial sought: to investigate the feasibility of installing ditch blocks at the BMB site, clarify the temporal and spatial response of shallow groundwater to the installation of ditch blocks and investigate questions concerning the spacing and placement of ditch blocks to achieve the desired water level outcomes. ECA proposed that water retention structures be placed within the ditches and the response of shallow groundwater to these actions monitored.

10.2.5.1 Installation of Trial Ditch Blocks

Steel plates and monitoring wells were installed in the north-draining ditch adjacent to the Avens BM1 site in late August 2015, after an extended dry period (Fig. 10.7 and 10.8). Steel plates (approximate dimensions 90 cm × 240 cm × 1.3 cm) were used for the trial as they provided a reversible and quickly installable method to temporary block water flow to a precise elevation with minimal ground disturbance impacts. The top of plates blocking the ditches were installed to an elevation 4–7 cm below the immediate adjacent ground level at each site, and approximately at the elevation that would be desirable for groundwater levels based on previous project team monitoring of productive *G. peckii* sites (< 20 cm below the ground during the dry period). The spacing of plates was further set such that, based on drainage slope, one ditch block would backflooded the next upstream ditchblock to within approximately 25 cm freeboard from the top of the upslope ditchblock and thereby achieve an average shallow water table depth of < 20 cm across the distance in between the two consecutive plates. This target was based in part on the preferred range of water table for the Eastern Mountain Avens as identified by the Recovery Team and Landry and Rochefort (2012).

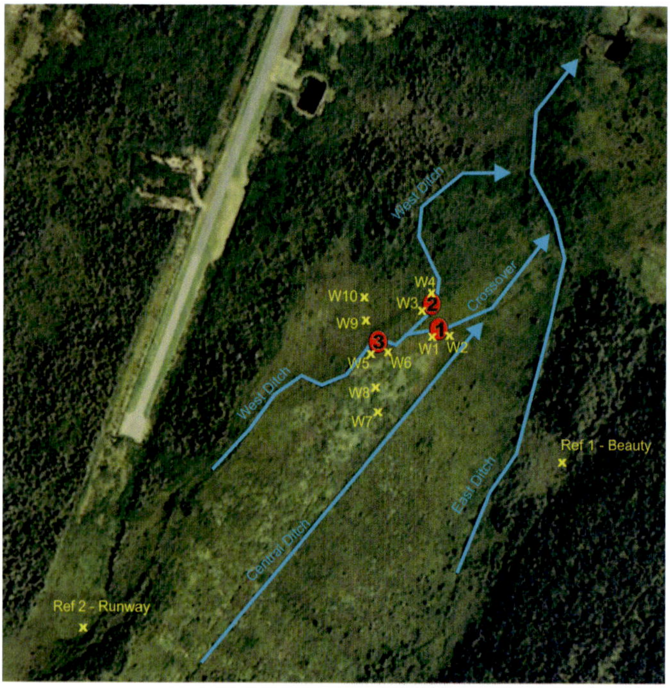

Fig. 10.7 Approximate location of Big Meadow Bog Trials features. Plates to regulate flow were placed at the red circle locations. Instrumented wells were installed adjacent to the plates and at two reference "Avens Sites" (*yellow* "*x*"). Base image: Google Earth

Fig. 10.8 Installation of Plate 3 on the northwest channel in Big Meadow Bog, August 28, 2015. Photo credit: Andy Sharpe

The ECA field-measured slope of the western channel water surface at the trial site was determined to be 0.54%, very similar to previously determined slopes for the north-draining ditches (Webster et al. 2015). Based on the slope, a linear separation of 50 m between the plates on the northwest channel was established to meet the < 20 cm shallow water table target elevation.

Shallow groundwater wells were installed 4 m upstream and downstream from each plate, and at 10 m, 25 and 50 m away from the ditch channel in a perpendicular direction to gauge the spatial effect of the plates on shallow groundwater level. All wells were equipped with Solinist dataloggers (pressure transducers), recording water levels and temperature on an hourly basis. All of the plates, well casings and datalogger positions were surveyed to a single local benchmark in order to examine relative elevations of monitored parameters and structures. A rain tipping gauge was installed so that the response of all wells to rainfall events could be assessed. Solinist dataloggers were also installed in existing wells as Beauty Bog and Runway Bog to provide simultaneous observations of shallow groundwater at nearby unaltered locations where *Geum peckii* was doing moderately well.

10.2.5.2 Hydrology Restoration Trial Findings

Placement of Ditch Blocks

Study Question *Was the linear spacing and horizontal placement of the plates suitable to create the desired channel water levels?*

The ditch blocks were installed with the goal of establishing no more than a 25 cm drop in channel water level across the plates, with one channel block being backflooded by the next downstream block.

Backflooding was found to influence both the water level within the stream channel and the adjacent shallow groundwater. Following installation, Ditchblock #2 was successfully found to backflood the upstream Ditchblock #3, with a water level drop of 0.26 m across #3 when channel flow reached the top of the trial plate. Ditchblock #2, positioned at the downstream end of the site and thus not backflooded, had a water level drop of 0.39 m across the barrier.

Response to Precipitation

Study Question *Were the plates effective at retaining shallow groundwater within Big Meadow Bog following precipitation?*

The ditch blocks were installed in late August 2015, following a moderately prolonged period with little precipitation. On September 10 Brier Island received an intense rain event, with 84 mm of precipitation falling over a two-day period. No significant rain was received again at Brier Island until September 30. Water levels in the shallow groundwater wells adjacent to the ditch blocks increased rapidly during and immediately after the September 10th rain event.

Shallow groundwater levels above and below the ditch blocks were examined over the period of September 11–29 to determine if the ditch blocks had an observable effect on retaining the September 10th precipitation longer within the wetland.

A consistent pattern was observed across the two ditch blocks installed on the Western watercourse at Big Meadow Bog. Following the heavy precipitation on September 10 and 11, shallow groundwater levels upstream of the ditch blocks decreased slowly over the next 19 days at a rate of 0.007 and 0.009 m/day, for Wells 3 and 5, respectively. Downslope of the ditch blocks, shallow groundwater levels were observed to decrease more rapidly over this period, falling by 0.016 m/day.

For comparison, water levels in the reference sites (Beauty Bog and Runway Bog) remained virtually unchanged, having had a rate of change of + 0.0036 and – 0.00037 m/day respectively over the same period. Thus, while the ditch blocks did not fully create the target bog hydrology, they did appear to be effective at retaining precipitation for longer within Big Meadow Bog.

Response of Shallow Groundwater Levels

Study Question *Did the ditch blocks increase local shallow groundwater levels at the trial site?*

A central question concerning the hydrology restoration trial was the ability and spatial magnitude to which the installation of ditch blocks would alter the local hydrographic profile. Due to constraints concerning the timing of project delivery and obtaining the necessary consents for the implementation of the trial, it was not possible to obtain an extended baseline of shallow groundwater levels prior to the installation of the plates. Monitoring wells were installed one day before the ditch blocks were put in place. In order to partially address the lack of pre-installation monitoring data, monitoring wells were established 25 and 50 m to the east and west of Ditch Block #3. It was felt that these wells, in particular the ones at 50 m, were at or beyond the likely zone of spatial influence of the ditch blocks and could provide for a comparison of hydrographs over the monitoring period.

Monitoring well data indicated a clear spatial progression from dryer ground surface conditions at the ends of the transect to wetter conditions nearer the ditch block. This provided support for the conclusion that ditch blocks were effective at raising shallow groundwater levels at BMB. Based on the trials, that effect appeared to have been limited to 25 m perpendicular from the channel. However, it is possible that a longer time frame would result in a greater spatial effect being observed as the soil matrix became saturated with water and vibrant recolonization of sphagnum began to influence hydrology through capillary action.

Additional Findings and Outcomes of Trial

1. The restoration trial provided confirmation that a rubber track-mounted 3.5 ton compact excavator, exerting a ground pressure of < 5 psi, could access and safely traverse the bog without breaking through the surface vegetation mat. The excavator was found to cause limited impact to the bog shrub community, with the transit route being almost impossible to identify six weeks post-installation. It was further found to be feasible for an individual on foot to successfully guide the excavator around concentrations of *Geum*.
2. Access to BMB for the installation of the ditch blocks for the restoration trial required the excavator to traverse approximately 280 m along an existing footpath and bog margin. The access route used was in places steep, with very soft ground conditions and standing shrub vegetation. The access route used was not suitable for multiple passes with a wheeled vehicle. These factors placed logistical and practical hurdles associated with movement of conventional construction materials from access points out onto the wetland surface.

10.2.6 Overview of Restoration Techniques

Based on a review of available scientific literature and site documents, consultation with project stakeholders, field surveys, the completion of the Autumn 2015 hydrology restoration trial, and consideration of design goals, principles and

Table 10.1 Summary of hydrology restoration approaches

Prescription	Brief description	Brief rationale
1. Ditch blocks	Construction of frequently spaced ditch blocks composed of peat (or peat and stone where applicable) along all excavated ditches, completely blocking the flow of water within the ditch	All surface water flows are re-directed into adjacent vegetation and natural channels, leading to a re-activation of the site's historic drainage network
2. Re-profiling of ditch edges	Steep, erodible ditch edges are re-profiled to a gradual slope and covered with native vegetation	Reduction of deep-water safety hazards, increased ditch stability, increased roughness to water flow and provision of a diversity of microhabitats to allow for native vegetation recolonization
3. Woody vegetation reduction	Cutting of deciduous and coniferous tree species and shrubs within the central raised bog	Reduction in evapotranspiration and ground shading, promoting the faster recovery of water levels and former vegetation communities

constraints, ECA developed a suit of restoration techniques for phased implementation at BMB. The three primary components of the restoration design approaches are summarized in Table 10.1.

10.2.6.1 Ditch Blocks

The network of ditches established in 1958/59 allowed surface water to be rapidly conveyed from the project site, leading to progressive desiccation of the surface. Analysis of historic aerial photography and field surveys indicate that the ditches were placed to maximize drainage of the site and to allow ease of construction, rather than follow the alignment of existing watercourses. Prior to 1958, natural water features existed at the site and were of sufficient size to convey surface flows. Reviews of historic air photos identified many pre-ditching features at BMB, including watercourses, ponds, and footpaths. Many of these features are also referenced in the oral history of Brier Island (Goodale 2015). The pre-alteration water features were numerous, of small but highly variable cross section, meandering, and of low gradient. These characteristics allowed retention of surface flows and maintenance of ombrotrophic bog conditions. Field surveys conducted by ECA confirmed that remnants of these water features still existed at BMB, including stream channels and depressions, which, if natural hydrology was returned to the site, would become reactivated, functional streams and ponds.

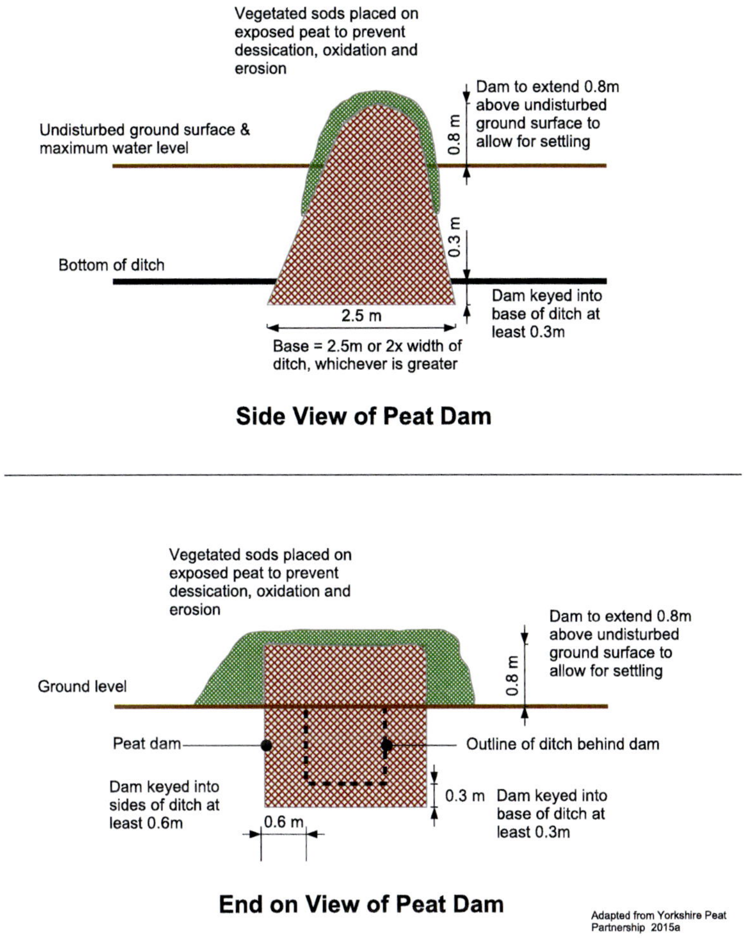

Fig. 10.9 Side and end view of peat ditch block

Summary of Methods

The methods presented here were adapted from Yorkshire Peat Partnership (2015a, 2015b, 2015c) and Scottish Nature, as appropriate for the Big Meadow Bog project site characteristics. Numerous blocks in all ditches excavated in the late 1950s time period were established such that the ditches no longer convey surface waters, but instead diverted water away from the ditchline to the former natural channels. The tops of the ditch blocks extended ~ 0.8 m above the adjacent ground surface at the time of construction to allow for settlement, ensuring that the blocks would not be overtopped and all surface water being re-directed into adjacent vegetation as sheet flow. Ditch blocks were constructed of locally available catotelmic peat, dug from adjacent borrow pits (Figs. 10.9 and 10.10).

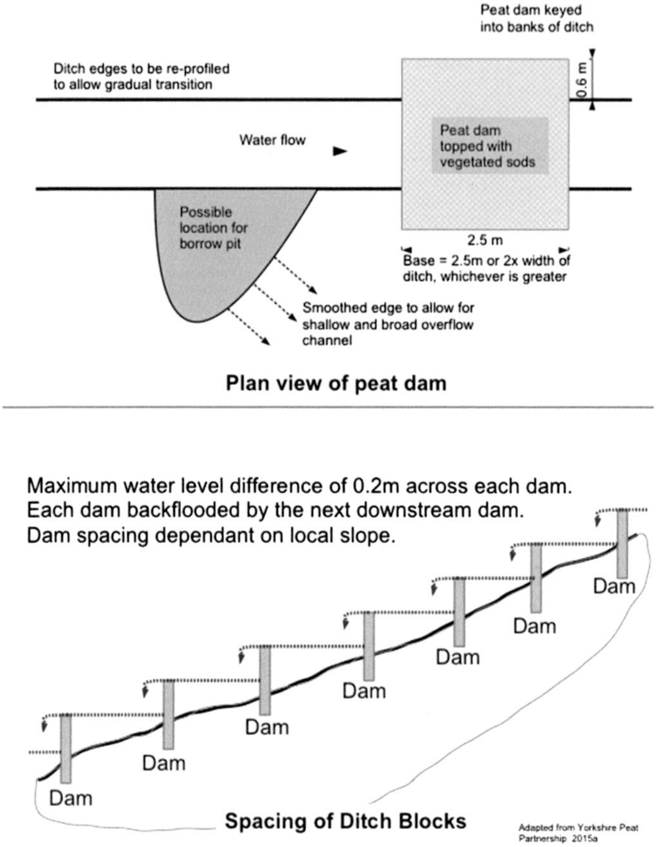

Fig. 10.10 Plan view of peat dam and ditch block spacing schematic

The horizontal spacing between ditch blocks was based on the local slope, being closer together at the northern and southern extents of Big Meadow Bog where slopes were the steepest. On average, ditch block spacing was every 15–20 m, and as close as 10 m on steeper segments of the central ditch, with spacing never exceeding 50 m. Ditchblocks were spaced horizontally to maintain a maximum 0.2 m difference in the water surface on the up and downstream sides of the ditch block (Price et al. 2003) and to ensure that each ditch block barrier was partially backflooded by the next downstream ditch block (Landry and Rochefort 2012).

Following an evaluation of alternatives, ECA concluded that ditch blocks should be constructed of available materials at each site. It is estimated that for most of the project area (> 75%), ditch blocks could be constructed exclusively of catotelmic peat, excavated immediately adjacent to the construction site. At the north and south ends of the BMB, where the overlying peat thins and was underlain by cobble, rock and marine-derived gravel, ditch blocks were constructed with a cobble and rock core, with a peat covering.

10.2.6.2 Re-profiling of Ditch Slopes

Following > 50 years of lowered groundwater levels, the surficial peat (acrotelm) at BMB was degraded, as a result of oxidation due to increased aeration, desiccation, and compaction. Such changes to peat pore structure led to an impairment of its natural water regulation functions, such as water storage, which may not automatically re-establish with rewetting (Price et al. 2003). Post-restoration, an undulating surface topography would provide a range of microhabitats and aid in the re-establishment of a healthy and diverse sphagnum community (Shepherd et al. 2013; Price et al. 2003).

Following restoration of the site and construction of ditch blocks, water depths in some ditches would exceed 1.5 m and represent a safety hazard for researchers, visitors, and residents of Brier Island. The abrupt, steep-sided ditches would further provide little in terms of diversity of wetland habitats. Re-profiling of ditch walls was completed at the same time as ditch blocks were constructed to simplify logistics. Re-profiling ditch walls had the advantage of addressing the hazard of deep-water while providing a range of aquatic microhabitats (~0.5–0 m water depths) and enhancing ditch wall stability. The Yorkshire Peat Partnership has found that re-profiling, when completed in conjunction with ditch blocking, allows for the rapid establishment of native ombrotrophic bog floral species (Matt Cross, personal communication, October 23, 2015).

Re-profiling of ditch slopes involved the following:

(a) The sides of ditches should be re-profiled where the sides are steep or undercut with overhanging vegetation (Figs. 10.11 and 10.12).

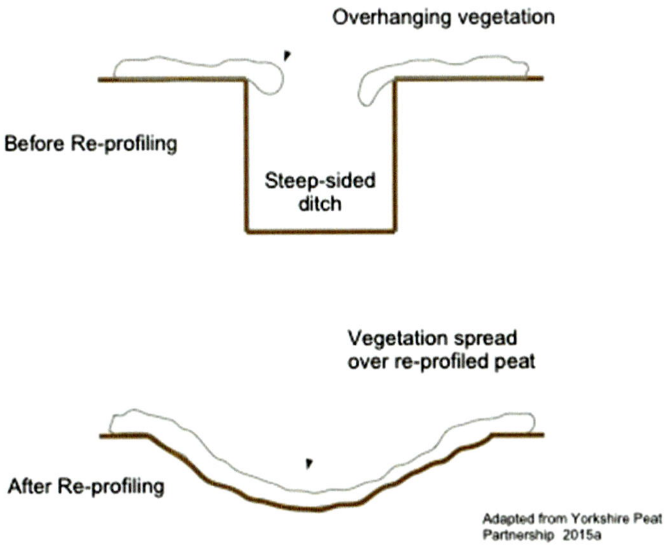

Fig. 10.11 Schematic for re-profiling of ditch edges

Fig. 10.12 Re-profiling of ditch edges, with photo re-profiled ditch and ditch blocks from Yorkshire Dales National Park, England. Reproduced from Yorkshire Peat Partnership (2015a)

(b) Using an excavator with a trenching bucket, a 2 m length of vegetation on the top of the ditch should be "rolled" back or undermined (to a depth that retains the root structure of the vegetation) far enough to enable the underlying peat to be removed to create a sloping bank from the top of the ditch to the base.
(c) The vegetation should then be rolled back and compacted to cover the newly profiled slope.
(d) As necessary, ditch spoils may be placed and compacted over any remaining exposed areas.

10.2.6.3 Woody Vegetation Removal

With progressive drying over the past 55 years, Big Meadow Bog had undergone numerous changes at a landscape scale, including colonization by coniferous and deciduous trees. The establishment of trees and woody shrubs very likely further contributed to the drying of the site through increased evapotranspiration (Shepherd et al. 2013; Fay and Lavoie 2009; Van Seters and Price 2001; Gonzalez et al. 2013), as well as altering the herbaceous plant community through shading and competition.

Based on a review of available evidence, Shepherd et al. (2013) concluded that tree removal contributes to the recovery of blanket bog vegetation. The UK Forestry Commission found that "felling is essential for restoring blanket and raised bogs altered by afforestation" (Anderson 2010). Landry and Rochefort (2012) recommend the trees be cut in peatlands undergoing restoration to facilitate optimal rewetting. However, Gonzalez et al. (2013) found that rewetting alone was sufficient to alter the colonization trajectory of herbaceous vegetation and limit the spread of invading tree and shrub species into a restored bog.

Concerning the question of how best to dispose of the forestry waste after felling, Shephard et al. (2013) has concluded that "the practice of felling trees to waste and disposing of waste on site need not prevent recovery towards blanket bog vegetation". Anderson (2010) found that the extra cost of removing felled tree waste is not justified in terms of the improved ecological conditions with the bog. The nutrient enriched context of Big Meadow Bog may alter the priority of removing slash material, but from a hydrological recovery standpoint removal would seem an unnecessary step. In principle, trees can be cut at any time of the year, although Landry and Rochefort (2012) recommend that best results are obtained with competitive species such as birch if cutting occurs during the summer growing season. Winter cutting has the advantage of limiting impacts to herbaceous vegetation.

Based on the available evidence, ECA concluded that, at a minimum, coniferous and deciduous trees within the raised bog portion of BMB would be felled as part of the hydrology restoration. Tree and shrub felling would seek to remove woody plants from BMB back to the circa 1955 boundary, as determined by air photo analysis. Pre-restoration surveys identified that Herring gulls nested and roosted on and under coniferous trees across the site. Removal of these trees was felt to make the site less attractive for Herring gull occupation. Care was taken during felling operations to ensure that tree trunks were limbed, allowing the trunks to be contact with the ground to facilitate decomposition.

10.3 Implementation of Design

10.3.1 Phasing of Restoration

From a practical perspective, the North/South divide in BMB drainage allowed the two ends of the wetland to be addressed separately. Initial landownership constraints dictated that 2016 construction could only take place in the southwest on lands owned by the Nature Conservancy of Canada. This allowed restoration of all western ditching to be completed during April 2016. By the spring of 2017 access had been secured for the remaining restoration areas on the central and eastern ditches. However, given that the most feasible machinery access point was in the north, and the desire to not have to travel over restored areas to access un-restored areas, work was initiated in the south half of BMB in April of 2017. A period of shutdown was then followed to allow for nesting of Herring gulls on the site, an activity that was occurring more predominantly in the north at the time. Following fledging, crews returned in late August 2017 to complete the restoration of Jimmy's Pond in the most northerly extent of the project area, followed immediately by completion of all ditch restoration in the northern half of BMB during September 2017. Following completion of all heavy machinery restoration activities, tall shrubs (> 2 m) and trees that had encroached the bog surface were cut across the entire site during several visits over the period of September 2017 to February 2018.

This overall sequencing is summarized in Table 10.2 and graphically shown in Fig. 10.13. Over 3.7 km of ditches were re-profiled following the installation of 123 ditchblocks. This was approximately 40% more than proposed (87 ditchblocks), and a result of additional structures being installed at key confluences of spur and main ditches, and along steeper areas at the north and south extremes of the project area. As anticipated, additional structures were required once tree cover was removed, and each individual site could be carefully assessed during construction. ECA utilized a conservative approach, erring on the side of redundancy during construction to ensure the long-term stability and minimal risk of failure of the restoration.

Table 10.2 Number of constructed ditchblocks across BMB during the restoration activities

Drainage	Ditch	Slope of ground surface	Ditched length (m)	Vertical elevation change (m)	Average spacing between blocks (m)	Constructed number of ditchblocks
North	East	0.37	991	3.68	50	27
	Central	0.48	1011	4.86	40	32
South	East	0.61	530	3.23	30	34
	Central	0.49	690	3.38	20	18
	West	1.2	176	2.18	15	12
					Total	123

Fig. 10.13 A pictorial representation of ditchblock construction and re-profiling sequencing at Big Meadow Bog. Each dot represents the specific location of a constructed ditchblock. Also shown is the location of Jimmy's Pond restoration site in the extreme north, and the post-restoration boundary of the Big Lily Pond (*blue line*). Base image: Google Earth

10.4 2023 Post-installation Assessment of Restoration Measures

10.4.1 Introduction

In February 2023 East Coast Aquatics conducted a self-directed review of the restoration structures for the purpose of self-learning and improvement of practice. The results of the qualitative assessment are included here to provide a longitudinal understanding of the restoration process from April 2016 to present. Seasonal timing of the assessment was chosen so that moderate water levels would be encountered, and herbaceous and seasonal foliage cover would be minimal and therefore permit good viewing conditions for the assessment of the ground level structures. The assessment was intended to provide a post-construction (64–82 months) review of stability and function for 110 ditch block structures. The review examined the condition of the former drainage ditches, following the installation of the ditch blocks and reprofiling of the ditch alignments. Additionally, a GPS boundary of the extent of standing water at the Big Lily Pond and the Southeast Lily Pond was captured. These freshwater pond habitats were two of the six that had been historically noted within Big Meadow Bog (LaRue and Toms 2013) and used for hunting and skating (Goodale 2015), but which had been fully drained by ditching prior to the restoration. Lastly, thirteen staked weirs installed along the northwest lagg channel were assessed and upslope drainage realignment toward the bog reviewed. Other restoration activities such as woody vegetation removal were not explicitly assessed.

10.4.2 Methods

As stated in the restoration design completed by ECA in 2015, the intended functions of the ditch blocks and ditch re-profiling were:

1. To retain water to an elevation "…target of water within 0.0–0.2 m below the ground surface.
2. To have "…all surface water flows within the existing ditches…re-directed into adjacent vegetation, leading to a re-activation of the site's historic drainage network".

A series of Assessment Categories (see Table 10.3) were developed and into which each surveyed ditch block was placed following field assessment. The category chosen reflected both the observed stability of the structure and whether it was functioning as intended. The descriptive text developed for each Assessment Category was used to maintain consistency in application during the qualitative assessment. During assessment visual observations were made that would allow selection of an Assessment Category. Observations were made of:

Table 10.3 2023 Post construction assessment categories for stability and function of installed ditch blocks at Big Meadow Bog

Assessment category	Structure stability	Structure function	Required maintenance
S—Stable *Green*	**Fully stable** with recolonization of vegetation	**Fully functioning** (100%) as intended, both water retention and flow path	None required
PS—Partial Stable *Green*	**Mostly stable** with good recolonization and only small erosion if any. Unlikely to become more unstable	**Fully functioning** (90+ %) as intended, although water pathway may not be as intended/anticipated. Water retention at or near intended	None required currently
PU—Partially Unstable *Yellow*	**Mostly stable** with good recolonization and some erosion that will likely continue	**Mostly functioning** (65–90%) as intended, although water pathway not as intended and retention likely slightly lower because of erosion and/or flow pathway	Should have preventative maintenance. Staked weir and/or sod transplant likely adequate
U—Unstable *Red*	**Mostly unstable** with poor recolonization and both obvious and ongoing erosion. Further degradation the likely trajectory	**Poorly functioning**, providing < 65% of intended function. Flow pathway and/or water retention significantly different than intended	Needs immediate maintenance to prevent complete failure. Might be done by hand but with considerable effort
CF—Complete failure *Red*	Structure is **not recognizable** as having been built. Significant erosion and failure of structure	**None (< 30%) of intended function is provided** by structure	Full new structure would be required. Would require machinery to implement
Not built *Blue*	Several structures were not built due to water levels encountered at the time of construction		
Not found *Blue*	Several structures were not relocated during the 2023 assessment. This may have been because they were under water or dense vegetation and therefore not visible		

Colors indicated under the Assessment Category correspond to the point icons presented in subsequent Figs. 10.16 and 10.17

- Whether the surface water table upslope of the structure was near ground level.
- Whether there was a similar or greater than anticipated water table elevation drop between the upslope and downslope sides of a ditch block (> 0.20 m).
- Whether water could be observed flowing as preferred in a direction away from the ditch line in a manner that it was unlikely to re-enter the ditch line within a short distance (< 6–8 m).
- Whether the ditch block had become fully covered with vegetation as anticipated.
- Whether water was moving along a vegetated path or along an undesired path of exposed bare peat/mineral soil.

Higher hydrological function was inferred at structures where the surface water table was < 0.2 m from the adjacent ground elevation on the upslope side of the ditch block and/or where surface water flow was being directed away from the ditch line path. Furthermore, if the downstream ditch block was functioning properly, it was anticipated that difference in surface water elevation on either side of a ditch block should be approximately 0.2 m or less. Higher stability was inferred at locations with dense vegetation cover over the ditch block and along any flow path away from the upstream side of the ditch block along which surface water was currently or likely to move during high water events.

It was anticipated that not all structures installed during the restoration would remain entirely stable and functional over time, and that weather conditions such as precipitation events within one year of construction could have a measurable effect on the stability and function of restoration structures. The 2023 post-construction assessment considered that while individual ditch blocks structures may have failed, the restoration design incorporated an element of over-building and redundancy, with additional ditch blocks constructed at the north and south ends of BMB where slope angle was higher.

Lastly, the assessment provided an opportunity to identify maintenance needs at individual structures. Proactively addressing areas of instability can positively affect long-term functional outcomes for the site. In assessing for required maintenance, a key factor was whether successful/beneficial maintenance could be done by hand, or whether machinery would be required. This resulted in the definition of five maintenance categories, ranging from structures needing no maintenance, to hand maintenance, to machinery maintenance.

10.4.3 Results

Of the 110 ditch block structure examined in February 2023, 88% were found to be Mostly to Fully Functioning as intended (Fig. 10.14). The restoration dataset identified GPS coordinates for 118 structures, although eight ditch blocks could not be relocated during the 2023 assessment and were thus not included the following analysis. A structures' stability need not fully correspond with its function. An unstable structure can still be high functioning, although such function is at risk of loss when structure stability is diminished. Therefore, restoration structure stability was also considered independently of structure function.

Although 88% of the ditch block structures were assessed as Mostly Functioning to Fully Functioning, a lesser 66% of those ditch blocks evaluated were categorized as either Stable or Partially Stable (Fig. 10.15). Stable or Partially Stable ditch blocks were those that require no maintenance. The evaluation identified a total of another 22 structures as Partially Unstable, which could be improved with hand tool maintenance. Examples of hand maintenance could include the addition of stake weirs or sods to existing flow paths around ditch blocks in cases where flow was not moving in a perpendicular direction away from the former ditch line. The objective

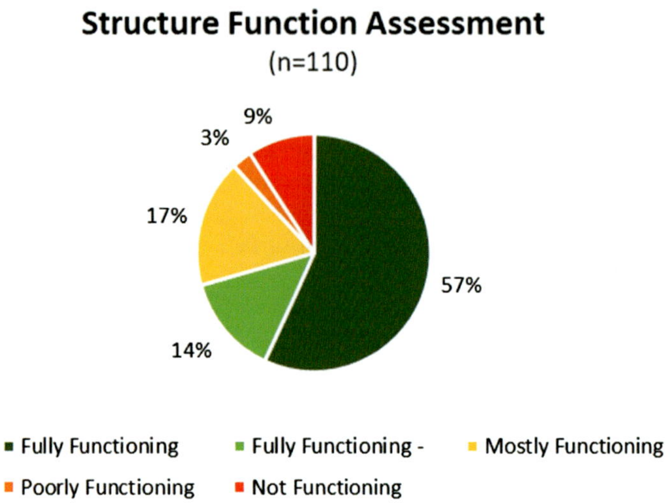

Fig. 10.14 Functional assessment for BMB ditch blocks, based on February 2023 post-construction qualitative assessment

of such maintenance would be to enhance stability and facilitate vegetation colonization along the flow paths. The February 2023 assessment identified 10% of the ditch block structures as Complete Failures, indicating a requirement for machinery-based work to re-establish the ditch block and its intended functions. This number of failed structures is not considered critical given the level of redundancy within the original restoration design, and, significantly, the spatial sparseness of blocks assessed as Complete Failures (Fig. 10.16).

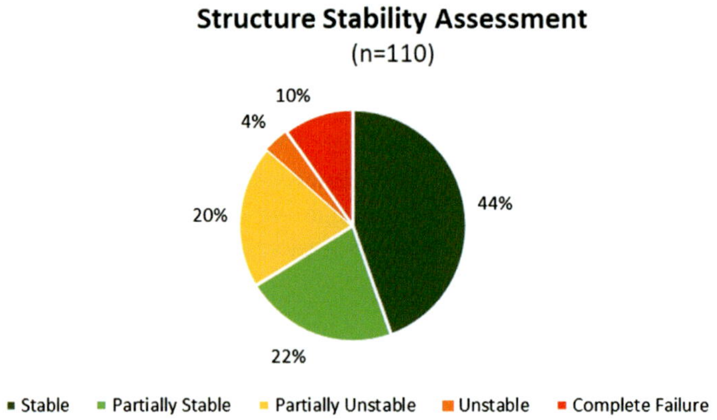

Fig. 10.15 Structural stability for BMB ditch blocks, based on February 2023 post-construction qualitative assessment

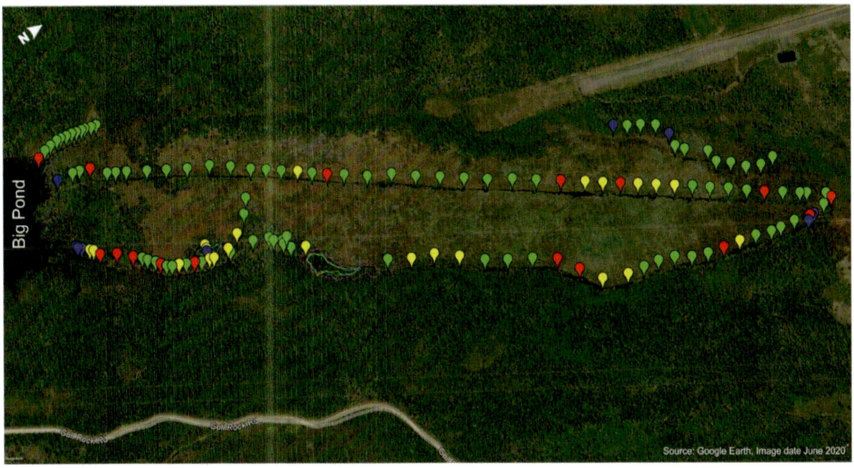

Fig. 10.16 Spatial layout of restoration structures at BMB and their respective post-construction status, based on a February 2023 survey. Ditch block coding key: Stable and Partially Stable (*green*), Partially Unstable (*yellow*), Unstable and Complete Failures (*red*), and non-built/not located ditch blocks (*blue*). Base image: Google Earth June 2020

The February 2023 post-construction assessment noted the complimentary and supportive role of ditch reprofiling as part of the overall restoration design. The reprofiling resulted in an overall reduction in the cross-sectional area of the drainage ditches, with a corresponding reduction in water velocity, frequently to zero. The addition of roughness to the former ditches, through the placement of vegetation mats, cut trees and shrubs, slowed the longitudinal movement of water along the ditches, allowing for additional infiltration into the adjacent bog. The reduction in the water velocity along the ditches is thought to have decreased the erosive forces on individual ditch blocks and contributing to their long-term stability.

Qualitatively, it appears that the structures assessed as Partially Unstable to Complete Failure occur in areas where surface water accumulation has occurred post-restoration. They are not concentrated in the steeper gradient areas in the northern and southern extremes of the restoration site where water velocities would be predicted to be greatest (Fig. 10.16). The more centrally located structures have been inundated with water, but very low flow velocities around these structures occur due to the low landscape slope. As a result, overall ditchblock functions are not significantly impaired despite some observed structure instability in the north central and east central areas of BMB. Instability was categorized as lack of vegetation colonization, and poor plant colonization around individual structures, resulting in part from the restoration induced surface water inundation around the structures.

The first ditch block structures constructed at Big Meadow Bog were in the south-west of the project site in the spring of 2016. As shown in Fig. 10.17 the natural lagg stream channel in this area was reactivated following the restoration, and surface water again began to flow along the former channel that had long been dry following the ditching of Big Meadow Bog in 1958–59. During late summer 2017 a storm caused accretion of the barrier beach at Big Pond, with a significant but unmeasured increase in water impoundment at Big Pond. This impoundment flooded the lower elevation ditch block structures (Fig. 10.17) and intended construction locations along the south end of BMB. These high-water conditions influenced both September 2017 construction and hydrological conditions around numerous south central and southeastern ditch blocks for a period of years until a subsequent storm in the winter of 2020/21 caused a break in the barrier beach that resulted in large drop in water table elevation of Big Pond and the south end of Big Meadow Bog. The impact on structure function and stability of southern ditch blocks was not assessed. However, high water was observed for an extended period along the relatively lower gradient southeast ditch line, and several ditch blocks there were subsequently categorized as Unstable or Complete Failures.

Along the southeastern ditch line of Big Meadow Bog, a series of four "Lily" ponds are visible in air imagery between 1929 and 1955 prior to the ditching of the bog (Fig. 10.18). Like the southwest lagg channel (Fig. 10.17), the shallow open water wetland areas have rewetted in each of the former Lily Pond locations since restoration. Dense shrub, sapling and tree stratum vegetation that has colonized the portions of bog since ditching makes observation of the rewetted ponds through current air imagery difficult, and site visitation was necessary to clearly see the locations. During the February 2023 field assessment, the boundaries of the Big Lily Pond (#1) and Southeast Lily Pond (#3) were captured with GPS and cover 2120 m^2

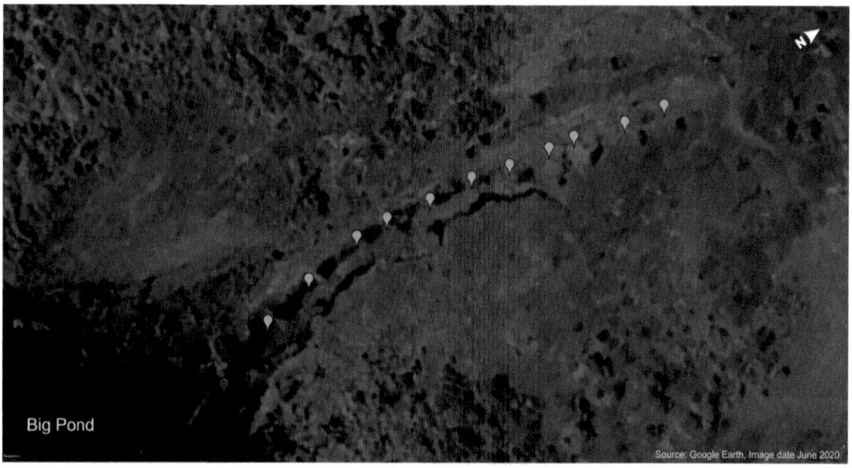

Fig. 10.17 The southwest ditch at BMB, along with 12 ditch blocks and adjacent borrow pits are visible in this June 2020 air image. The structures installed in the spring of 2017 were partially inundated with water following a 2018 storm that held additional water in Big Pond until 2020/21. In this June 2020 air image, the lower ditch blocks are flooded. Water can also be seen in the reactivated lagg channel east of the ditch blocks and adjacent ditch line. Ditch block coding key: Stable and Partially Stable (*green*), Unstable and Complete Failures (*red*). Base image: Google Earth June 2020

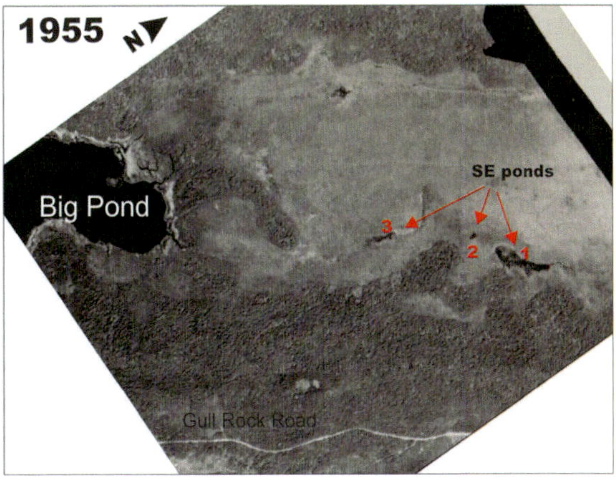

Fig. 10.18 1955 air photo of the southeast portion of Big Meadow Bog, showing the pre-disturbance boundaries of the three Lily ponds. Following the 1958/59 ditching, the ponds were consistently dry until the 2017/18 restoration

and 487 m^2, respectively. The boundary of the Big Lily Pond has been documented four times since May 2017 and has ranged from 1853 to 4551 m^2 in area depending on seasonal site conditions. These ponds add to the embedded habitat diversity within Big Meadow Bog and provide further evidence of restored hydrological functions.

Several additional qualitative observations were made during the February 2023 post-restoration assessment, including:

- Re-locating and observing 11 functional and stable staked weir structures along the northwest lagg channel (shown in Fig. 10.16) in a location referred to as BM1. Two staked weirs could not be located.
- Confirmation of stability at the Jimmy's Pond (Fig. 10.13) outlet rock weir and recolonization of the excavation spoil pile.
- Observation of Big Meadow Bog ditch line reprofiling, and sphagnum moss colonization along some sections of the reprofiled ditch.
- Observation of stable drainage channels and "seep away" structure that had been constructed to redirect upslope water toward the BM1 portion of Big Meadow Bog below a private runway.

It is noted that, at the time of this reporting, no maintenance requiring machinery is recommended as the potential damage to recovering areas of BMB that might result from accessing the site with machinery outweighs the anticipated benefits of repairing lost function of restoration structures with machinery. It is recommended that maintenance of individual structures by "hand" methods be undertaken to help ensure restored wetland hydrology functions are not adversely impacted by further destabilization of structures currently identified as Partially Unstable. The 22 structures are indicated by the yellow icons in Fig. 10.16.

References

Anderson R (2010) Restoring afforested peat bogs: results of current research. Forestry Commission. United Kingdom

Armstrong A, Holden J, Kay P, Foulger M, Gledhill S, McDonald TA, Walker A (2009) Drain-blocking techniques on blanket peat: a framework for best practice. J Environ Manage 90:3512–3519

Blawhorn Moss National Nature Reserve (2024) https://www.nature.scot/enjoying-outdoors/visit-our-nature-reserves/blawhorn-moss-national-nature-reserve. Accessed 25 May 2024

Environment Canada (2011) Action Plan for the Eastern Mountain Avens (Geum Peckii) in Canada {Proposed}. Species at Risk Act Action Plan Series. Environment Canada. Ottawa

Evans M, Allott T, Holden J, Flitcroft C, Bonn A (2005) Understanding gully blocking in deep peat. Moors for the Future Report No 4. Castleton, Derbyshire

Fay E, Lavoie E (2009) The impact of birch seedlings on evapotranspiration from mined peatland: an experimental study in southern Quebec. Canada. Mires and Peat 5(3):1–7

Gonzalez E, Henstra SW, Rochefort L, Bradfield GE, Poulin M (2013) Is rewetting enough to recover Sphagnum and associated peat-accumulating species in traditionally exploited bogs? Wetlands Ecological Management. https://doi.org/10.1007/s11273-013-9322-6

Goodale K (2015) Historical changes in the big meadow bog. Brier Island Interviews. Prepared for the Nature Conservancy of Canada

Grosvernier P, Staubli P (2009) Régénération des hauts-marais, bases et mesures techniques, L'environnement pratique n°0918. Office fédéral de l'environnement, Berne, Switzerland

Hill N, Kennedy G, Smith C, Horne D, Brazner J, Drage J, Boates S, Eaton S (2015) A conceptual restoration plan for the Big Meadow Bog, Brier Island, Nova Scotia

Kennedy G, Drage J (2015) Preliminary results of baseline hydrological monitoring at Big Meadow Bog, Brier Island, Digby County, Nova Scotia (Open File Report ME 2015–001). Halifax, NS: Nova Scotia Department of Natural Resources

Kozulin AV, Tanovitskaya NI, Vershitskaya IN (2010) Methodical Recommendations for ecological rehabilitation of damaged mires and prevention of disturbances to the hydrological regime of mire ecosystems in the process of drainage. Scientific and Practical Center for Bio Resources Institute for Nature Management of the National Academy of Sciences of Belarus, Belarus

LaRue D, Toms B (2013) Building for the future: establishing a foundation for the recovery of Eastern Mountain Avens: Final Report. HSP Project: 2012HSP6304. Mersey Tobeatic Research Institute for Environment Canada and the Eastern Mountain Avens Recovery Team. Kempt, Nova Scotia

Landry J, Marcoux O (2010) Rapport d'activites 2010. Remouillage de la section sud de la tourbiere Grande plee Bleue. Mars 2011. Groupe de recherché en ecologie des tourbieres, Universite Laval, Quebec

Landry J, Rochefort L (2012) The drainage of peatlands: impacts and rewetting techniques (trans by Graf MD. Peatland Ecology Research Group)

McCarter CPR, Price JS (2013) The hydrology of the Bois-des-Bel bog peatland restoration: 10 years post-restoration. Ecol Eng 55:73–81

Mills E, Laviolette L (2011) Birds of Brier Island. In: Proceedings of the Nova Scotia Institute of Science, vol 46, part 1

Price JS, Heathwaite AL, Baird JA (2003) Hydrological processes in abandoned and restored peatlands: an overview of management approaches. Wetlands Ecol Manage 11:65–83

Shepherd MJ, Labadz J, Caporn SJ, Crowle A, Goodison R, Rebane M, Waters R (2013) Natural England review of upland evidence—restoration of Degraded Blanket Bog. Natural England Evidence Review

Van Seters T, Price JS (2001) The impact of peat harvesting and natural regeneration on the water balance of an abandoned cutover bog, Quebec. Hydrol Process 15:233–248

Webster T, Collins K, Crowell N, MacDonald C, McGuigan K (2015) Hydrological modelling of surface Runoff on Brier Island, Nova Scotia. Technical report, Applied Geomatics Research Group, NSCC Middleton, Nova Scotia

Yorkshire Dales National Park (2015) Nature in the Dales—Yorkshire Peat Partnership. http://www.natureinthedales.org.uk/projects/the-yorkshire-peat-partnership. Accessed 6 Nov 2015

Yorkshire Peat Partnership (2015a) Technical specification 1: grip blocking using peat dams

Yorkshire Peat Partnership (2015b) Technical specification 3: larger grip and gully blocking using stone dams

Yorkshire Peat Partnership (2015c). Technical Specification 5: Gully and Hag Reprofiling

Chapter 11
Baseline and 3-Year Post-restoration Bird Communities at Big Meadow Bog, Brier Island, Nova Scotia

John C. Brazner

Abstract Anticipating the opportunity to restore the historic hydrologic conditions in Big Meadow Bog (BMB), it was important to assess the current status of key faunal taxa to provide a baseline for evaluating potential changes following restoration and assessing success. To accomplish this, bird communities were surveyed in swamp, lagg, and bog habitats along two transects associated with BMB—northwest Big Meadow (NWBMB) and southeast Big Meadow (SEBMB), and one transect at a reference area associated with Lost Bog. Transect points were surveyed during the breeding season in 2015–2018 and 2021. Ditches were plugged and water levels were raised in BMB during the fall of 2017, so bird surveys in 2015–2017 preceded water level manipulation, while those in 2018 and 2021 followed the manipulation. Overall, the bird communities were dominated by herring gulls (*Larus smithsonianus*), regardless of general transect location or habitat type. This result was driven mainly by high numbers (n = 3162) in the bog at NWBMB but to a lesser extent moderate numbers (n = 723) at SWBMB. Common yellowthroat (*Geothlypis trichas*; n = 86), American redstart (*Setophaga ruticilla*; n – 49), song sparrow (*Melospiza melodia*; n = 40), and black-throated green warbler (*Setophaga virens*; n = 33) were the next most common species overall. A total of 46 of the 83 breeding species known from Brier Island were detected across all surveys. Lost Bog and SEBMB had the highest number of species overall (n = 38) and NWBMB the lowest (n = 16). Ordinations revealed clear differences in the bird communities associated with BMB and the reference system at Lost Bog. Differences were most pronounced at NWBMB, where there were the fewest species, the lowest diversity and evenness, the lowest numbers of individuals apart from herring gulls and the fewest species of conservation concern, reflecting the highly degraded nature of the habitat around NWBMB. Based on their similarity to those at the Lost Bog reference site and lower numbers of gulls in the southern part of BMB, it is expected that bird communities at

---

J. C. Brazner (✉)
Wildlife Division, Nova Scotia Department of Natural Resources and Renewables, Kentville, NS B4N 4E5, Canada
e-mail: john.brazner@novascotia.ca

Brazner Ecological Services, Black Rock, NS, Canada

N. M. Hill et al. (eds.), *Wetland Restoration for Endangered Species Recovery*, Environmental Contamination Remediation and Management,
https://doi.org/10.1007/978-3-031-71344-6_11

SEBMB will be first to return to a full array of typical peatland species, possibly as quickly as 10–15 years. Recovery of bird communities in the northern part of BMB is likely to take several more decades and initiation of recovery will not progress until there is substantial reduction in the herring gull population. Continued systematic monitoring of bird and vegetation communities at the site will be necessary into the indefinite future to fully assess success of the restoration from an avian perspective.

Keywords Herring gulls · Bird communities · Population · Bog

11.1 Introduction

One of the keys to evaluating the success of any wetland restoration effort is selecting appropriate response variables (Cristescu et al. 2012; Prach et al. 2019) and using clear standards when assessing success (Gann et al. 2019). Selecting appropriate response variables is dependent upon an accurate *apriori* understanding of the typical structure and function of the type of ecosystem being restored when it is in a relatively undisturbed condition (e.g., Kentula 1989; Stelk et al. 2017; Andersen et al. 2023). This sort of understanding can be obtained through systematic study of appropriate reference ecosystems, but because this is not always feasible, selection of response variables is often based on extensive literature review and expert opinion (e.g., Moreno-Mateos et al. 2015).

The literature related to peatland restoration has grown substantially in the last three decades (Chimner et al. 2017). Most of these studies have focused on physical–chemical responses such as changes in groundwater levels (e.g., McCarter and Price 2013; Haapalehto et al. 2014; D'Acunha et al. 2018) and carbon dynamics (Tuittila et al. 1999; Bernard-Jannin et al. 2018; Gaffney et al. 2020) or water and soil chemistry (e.g., Andersen et al. 2010; Lundin et al. 2017). Fewer have focused on assessing biodiversity responses and these have primarily been studies on changes in the vegetation community over time (e.g., Rochefort et al. 2013; Maanavilja et al. 2014; Yepsen et al. 2014), whereas some have incorporated aspects of both vegetation and physical–chemical responses (e.g., Howe et al. 2007; Dimick et al. 2010; Goud et al. 2022). Much less has been done to track responses of wildlife communities (e.g., Prach et al. 2019; Hancock et al. 2020), such as birds (e.g., Desrochers and Rochefort 2021), to peatland restoration even though assessing a wide range of biodiversity responses is considered essential for fully evaluating the success of these sorts of projects (Stelk et al. 2017; Cross et al. 2019; Andersen et al. 2023).

Anticipating the opportunity to restore the historic hydrologic conditions in Big Meadow Bog (BMB; details in Chap. 1, this volume), we felt it would be important to assess the current status of certain key taxa to provide a baseline for evaluating potential changes following restoration and assessing success. We chose to focus on birds because the responses of bird communities are excellent indicators of ecological condition in a variety of ecosystems, including wetlands (e.g., Mensing et al. 1998; Bryce et al. 2002; Howe et al. 2007). Birds are also known to respond to habitat conditions at a variety of spatial scales (e.g., Crozier and Niemi 2003; Cunningham and Johnson 2016), they are relatively easy to survey reliably during the breeding season (e.g., Koskimies 1989; Howe et al. 2007; Gregory and van Strien 2010) and avian diversity is high in peatland ecosystems (e.g., Calme et al. 2002; Brazner and Achenbach 2020).

Although birds have seldom been used to track and evaluate the success of peatland restoration projects, I expected that they would be highly responsive to the sorts of habitat changes that were anticipated in BMB post-restoration (Hill et al. 2016). We believed the wetter conditions would lead to fewer trees and tall shrubs, a reduction in the large patches of non-native and aggressive early successional vegetation species (e.g., European grasses, blackberry and raspberry, *Rubus* spp.) and an increase in native vegetation (e.g., sphagnum mosses and ericaceous shrubs) typical of ombrotrophic bogs (e.g., Bellamy et al., 2012; Hill et al. 2019). Bird communities respond to these kinds of differences in habitat structure (e.g., MacArthur and MacArthur 1961; Roth 1976; Whelan 2001), but given these structural changes were expected to occur over a decadal or longer time scale (e.g., Price et al. 2003; Shepherd et al. 2013), detecting change in bird communities (or other aspects of biodiversity) at BMB was expected to require a long-term monitoring effort (Eastern Mountain Avens Recovery Team 2016; Hill et al. 2016).

An additional factor that we expected would have a cascading influence on the bird communities at BMB was the anticipated reduction in the size of the herring gull (*Larus smithsoniansus*) breeding colony over time. As shallow groundwater levels in BMB were restored to historic conditions the suitability of the bog for herring gulls was expected to decrease as preferred drier breeding conditions deteriorated (Billerman et al. 2020). In turn, this was expected to reduce the size of the herring gull colony over time and allow for a greater diversity of birds as competition for resources and predation risk were reduced (Eastern Mountain Avens Recovery Team 2016). Ultimately, all of these changes are expected to result in the return of a bird community in the coming decades that is more typical of bogs in eastern Canada (Calme et al. 2002; Brazner and Mackinnon 2020; Desrochers and Rochefort 2021), with Palm warbler (*Setophaga palmarum*), alder flycatcher (*Empidonax alnorum*), American redstart, common yellowthroat, swamp sparrow (*Melospiza georgiana*), yellow-rumped warbler (*Setophaga coronata*), and white-throated sparrow (*Zonotrichia albicollis*) among the most prevalent species.

11.2 Methods

11.2.1 Study Sites

Three transects were established in 2015 to characterize the baseline bird community in and adjacent to BMB (Fig. 11.1). The general character of BMB and the process used to restore historic water levels to BMB in the fall of 2017 was described by

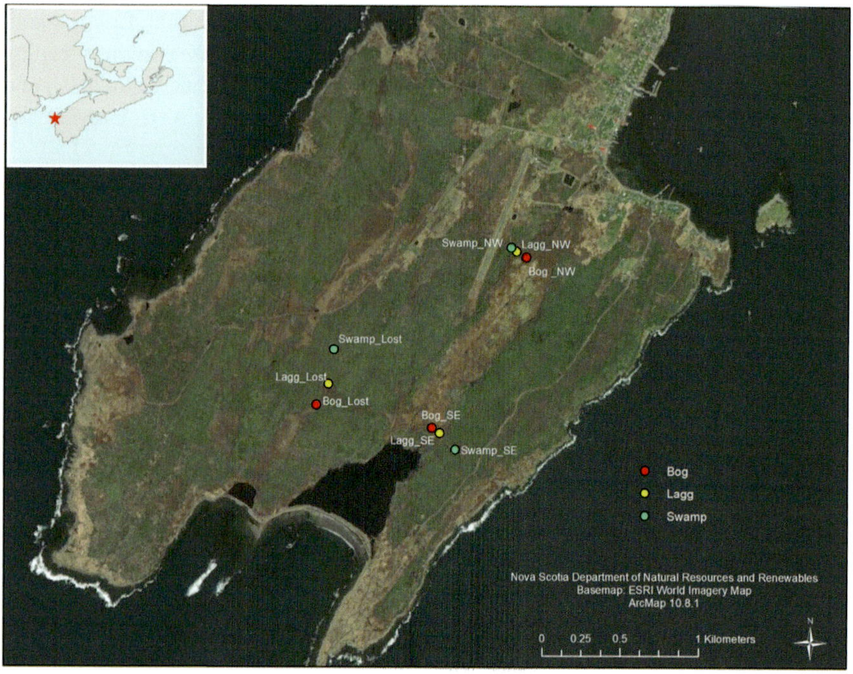

Fig. 11.1 Map of bird survey locations in Big Meadow Bog and Lost Bog

Hill et al. (2019), so that will not be repeated here. However, it is important to point out that each transect consisted of one point in swamp habitat immediately adjacent to the bog, one in the lagg at the bog margin and one in the bog interior. One transect was located in the southeast portion of BMB (SEBMB), one in the northwest portion of BMB (NWBMB) and one in the nearby Lost Bog, which we included as a relatively undisturbed reference site. It is similar to BMB in vegetation character and the adjacency of swamp habitat, although it is smaller (only ~ 11 ha in size compared to ~ 55 ha at BMB). Swamp points were dominated by stunted trees (e.g., black spruce, *Picea mariana*) and tall shrubs (e.g., speckled alder, *Alnus incana*), whereas lagg points were mainly dominated by ericaceous shrubs and bogs more by herbaceous cover and sphagnum mosses.

11.2.2 Bird Surveys

Bird communities at each point along the 3 transects were surveyed with fixed-distance (50-m radius), 10-min point counts (Howe et al. 2007) once during the breeding season in 2015–2018 and 2021 except for the Lost Bog transect which was not sampled in 2016. All surveys were conducted between 5:30 and 10:00 am and between 16 June and 5 July and on the same day in each year. Individual species abundances for each habitat type (swamp, lagg, bog) at each transect location were estimated as the number of individuals detected for a given species at a given point in each year. Birds deemed to be flyover transients were excluded, whereas aerial foragers and hovering gulls were included in the counts. Flyover transients made up $< 1\%$ of the total number of birds detected at any site, but hovering gulls were quite prevalent, particularly at bog points within BMB. Variation in detectability was minimized by conducting surveys only on days with optimal conditions (e.g., no rain, wind speeds below 20 km/h), ensuring all point counts were completed by a single observer (the author) and limiting data to birds detected within 50 m of the observer.

11.2.3 Habitat Characterization

Because in-depth vegetation surveys were being conducted as part of other parallel studies associated with the restoration of BMB (see Hill et. al. this volume) only a rapid assessment of habitat characteristics was completed at transect points in 2021 to provide a general sense of the differences in the structural character of the points where we surveyed birds. The total percentage cover in 5 vegetation strata—canopy (> 7 m high), tall shrub (2–7 m), low shrub (< 2 m), herb (ferns, graminoids, and forbs), and sphagnum moss layers—as well as the percentage of surface water were visually estimated within a 10 m radius plot at the bird survey locations. The number of dead snags (≥ 10 cm dbh and ≥ 3 m high) was also counted within each plot.

11.2.4 Data Analysis

Non-metric multidimensional scaling (NMDS; Clarke and Ainsworth 1993) was used to ordinate bird communities at different transect points based on untransformed abundances of all bird counts (PC-ORD version 7; McCune and Mefford 2016). Ordination was used to test whether patterns in overall community composition were related to specific locations along transects (i.e., swamp, lagg, bog) or to the general location of the 3 transects (e.g., NWBMB, SEBMB, Lost Bog). Methods employed for NMDS and related analyses are described in Brazner and Achenbach (2020), so are only reiterated briefly here. The method used a steepest-descent minimization algorithm and a Monte Carlo randomization technique to determine the optimal number of axes. The percentage of variance in the original data matrix explained by each axis in the ordination was calculated as the coefficient of determination (R^2) between distances among sites in the ordination space and distances in the original matrix and a nonparametric multiple response permutation procedure (MRPP) to test for differences in the centroids of site distributions among general transect locations in the final NMDS result to characterize overall patterns in community composition (McCune and Grace 2002). To examine the relationship between species abundances and NMDS axes, we plotted joint-plot vectors (McCune and Mefford 2016) radiating from the ordination centroid based on the combined correlation (r) between each variable and the axes. The angle and length of the vectors reflect the strength of the correlations (only species with $r \geq 0.4$ included).

Because data distributions for bird species were often not normal or homogeneous, we used two-way mixed-model nonparametric analysis of variance (npANOVA) to examine species richness, Shannon's diversity index (H'; Shannon and Weaver 1949), evenness (H'/ln[species richness]; Pielou 1969), and total number of individuals among general transect locations and specific transect points (Anderson 2001). Data from 2016 were eliminated from these analyses because there were no data from Lost Bog that year and npANOVA requires a balanced design (Anderson 2001). For each model, we included general transect locations (NWBMB, SEBMB, Lost Bog) as a random effect and specific transect point location (swamp, lagg, bog) as a fixed effect. Where the resulting model was significant for either of the main factors, pairwise comparisons for all levels of each factor were completed using the method described by McCune and Mefford (2016). Potential interaction effects were included for each model but partitioned out before paired comparisons were examined.

11.3 Results

11.3.1 Habitat Conditions

Swamp points were typified by stunted black spruce and white birch (*Betula papyrifera*) as well as witherod (*Viburnum nudum*), speckled alder and false holly (*Ilex mucronata*) as the primary woody species. There was relatively sparse herb cover in the swamps (< 20%) at all 3 transects and dense sphagnum cover (> 50%) on the ground at SEBMB and Lost Bog swamps (NWBMB had no sphagnum; Fig. 11.2). Lagg points were dominated by dense ericaceous shrub cover (e.g., lambkill, *Kalmia angustifolia*, Labrador tea, *Rhododendron groenlandicum*, leather-leaf, *Chamaedaphne calyculata*) and other low woody species (e.g., dwarf birch, *Betula michauxii*, steeplebush, *Spiraea tomentosa*, huckleberry, *Gaylussacia* spp.) at all three transects (90%) and minimal tall shrub (~ 10%) and herb cover (~ 5%). Bog points varied considerably among the 3 transects (Fig. 11.2), especially in their herb and sphagnum cover, but all had dense low shrub cover (~ 70%) similar to the laggs. Herb cover at bog points was highest at SEBMB and Lost Bog (Fig. 11.2) with cinnamon fern, *Carex* spp. sedges, and small cranberry (*Vaccinium oxycoccos*) common, and extremely sparse at NWBMB (~ 5%). Sphagnum cover was moderate to high at all SEBMB and Lost Bog transect points (> 50%) and nearly absent from the NWBMB points, with no sphagnum at the NWBMB bog point (Fig. 11.2).

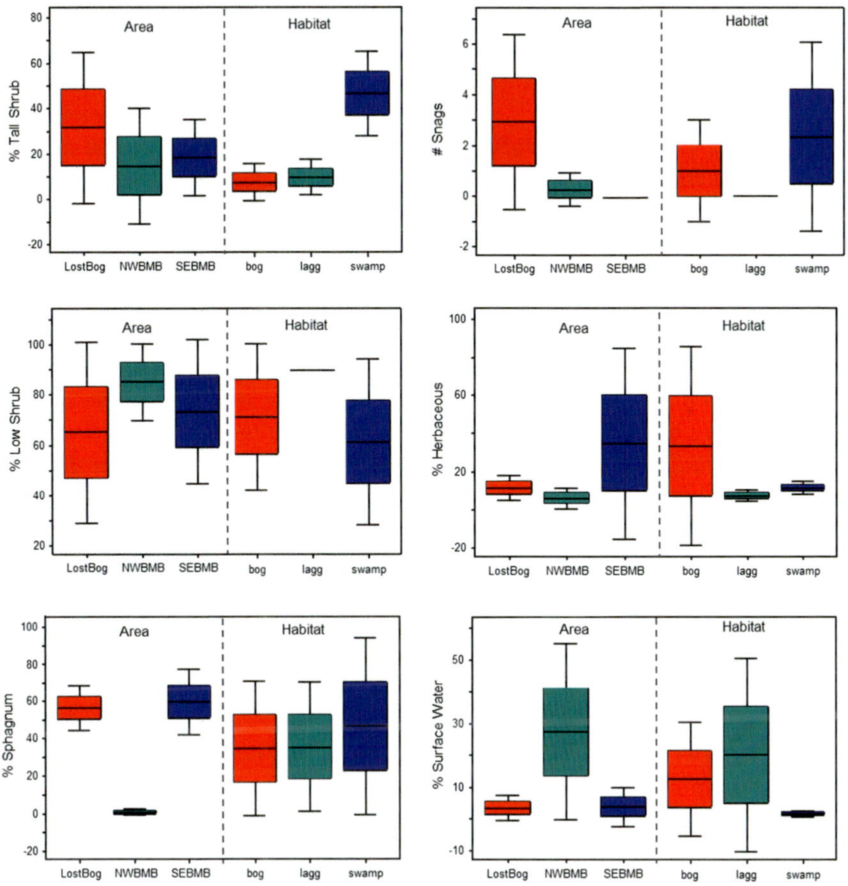

Fig. 11.2 Boxplots of mean habitat conditions by area and habitat for all bird survey locations in 2021 (box length represents one standard error above and below the mean; whiskers represent plus or minus two standard errors)

11.3.2 General Composition and Structure of Bird Communities

Overall, the bird communities were dominated by herring gulls, regardless of general transect location or habitat type (Table 11.1). With over 5500 individuals detected (out of a total of 6128 birds overall), herring gulls were over 50 times more abundant than the second most abundant species ($n = 86$), Common yellowthroat. In fact, herring gull abundance was highest at bog locations overall ($n = 3977$) compared to swamps and laggs ($p = 0.0004$), with the result driven mainly by high numbers ($n = 3162$) in the bog at NWBMB but to a lesser extent moderate numbers ($n = 723$) in the bog at SWBMB (Table 11.1, Fig. 11.3a). American redstart ($n - 49$), Song sparrow (n

= 40), and black-throated green warbler ($n = 33$) rounded out the five most common species across all locations (Table 11.1). These same species were generally among the five most common species at the individual transect locations and habitat types along those transects, but there were a few exceptions. Winter wren (*Troglodytes hiemalis*) made the top-5 at SEBMB and in swamp habitats, white-throated sparrow was fifth most abundant at both SEBMB and Lost Bog, alder flycatcher was fourth most abundant in bog habitats and at SEBMB and fifth most abundant in laggs, swamp sparrow was fourth most abundant at lagg points, and yellow warbler (*Setophaga petechia*) was fourth most abundant at NWBMB.

It is informative to examine the patterns in the 10 most abundant species a bit more closely (Fig. 11.4). In many ways, it is most interesting to note where particular bird species were absent, particularly in NWBMB. Winter wrens were never detected at NWBMB even though they were the 8th most abundant species overall. Alder flycatcher, black-throated green warbler, and white-throated sparrow were all also nearly absent altogether from NWBMB, except for 1 alder flycatcher that was detected in the bog in 2021, one black-throated green warbler observed in 2015, and one white-throated sparrow in the swamp in 2017. In addition, American redstarts were common at almost all transect locations except the lagg and bog habitats at NWBMB where only 1 individual was detected. It is also interesting that only one song sparrow (*Melospiza melodia*), a disturbance tolerant species, was detected in the swamp at Lost Bog while 39 were detected across all 3 habitats at the NWBMB and SEBMB transects (Table 11.1; Fig. 11.4). Although we looked for trends in the abundances of our most common species over all survey periods, the data were quite noisy and no obvious patterns were apparent (Figs. 11.3b and 11.5), suggesting responses to rewetting of Big Meadow Bog have been minimal as of 2021.

Examining the community structure during the study, a total of 46 of the 83 (55%) known breeding species from the island (Mills and Laviollette 2022) were detected across all surveys between 2015 and 2021. Lost Bog and SEBMB had the highest number of species overall ($n = 38$) and NWBMB the lowest ($n = 16$; Table 11.2). Abundance followed an inverse trend with highest numbers at NWBMB ($n = 4794$), lowest at Lost Bog ($n = 255$; note no data collected in 2016) and intermediate numbers at SEBMB ($n = 1079$). The highest species total at any particular transect point was at the Lost Bog lagg ($n = 31$), followed closely by the swamp at SEBMB ($n = 29$). The lowest number of species was at the lagg ($n = 8$) and bog ($n = 7$) points in NWBMB (Table 11.2). As with individual species abundances, we found no obvious pattern of increasing or decreasing overall abundances or number of species across years for any given site and habitat apart from a decreasing trend in the number of individuals in the swamp at NWBMB (Table 11.3). This trend was almost entirely due to decreasing numbers of herring gulls detected at this site from 2015 to 2021.

Comparing mean differences in overall community structure, bird communities at Lost Bog had the highest species richness, diversity, and evenness ($p < 0.001$) compared to NWBMB, with values at SEBMB intermediate (Fig. 11.6). In contrast, the mean number of individuals was lowest at Lost Bog and highest at NWBMB ($p = 0.0002$) due to the extremely high density of gulls breeding in the northern portion of the bog at NWBMB. Species richness, diversity, and evenness were all highest in

Table 11.1 All species observed during the study, their four letter species codes, scientific nomenclature, total abundances across all sites, abundances by habitat (bog, lagg, and swamp) and transect (NWBMB, SEBMB, and Lost Bog) as well as rank abundance for the 10 most abundant species within each class (as superscripts)

Species	4 Letter Species Code	Scientific name	Total	Bog	Lagg	Swamp	NW BMB	SE BMB	Lost Bog
Alder flycatcher	ALFL	*Empidonax alnorum*	30^{7}	13^{4}	12^{5}	5	1^{10}	17^{4}	12^{8}
American crow	AMCR	*Corvus brachyrhynchos*	21	7^{9}	9^{9}	5	2^{9}	15	4
American goldfinch	AMGO	*Spinus tristis*	13	3	1	9^{7}	6^{6}	2	5
American redstart	AMRE	*Setophaga ruticilla*	49^{3}	11^{5}	18^{5}	20^{5}	7^{5}	25^{3}	17^{4}
American robin	AMRO	*Turdus migratorius*	2	1	0	1	0	1	1
American woodcock	AMWO	*Scolopax minor*	1	0	0	1	0	1	0
Bald eagle	BAEA	*Haliaeetus leucocephalus*	1	1	0	0	0	0	1
Blackpoll warbler	BLPW	*Setophaga striata*	7	0	5	2	0	4	3
Black and white warbler	BAWW	*Mniotilta varia*	20	4	7	9^{7}	3^{7}	4	13^{7}
Black-capped chickadee	BCCH	*Poecile atricapillus*	7	3	2	2	0	3	4
Black-throated green warbler	BTNW	*Setophaga virens*	33^{5}	7^{9}	11^{7}	15^{4}	1^{10}	13^{8}	19^{3}
Boreal chickadee	BOCH	*Poecile hudsonicus*	5	0	4	1	0	1	4
Canada warbler	CAWA	*Cardellina canadensis*	2	0	0	2	0	2	0
Cedar waxwing	CEWA	*Bombycilla cedrorum*	3	0	3	0	0	2	1
Chestnut-sided warbler	CSWA	*Setophaga pensylvanica*	2	0	2	0	0	1	1
Common grackle	COGR	*Quiscalus quiscula*	3	2	1	0	1	1	1
Common raven	CORA	*Corvus corax*	3	1	2	0	0	1	2
Common yellowthroat	COYE	*Geothlypis trichas*	86^{2}	26^{2}	25^{2}	35^{2}	26^{3}	35^{2}	25^{2}
Dark-eyed junco	DEJU	*Junco hyemalis*	4	1	2	1	0	2	2
Golden-crowned kinglet	GCKI	*Regulus satrapa*	10	2	5	3	1	7	2

(continued)

Table 11.1 (continued)

Species	4 Letter Species Code	Scientific name	Total	Bog	Lagg	Swamp	NW BMB	SE BMB	Lost Bog
Gray catbird	GRCA	*Dumetella carolinensis*	6	1	4	1	1	5	0
Greater yellowlegs	GRYE	*Tringa melanoleuca*	1	0	1	0	0	1	0
Hermit thrush	HETH	*Catharus guttatus*	2	0	2	0	0	1	1
American herring gull	HERG	*Larus smithsonianus*	5572[1]	3817[1]	1394[1]	361[1]	4704[1]	833[1]	35[1]
Magnolia warbler	MAWA	*Setophaga magnolia*	12	1	4	7	0	4	8
Mourning dove	MODO	*Zenaida macroura*	7	2	1	4	0	2	5
Northern flicker	NOFL	*Colaptes auratus*	1	1	0	0	0	0	1
Northern parula	NOPA	*Setophaga americana*	5	1	3	1	0	5	0
Palm warbler	PAWA	*Setophaga palmarum*	3	3	0	0	0	0	3
Pine grosbeak	PIGR	*Pinicola enucleator*	2	0	1	1	0	1	1
Purple finch	PUFI	*Haemorhous purpureus*	11	3	5	3	0	2	9
Red crossbill	RECR	*Loxia curvirostra*	2	0	0	2	0	0	2
Red-breasted nuthatch	RBNU	*Sitta canadensis*	2	0	2	0	0	0	2
Red-eyed vireo	REVI	*Vireo olivaceus*	2	0	1	1	1[10]	0	1
Savannah sparrow	SAVS	*Passerculus sandwichensis*	1	1	0	0	0	1	0
Song sparrow	SOSP	*Melospiza melodia*	40[3]	22[3]	12[5]	6	27[2]	12[10]	1
Swainso"s thrush	SWTH	*Catharus ustulatus*	21	5	7	9	0	10	11
Swamp sparrow	SWSP	*Melospiza georgiana*	24[10]	9[6]	13[4]	2	3[7]	13[8]	8
Tree swallow	TRSP	*Tachycineta bicolor*	4	0	4	0	0	0	4

(continued)

Table 11.1 (continued)

Species	4 Letter Species Code	Scientific name	Total	Bog	Lagg	Swamp	NW BMB	SE BMB	Lost Bog
White-throated sparrow	WTSP	*Zonotrichia albicollis*	32^{6}	8^{8}	11^{7}	13^{6}	1^{10}	15^{5}	16^{5}
Wilson's snipe	COSN	*Gallinago gallinago*	2	0	0	2	0	1	1
Wilson's warbler	WIWA	*Cardellina pusilla*	2	2	0	0	0	0	2
Winter wren	WIWR	*Troglodytes hiemalis*	30^{8}	7^{9}	9^{9}	14^{5}	0	15^{5}	15^{6}
Yellow-bellied flycatcher	YBFL	*Empidonax flaviventris*	11	4	1	6	0	1	10^{10}
Yellow-rumped warbler	YRWA	*Setophaga coronata*	6	0	1	5	0	5	1
Yellow warbler	YWAR	*Setophaga aestiva*	25^{8}	9^{6}	8	8^{8}	9^{4}	14^{7}	2
Total (with herring gull)			6128	3977	1595	556	4794	1079	255
Total (without herring gull)			556	160	201	195	90	246	220

swamps, lowest in bogs, and intermediate in laggs. All differences in these variables among the habitats were statistically clear ($p \leq 0.001$), except for species richness ($p \geq 0.12$; Fig. 11.6).

11.3.3 Ordinations of Sites Based on Bird Communities

The ordination of sites based on the presence of all bird species (Fig. 11.7) revealed clear and highly significant separation among the bird communities associated with the 3 transect locations ($p \leq 0.0001$), particularly along Axis 1 which explained 65% of the variation in the original data matrix. Pairwise comparisons among transect locations revealed significant differences among the centroids for all 3 transects in the two-dimensional solution ($p \leq 0.002$). Axis 2 only accounted for 13% of the variation in the original data matrix, but some differences in the bird communities associated with the 3 different habitat types were apparent along this axis based on the MRPP analysis we employed ($p = 0.003$). Pairwise comparisons among the three habitat types revealed the centroids for lagg and bog bird communities were not significantly different from each other, but swamp communities differed from both of those habitats ($p \leq 0.035$).

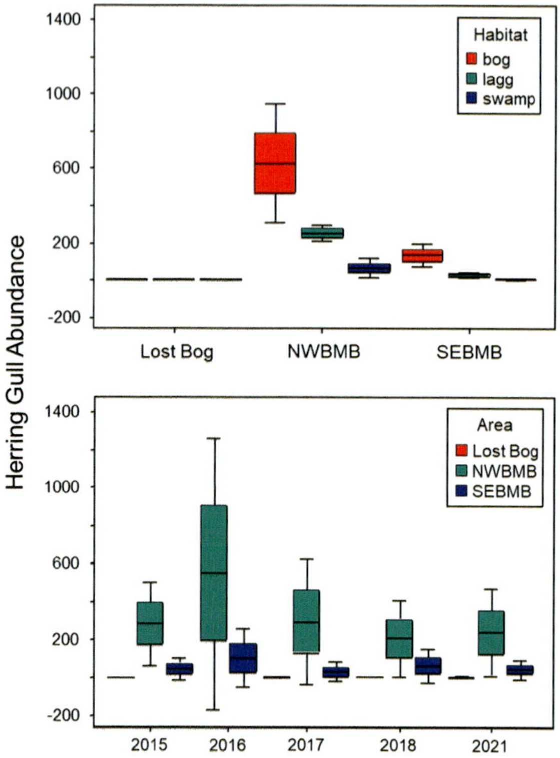

Fig. 11.3 Estimated mean herring gull abundance by survey area and habitat, as well as survey year and area (box length represents one standard error above and below the mean; whiskers represent plus or minus two standard errors)

Joint-plots (Fig. 11.7) revealed herring gull and Song sparrow were most strongly associated with NWBMB points in the ordination, gray catbird (*Dumetella carolinensis*) and golden-crowned kinglet (*Regulus satrapa*) were most strongly associated with SEBMB sites and black and white warbler (*Mniotilta varia*) with Lost Bog (particularly the Lost Bog swamp point). In addition, winter wren, white-throated sparrow, American redstart, Swainson's thrush (*Catharus ustulatus*), black-throated green warbler, and purple finch (*Haemorhous purpureus*) all had combined affinities for both Lost Bog and SEBMB sites.

11.3.4 Species of Conservation Concern

Twelve species of conservation concern (SOCC) based on Conservation Ranks less than "S5" from the Atlantic Canada Conservation Data Center (http://www.accdc.com/en/ranks.html) were observed across all survey points. These included, boreal chickadee (*Poecile hudsonicus*; S3S4), Canada warbler (*Cardellina canadensis*; S3S4)—the only species detected that is listed under the provincial Endangered Species Act, gray catbird (S4), pine grosbeak (*Pinicola enucleator*; S2), purple finch

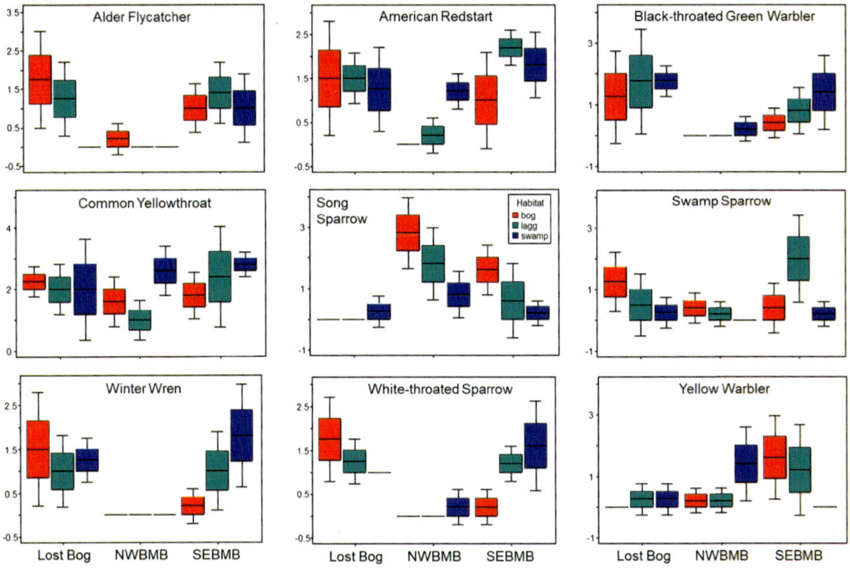

Fig. 11.4 Estimated mean abundance for the most common bird species (herring gulls plotted separately in Fig. 11.3) by survey area and habitat (box length represents one standard error above and below the mean; whiskers represent plus or minus two standard errors)

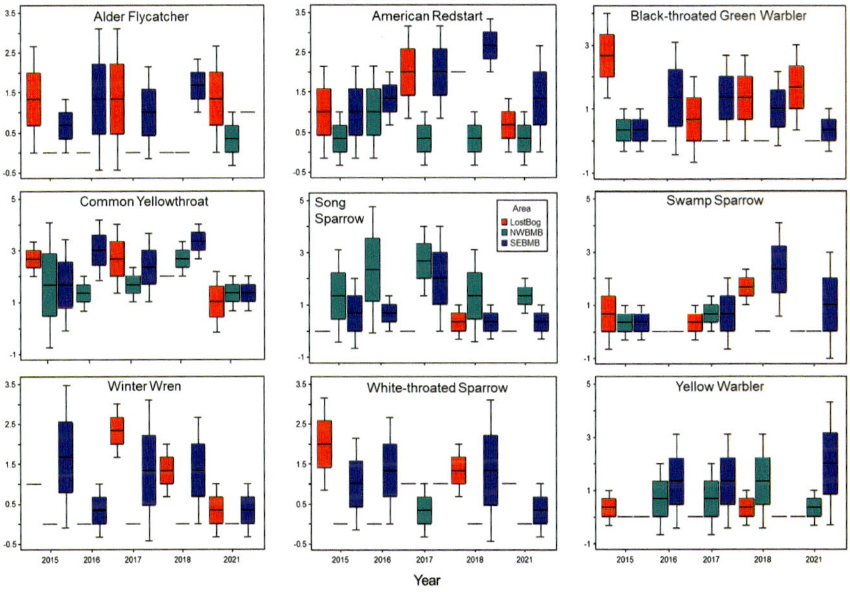

Fig. 11.5 Estimated mean abundance for the most common bird species (herring gulls plotted separately in Fig. 11.3) by survey year and area (box length represents one standard error above and below the mean; whiskers represent plus or minus two standard errors)

Table 11.2 Number of species and number of individual birds detected at the three transect areas and the three habitats sampled along each transect (2015–2018 and 2021)

Transect area	Habitat	# of Species	# of Individuals
NWBMB	Swamp	13	362
	Lagg	8	1270
	Bog	7	3162
	Total	16	4794
SEBMB	Swamp	29	124
	Lagg	24	232
	Bog	18	723
	Total	38	1079
Lost Bog[a]	Swamp	20	70
	Lagg	31	93
	Bog	24	92
	Total	38	255

[a] No data for 2016

Table 11.3 Number of species and number of individual birds detected at the three transect areas by year and habitat

Transect area	Year	Swamp		Lagg		Bog	
		# of Species	# of Individuals	# of Species	# of Individuals	# of Species	# of Individuals
NWBMB	2015	7	159	1	200	4	505
	2016	5	97	4	305	3	1255
	2017	8	42	6	256	4	607
	2018	7	44	4	206	5	391
	2021	5	20	4	303	4	404
SEBMB	2015	10	22	10	40	7	112
	2016	14	25	14	71	7	259
	2017	16	26	14	35	11	94
	2018	14	28	11	49	7	162
	2021	8	23	11	37	6	96
Lost Bog	2015	9	14	12	20	10	19
	2017	13	25	16	31	15	28
	2018	14	23	20	30	15	24
	2021	6	8	10	12	10	21

(S4S5), red crossbill (*Loxia curvirostra*; S3), Savannah sparrow (*Passerculus sandwichensis*; S4S5), Swainson's thrush (S4S5), tree swallow (*Tachycineta bicolor*; S4), Wilson's snipe (*Gallinago delicata*; S3S4), Wilson's warbler (*Cardellina pusilla*; S4), and yellow-bellied flycatcher (*Empidonax flaviventris*; S4S5). All but 3 (Canada

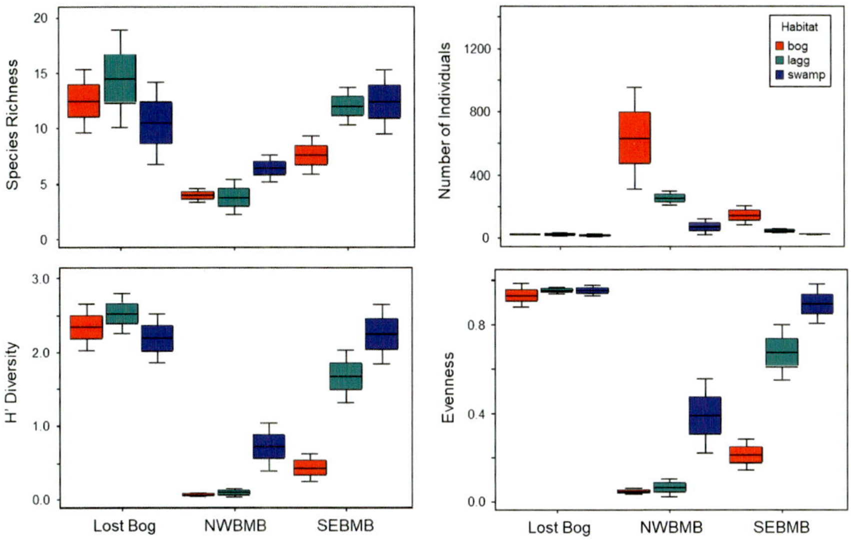

Fig. 11.6 Box plots of bird community characteristics (box length represents one standard error above and below the mean; whiskers represent plus or minus two standard errors)

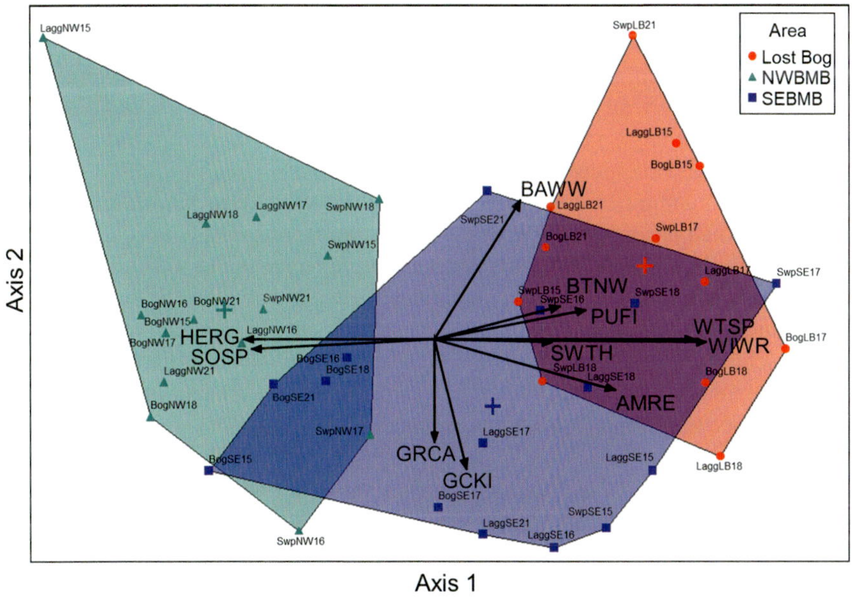

Fig. 11.7 Non-metric multidimensional scaling ordination plot of bird survey locations based on the presence or absence of all bird species detected during surveys from 2015 to 2021. Colored polygons (convex hulls) outline the outermost points of each site type (centroids for each group denoted by "+"). The angle and length of bird species (4 letter species codes defined in Table 11.1) joint-plot vectors radiating from ordination centroid reflect the strength of the combined correlation (r) of each species site abundances with NMDS axes 1 and 2 (only vectors with $r \geq 0.4$ included)

warbler, gray catbird and savannah sparrow) of the SOCC were observed at Lost Bog and three SOCC (boreal chickadee, red crossbill, tree swallow, and Wilson's warbler) were *only* observed at Lost Bog. Nine SOCC were also observed at SEBMB, the only transect where Canada warbler (swamp) and savannah sparrow (bog) were detected throughout the study. A single gray catbird comprised the only SOCC that was detected at NWBMB (Table 11.1).

11.4 Discussion

There were clear differences in the bird communities associated with BMB and the reference system at Lost Bog. The differences were most pronounced at NWBMB, where there were the fewest species, the lowest diversity and evenness, the lowest numbers of individuals apart from herring gulls, by far the highest abundance of herring gulls and the fewest SOCC. This community structure and species composition reflects the highly degraded nature of the habitat around NWBMB. The lack of substantive sphagnum cover that was observed at this site is reflective of these degraded conditions. Hill et al. (2019 and this volume) describe these degraded vegetation conditions in much greater detail, but these include a predominance of exotic grasses (e.g., *Holcus lanatus*), annuals, raspberry and blackberry (Rubus spp.) canes, and rose shrubs (*Rosa* spp.) in contrast to the native heathland community (e.g., lambkill, Labrador tea, and dwarf huckleberry, *Gaylussacia dumosa*) that would have occurred there historically along with other plant species that typify raised bog habitats in this region (e.g., small cranberry, *Vaccinium oxycoccos*; cloudberry, *Rubus chamaemorus,* and deergrass, *Trichophorum cespitosum*).

The bird communities associated with SEBMB were most similar to those in Lost Bog, which has a bird community that is typical of healthy wooded peatland ecosystems in Nova Scotia (Desrochers et al. 1998; Brazner and Achenbach 2020). Even so, the birds at SEBMB were dominated by an abundance of herring gulls, mainly due to high numbers in the bog and so, not surprisingly, had relatively low species richness, diversity and evenness at this bog point. The fact that Lost Bog and SEBMB supported a similar community of bird species overall, had similar diversity and abundance of birds at all habitats combined and a similar compliment of SOCC, suggests SEBMB is much closer to a recovered state than NWBMB. SEBMB has a vegetation community in the lagg area with many of the species typical of a native heathland for this area (e.g., low ericaceous shrubs, dwarf birch, certain sedges such as, *Carex michauxiana,* bog goldenrod, *Solidago uliginosa*, and small cranberry) that is similar to the vegetation present in the lagg at Lost Bog, even though SEBMB the vegetation community is still relatively degraded in the bog (e.g., dominated by exotic grasses, *Rubus* spp. and rose shrubs; LaRue and Toms 2013; Hill et al. 2016; Brazner, unpublished).

The fact that there were no obvious patterns of increasing or decreasing numbers of individuals or species across years apart from the decreasing trend in herring gulls in the swamp at NWBMB suggests the bird community in the northern portion of

BMB has yet to respond in any substantive way to the manipulation of water levels that was completed in 2017. Although this decreasing trend in herring gulls was not detected in recent nest surveys (Brad Toms, Mersey Tobeatic Research Institute, personal communication), Mills and Laviolette (2022) report that "as of 2021, it appears that nesting in the Big Meadow Bog is decreasing, and that recruitment from that area is declining". Declining herring gull numbers at the NWBMB swamp point lend at least some support for this idea and if borne out by additional evidence in coming years would suggest one of the key steps necessary for the recovery of a healthy bog ecosystem at BMB is underway. As explained by Hill et al. (2019), vegetation recovery to typical ombrotrophic bog conditions with healthy associated lagg in BMB cannot fully proceed without a dramatic reduction in the nesting herring gull population. In addition, based on the idea that if you restore the vegetation structure the appropriate biotic communities will follow (so-called "Field of Dreams" hypothesis; Palmer et al. 1997), the expectation is that once healthier vegetation conditions are present, birds that typify those conditions will recolonize the area. However, in a recent study of drained peatlands in western Finland, bird communities in eleven peatlands where historic water levels had been restored had not converged with birds typical of nearby areas that were in "pristine" condition after 8 years (Alsila et al. 2021) This suggests convergence of bird communities is not guaranteed even if the habitat is restored, at least not on shorter (less than decadal) time scales. Unfortunately, there is a paucity of studies focused on faunal recovery in restored peatlands, so it is difficult to draw firmer conclusions.

Regardless, I am optimistic that bird communities in the southern portion of BMB will return to a relatively full array of typical bird species much more quickly than in the northern portion. This could occur in as few as 10–15 years, given they are already well on their way based on their current similarity to the relatively undisturbed Lost Bog bird communities and low numbers of gulls in this part of BMB. On the other hand, the recovery of bird communities in the northern part of BMB is likely to take several decades based on observations of the recovery of bird communities following peat harvesting in Quebec (Desrochers and Rochefort 2021) and the assumption that the recovery of bird communities will not proceed faster than the recovery of vegetation which has been estimated to be about 30 years once gull populations are reduced to more natural levels (Hill et al. 2019). Finally, it is important to note that it will only be possible to predict the specific timing of recovery once this reduction in herring gull numbers is clearly underway. Continued systematic monitoring (ideally at a higher level of intensity than was possible for this study) of bird and vegetation communities at the site will be necessary into the indefinite future to fully assess success of the restoration from an avian perspective.

Acknowledgements I acknowledge that we conducted this study in Mi'kma'ki, the ancestral and unceded territory of the Mi'kmaq people, that we are all treaty people and bound by mutual obligations under the Peace and Friendship Treaties of 1760–61. I thank Nick Hill and Randy Milton for ongoing advice and assistance during the development and completion of the fieldwork for this project. Thanks also to Nova Scotia Department of Natural Resources and Renewables Librarian, Tracy Lenfesty help locating supporting references, Maggie Davidson for data management and Frances MacKinnon for GIS support. This project has been funded wholly by the NSDNRR. It has been reviewed by the Wildlife Division of NSDNRR and approved for publication. Approval does not signify that the contents reflect the views of the Department, nor does mention of trade names or commercial products constitute endorsement or recommendation for use.

References

Alsila T, Elo M, Hakkari T, Kotiaho JS (2021) Effects of habitat restoration on peatland bird communities. Restor Ecol 29(1):e13304

Andersen R, Rochefort L, Poulin M (2010) Peat, water and plant tissue chemistry monitoring: a seven-year case-study in a restored peatland. Wetlands 30:159–170

Andersen AN, Einoder LD, Fisher A, Hill B, Oberprieler SK (2023) Faunal standards for the restoration of terrestrial ecosystems: a framework and its application to a high-profile case study. Restor Ecol 31(1):e13735

Anderson MJ (2001) A new method for non-parametric multivariate analysis of variance. Austral Ecol 26(1):32–46

Bellamy PE, Stephen L, Grant MC, MacLean IS (2012) Response of blanket bog vegetation to drain-blocking. Appl Veg Sci 15:129–135

Bernard-Jannin L, Binet S, Gogo S, Leroy F, Défarge C, Jozja N, Zocatelli R, Perdereau L, Laggoun-Défarge F (2018) Hydrological control of dissolved organic carbon dynamics in a rehabilitated Sphagnum-dominated peatland: a water-table based modelling approach. Hydrol Earth Syst Sci 22(9):4907–4920

Billerman SM, Keeney BK, Rodewald PG, Schulenberg TS (eds) (2020) Birds of the world. Cornell Laboratory of Ornithology, Ithaca, NY, USA. https://birdsoftheworld.org/bow/home

Brazner J, Achenbach L (2020) Do breeding bird communities or conservation value differ among forested wetland types or ecoregions in Nova Scotia? Wetlands 40(4):811–823

Brazner J, MacKinnon F (2020) Relative conservation value of Nova Scotia's forests: forested wetlands as avian diversity hotspots. Can J for Res 50(12):1307–1322

Bryce SA, Hughes RM, Kaufmann PR (2002) Development of a bird integrity index: using bird assemblages as indicators of riparian condition. Environ Manage 30:294–310

Calme S, Desrochers A, Savard J-PL (2002) Regional significance of peatlands for avifaunal diversity in southern Québec. Biol Cons 107:273–281. https://doi.org/10.1016/S0006-3207(02)00063-0

Chimner RA, Cooper DJ, Wurster FC, Rochefort L (2017) An overview of peatland restoration in North America: where are we after 25 years? Restor Ecol 25(2):283–292

Clarke KR, Ainsworth M (1993) A method of linking multivariate community structure to environmental variables. Marine Ecology Progress Series

Cristescu RH, Frère C, Banks PB (2012) A review of fauna in mine rehabilitation in Australia: current state and future directions. Biol Cons 149(1):60–72

Cross SL, Tomlinson S, Craig MD, Dixon KW, Bateman PW (2019) Overlooked and undervalued: the neglected role of fauna and a global bias in ecological restoration assessments. Pac Conserv Biol 25(4):331–341

Crozier GE, Niemi GJ (2003) Using patch and landscape variables to model bird abundance in a naturally heterogeneous landscape. Can J Zool 81(3):441–452

Cunningham M, Johnson DH (2016) What you find depends on where you look: responses to proximate habitat vary with landscape context. Avian Conserv Ecol 11(2):1

D'Acunha B, Lee SC, Johnson MS (2018) Ecohydrological responses to rewetting of a highly impacted raised bog ecosystem. Ecohydrology 11(1):1922

Desrochers A, Rochefort L, Savard JPL (1998) Avian recolonization of eastern Canadian bogs after peat mining. Can J Zool 76(6):989–997

Desrochers A, Rochefort L (2021) Avian recolonization of unrestored and restored bogs in Eastern Canada. bioRxiv 2021:11

Dimick BP, Stucky JM, Wall W, Vepraskas MJ, Wentworth TR, Arellano C (2010) Plant-soil-hydrology relationships in three Carolina bays in Bladen County, North Carolina. Castanea 75(4):407–420

Eastern Mountain Avens Recovery Team (2016) A restoration plan for big meadow bog: background, baseline studies and overall goals. Application to Nova Scotia Environment, Yarmouth Regional Office

Gaffney PP, Hancock MH, Taggart MA, Andersen R (2020) Restoration of afforested peatland: immediate effects on aquatic carbon loss. Sci Total Environ 742:140594

Gann GD, McDonald T, Walder B, Aronson J, Nelson CR, Jonson J, Hallett JG, Eisenberg C, Guariguata MR, Liu J, Hua F (2019) International principles and standards for the practice of ecological restoration. Restor Ecol 27(S1):S1–S46

Goud EM, Touchette S, Strachan IB, Strack M (2022) Graminoids vary in functional traits, carbon dioxide and methane fluxes in a restored peatland: implications for modelling carbon storage. J Ecol 110(9):2105–2117

Gregory RD, van Strien A (2010) Wild bird indicators: using composite population trends of birds as measures of environmental health. Ornithol Sci 9(1):3–22

Haapalehto T, Kotiaho JS, Matilainen R, Tahvanainen T (2014) The effects of long-term drainage and subsequent restoration on water table level and pore water chemistry in boreal peatlands. J Hydrol 519:1493–1505

Hancock MH, Klein D, Cowie NR (2020) Guild-level responses by mammalian predators to afforestation and subsequent restoration in a formerly treeless peatland landscape. Restor Ecol 28(5):1113–1123

Hill N, Kennedy G, Smith C, Horne D, Brazner J, Drage J, Boates S, Eaton S (2016) A restoration plan for Big Meadow Bog. Application to Nova Scotia Environment, Parts I and II

Hill NM, Chinner K, Toms B, Garbary DJ (2019) Recovery succession of native plant communities at Big Meadow Bog after disturbance and eutrophication from herring gulls. Wetlands 39(3):533–548

Howe RW, Regal RR, Hanowski J, Niemi GJ, Danz NP, Smith CR (2007) An index of ecological condition based on bird assemblages in Great Lakes coastal wetlands. J Great Lakes Res 33:93–105

Kentula ME (1989) Wetland creation and restoration: the status of the science. Corvallis Environmental Research Laboratory, Office of Research and Development, US Environmental Protection Agency

Koskimies P (1989) Birds as a tool in environmental monitoring. Ann Bot Fenn 26:153–166

LaRue D, Toms B (2013) Building for the future: establishing a foundation for the recovery of Eastern Mountain Avens. Final Report to Environment Canada and the Eastern Mountain Avens Recovery Team

Lundin L, Nilsson T, Jordan S, Lode E, Strömgren M (2017) Impacts of rewetting on peat, hydrology and water chemical composition over 15 years in two finished peat extraction areas in Sweden. Wetlands Ecol Manage 25:405–419

Maanavilja L, Aapala K, Haapalehto T, Kotiaho JS, Tuittila ES (2014) Impact of drainage and hydrological restoration on vegetation structure in boreal spruce swamp forests. For Ecol Manage 330:115–125

MacArthur RH, MacArthur JW (1961) On bird species diversity. Ecology 42:594–598. https://doi.org/10.2307/1932254

McCarter CP, Price JS (2013) The hydrology of the Bois-des-Bel bog peatland restoration: 10 years post-restoration. Ecol Eng 55:73–81

McCune B, Grace JB (2002) Analysis of ecological communities. MjM Software Design, Oregon, USA

McCune B, Mefford MJ (2016) PC-ORD. Multivariate analysis of ecological data. Version 7.0 for Windows. Wild Blueberry Media, Corvallis, Ore

Mensing DM, Galatowitsch SM, Tester JR (1998) Anthropogenic effects on the biodiversity of riparian wetlands of a northern temperate landscape. J Environ Manage 53(4):349–377
Mills EL, Laviolette L (2022) Birds of Brier Island, Nova Scotia, revised and updated. Proc N Sn Inst Sci (NSIS) 52(1):173
Moreno-Mateos D, Meli P, Vara-Rodríguez MI, Aronson J (2015) Ecosystem response to interventions: lessons from restored and created wetland ecosystems. J Appl Ecol 52(6):1528–1537
Palmer MA, Ambrose RF, Poff NL (1997) Ecological theory and community restoration ecology. Restor Ecol 5(4):291–300
Pielou EC (1969) An introduction to mathematical ecology. Wiley, New York
Prach K, Durigan G, Fennessy S, Overbeck GE, Torezan JM, Murphy SD (2019) A primer on choosing goals and indicators to evaluate ecological restoration success. Restor Ecol 27(5):917–923
Price JS, Heathwaite AL, Baird AJ (2003) Hydrological processes in abandoned and restored peatlands: an overview of management approaches. Wetlands Ecol Manage 11:65–83
Rochefort L, Isselin-Nondedeu F, Boudreau S, Poulin M (2013) Comparing survey methods for monitoring vegetation change through time in a restored peatland. Wetlands Ecol Manage 21:71–85
Roth RR (1976) Spatial heterogeneity and bird species diversity. Ecology 57:773–782. https://doi.org/10.2307/1936190
Shannon CE, Weaver W (1949) The mathematical theory of communication. University of Illinois Press, Urbana, Ill
Shepherd MJ, Labadz J, Caporn SJ, Crowle A, Goodison R, Rebane M, Waters R (2013) Natural England review of upland evidence—restoration of Degraded Blanket Bog. Natural England Evidence Review, Number 003
Stelk MJ, Christie J, Weber R, Lewis RRIII, Zedler J, Micacchion M, Merritt J (2017) Wetland restoration: contemporary issues and lessons learned. Association of State Wetland Managers, Windham, Maine
Tuittila E-S, Komulainen V-M, Vasander H, Laine J (1999) Restored cutaway peatland as a sink for atmospheric CO_2. Oecologia 120:563–574
Whelan CJ (2001) Foliage structure influences foraging of insectivorous forest birds: an experimental study. Ecology 82:219–231. https://doi.org/10.1890/0012-9658(2001)082[0219:FSIFOI]2.0.CO;2
Yepsen M, Baldwin AH, Whigham DF, McFarland E, LaForgia M, Lang M (2014) Agricultural wetland restorations on the USA Atlantic Coastal Plain achieve diverse native wetland plant communities but differ from natural wetlands. Agr Ecosyst Environ 197:11–20

Chapter 12
Effects of Avian Biotransport of Industrial Wastes to a Bog Ecosystem

Mark L. Mallory, Sarah E. Gutowsky, Jennifer F. Provencher, Ian Spooner, Nelson J. O'Driscoll, and John P. Smol

Abstract Big Meadow Bog, in the center of Brier Island, Nova Scotia, has long been the nesting site for American herring gulls (*Larus smithsonianus* Coues), with recent counts of approximately 3000 pairs of birds at the colony in the summer. Gulls move from the colony to fish-processing plants, mink farms, aquaculture sites, and marine coastlines to feed, and then return to the colony to incubate their eggs or feed their chicks. These gulls defecate at the colony, regurgitate, and drop pieces of food; in some cases eggs, chicks, or adults die at the colony. These actions provide large nutrient and contaminant subsidies to the bog. Here we review information on this process of avian biotransport, and we examine evidence for amounts of nutrients and of mercury that birds contribute to the habitat.

Keywords Herring gull · *Larus* · Mercury · Nutrients · Water table

12.1 Introduction

Seabirds are a ubiquitous feature of marine environments around the globe, found in colonies along coastlines from polar to tropical regions (Coulson 2002). Breeding in colonies provides advantages to many seabird species, including safety in numbers,

M. L. Mallory (✉) · S. E. Gutowsky
Biology, Acadia University, 15 University Drive, Wolfville, Nova Scotia B4P 2R6, Canada
e-mail: Mark.mallory@acadiau.ca

J. F. Provencher
Science and Technology Branch, Environment and Climate Change Canada, Carleton University, Raven Road, Ottawa, Ontario K1A 0H3, Canada

I. Spooner · N. J. O'Driscoll
Earth and Environmental Science, Acadia University, 15 University Drive, Wolfville, Nova Scotia B4P 2R6, Canada

J. P. Smol
Paleoecological Environmental Assessment and Research Lab (PEARL), Biology, Queen's University, Kingston, Ontario K7L 3N6, Canada

N. M. Hill et al. (eds.), *Wetland Restoration for Endangered Species Recovery*, Environmental Contamination Remediation and Management,
https://doi.org/10.1007/978-3-031-71344-6_12

and sharing information (directly or indirectly) to locate food sources (Kharitonov and Siegel-Causey 1988), but colonies must be located at sites that are relatively safe from mammalian predators, and as such are often located on cliffs or on islands (Gaston 2004). When such favorable locations are found, colonies may have millions of birds (e.g., Sklepkovych and Montevecchi 1989). Often single colonies can represent a large part of a species' population, and thus the many threats faced by colonial-nesting seabirds at sea and on land (Croxall et al. 2012) can lead to deleterious, population-level effects in some locations.

Multiple threats in the twenty-first century are severely affecting seabirds. Through scientific research and extensive communications and outreach from global conservation organizations and other knowledge holders, it is now recognized that many seabird species face threats year-round and are among the most at-risk group of birds (Dias et al. 2019). Changing ocean conditions and attendant perturbations to food webs due to climate change, deleterious interactions with fisheries, and introductions of invasive species to nesting colonies are among the major threats that most seabirds face, with many albatross, petrel, and penguin species recently experiencing significant and alarming population declines.

In contrast to most other seabirds, there is one group of species that exploit human-altered landscapes or activities and often are considered pests: these are gulls (among other common names, Family Laridae). Most gull species are "generalists" in an ecological sense; that is, daily, seasonally, and annually they rely on a suite of different habitats and food resources for survival in contrast to a more specialized species such as thick-billed murres *Uria lomvia*, that rely principally on pelagic fish year-round (Gaston and Hipfner 2020). While common along marine coastlines, some gull species have breeding colonies inland near agricultural and urban habitats, including on building rooftops (e.g., Blight et al. 2019; Spelt et al. 2019). Consequently, with foraging regions that are highly diverse, dietary items consumed by gulls can range from insects, earthworms, zooplankton, birds (eggs, chicks, adults), fish, and small mammals (i.e., direct predation), to fruit, carrion, and garbage (i.e., scavenging; Weseloh et al. 2020). The exploitation of human-altered habitats has led to considerable conflict between gulls and people for several reasons. First, when gull colonies form, especially near urban or industrial centers, numbers of other (often considered more desirable) bird species may decline if their eggs and chicks become prey for gulls (e.g., Scopel and Diamond 2017). Second, the presence of gulls at dumps, parking lots, food outlets, airports, golf courses, or urban parks, and the associated issues with noise and excrement in those locations, has meant gulls are often viewed with disdain and subject to population control measures (e.g., Thomas 1972; Belant 1997). Lastly, gulls can vector pathogens, and thus again are viewed as nuisance species, as in the recent case of SARS CoV-2 (Wright et al. 2009; Boklund et al. 2021) or in the case of highly pathogenic avian influenza where a gull was implicated as patient zero for North America (Caliendo et al. 2022).

Similar to all other colonial species, gulls spend several months of their breeding season at a single nesting site, sympatric with a few to thousands of conspecifics. At and around their nesting site, the remains of what they have eaten are deposited. Birds defecate near their nests (or on entry or exit from a colony), they drop bits of food, they regurgitate food for offspring, and of course some eggs, chicks, and adults die (Blais et al. 2007). These biological processes occur for any nesting bird, but in large colonial aggregations, the chemical subsidies left by breeding birds can lead to profound changes in the environmental compartments (water, soils, biota) at and near a colony (Duda et al. 2020), to the point where some seabirds are considered environmental engineers (Sánchez-Piñero and Polis 2000; González-Berganzoni et al. 2017). Unfortunately, in some situations, excess amounts of feces and associated nutrients can have deleterious effects on the recipient habitats, leading to a total change in, or loss of, vegetation (Klimaszyk and Rzymski 2016; Simpson et al. 2021), as well as having other repercussions.

In this chapter, we review how issues of seabird relationships with industry, safe nesting sites and movement of chemicals by avian vectors have been the focus of management options at Brier Island, Nova Scotia (44.2° N, 66.3° W). Here, habitat conservation efforts have been underway for years to save a rare plant (eastern mountain avens *Geum pecki* Pursh) that inhabits a bog system on the island. Due to human intervention at the bog (see Chap. 1 for details), gulls established a huge colony on the island, the effects of which may threaten the survival of these plants.

12.2 Avian Biotransport

It has been known for decades that avian vectors (here, colonial seabirds) move various elements and chemicals from feeding to nesting areas, generally across ecosystem boundaries (e.g., marine to terrestrial; Gillham 1961; Heatwole 1971; Polis et al. 1997), but this topic has experienced an increase in attention since the early 2000s (e.g., Blais et al. 2007; Bauer and Hoye 2014; Grant et al. 2022). Movement of nutrients and subsequent effects of, or reliance by, vegetation formed the initial focus (e.g., Polis et al. 1997; Ellis et al. 2006), including the findings that, in some situations, biological communities of receiving environmental subsidies can be markedly altered (e.g., Michelutti et al. 2009a, 2010). However, the revelation that seabirds were also moving contaminants from sea to land (Blais et al. 2005; Evenset et al. 2007), and that those contaminants were entering freshwater and terrestrial ecosystems and reaching potentially toxic levels in biota (Brimble et al. 2009a, b; Choy et al. 2010a, b), may have driven additional research. Recently, Bourdages et al. (2021) found that seabirds also vectored microplastic pollution to their colony. Hammer et al. (2016) demonstrated that, in addition to microplastic deposition, seabirds can also deposit mesoplastics in and around their terrestrial colonies, which can be incorporated into regurgitated pellets. In any case, using information from recent studies from around the world has allowed researchers to estimate contributions of avian biotransport to global nutrient cycles (Otero et al. 2018), and trace element transport (De La Peña-Lastra et al. 2021). It would seem that birds move all forms of human pollutants.

The principles of avian biotransport are not complicated, and the case of colonial seabirds can be used to illustrate this concept (Fig. 12.1). Breeding can be highly synchronous (i.e., all the birds breed in a tight temporal window as is the case for many species of Arctic seabirds), or asynchronous (i.e., the breeding occurs in a wide temporal window that may be year-round, such as occurs in some penguin colonies in South Africa). Regardless of the temporal patterns, the proclivity of many seabird species to nest close together makes their colonies central locations of concentrated activity. They leave their colony on foraging trips to feed in the ocean (or various land-based sites in the case of gulls) and, depending on the species, can cover immense areas around the colony (e.g., Pollet et al. 2014), consuming various types of prey. Once prey is captured, seabirds return to their colony (along with many other sympatrically-nesting conspecifics), and at the colony the food they have captured may get dropped or digested, after which it is released in the environment around the colony as feces (the main form of chemical subsidy). Feces are elevated in nitrogen and phosphorus and may also have elevated concentrations of contaminants such as mercury (Signa et al. 2013; Geizer et al. 2021), depending on the trophic position and diet of the seabird species (e.g., Michelutti et al. 2010). Thus, through deposition of feces at the breeding site (as well as dropped food, regurgitations, and mortality of eggs, chicks, and adults), nutrients and contaminants are deposited in higher concentrations at or near a colony (Blais et al. 2007) compared to nearby, similar sites without nesting birds (Michelutti et al. 2009a; Mallory et al. 2015). Note that in some extreme cases, such as seabird nesting areas of western South America, archives of guano from countless seabirds became so thick that they could be mined, and they are often used as fertilizer by local peoples (Santana-Sagredo et al. 2021).

Nutrient deposition may enhance local nutrient supplies, invigorating vegetation growth (Ellis et al. 2006; Clyde et al. 2021); this effect can be striking in areas with low nutrient availability such as Arctic coastal areas (e.g., Blais et al. 2005; González-Bergonzoni et al. 2017). Nonetheless, effects of fecal nutrient subsidies are also evident at lower latitudes (Adame et al. 2015; Guerra et al. 2022), although behaviors of birds at colonies can make relationships with vegetation more complex (i.e., sometimes enhancing growth, sometimes inhibiting growth, and sometimes both depending on species; Sobey and Kentworthy 1979; Young et al. 2010). At the same time, birds deliver markedly elevated amounts of contaminants (e.g., organochlorines, bioaccumulated trace elements) compared to sites receiving only long-range transport and atmospheric deposition (Michelutti et al. 2009b). These contaminants can enter local soils (Mallory et al. 2015, 2023; Santamans et al. 2017) and aquatic ecosystems (Evenset et al. 2007), and may enter and accumulate in local terrestrial food webs (Choy et al. 2010a, b; Shoji et al. 2019), completing a circuit of emission from land sources, entry into the sea, delivery back to land, and re-entry into terrestrial areas.

We note that the vast majority of biotransport studies have documented the effects of avian vectors on recipient lakes and ponds, island soils and vegetation, or coastal vegetation (Blais et al. 2007; Sánchez-Piñero and Polis 2000; Grant et al. 2021, 2022) but there are few studies that have shown these effects in wetlands. These latter habitats may be highly productive and may store large quantities of organic material (Mitsch et al. 2009). Moreover, they are often heterogeneous in vegetation and morphometry creating sometimes challenging assessments of flow regimes

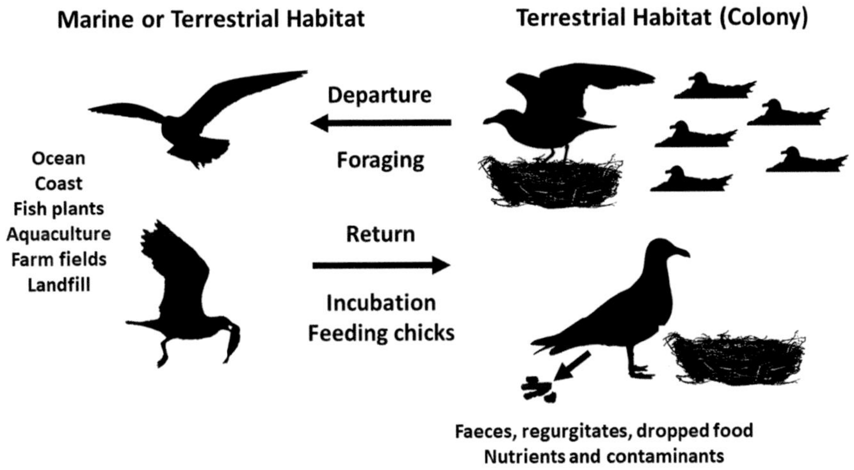

Fig. 12.1 Schematic representation of how avian biotransport by seabirds operates. Breeding birds (top right) depart the colony to forage, and may vary in what they ingest, depending on foraging location. After acquiring food, birds return to a colony, with their personal reserves restored, and/or to deliver food to their chicks (bottom right). At the colony, these birds digest their food and then defecate (in the air or on the ground), but they may also regurgitate (and lose) food or drop prey. In addition, eggs, chicks, and adults die on the colony. Collectively, this abundance of high trophic level biological material transfers chemicals (nutrients, contaminants) to the terrestrial and aquatic environment around colonies

(often through groundwater movement), and in some cases, these sites have high water tables with little movement (limited horizontal flow), depending on the type of wetland and its position in the local hydrogeological regime (Winter 1999; Mitsch et al. 2009). Collectively, these characteristics make it more difficult to locate in-situ environmental archives that might capture pollutant or nutrient signatures, in comparison to profundal zones of lakes or nesting soils on small islands (Ellis et al. 2006; Blais et al. 2007; Hargan et al. 2017).

12.2.1 Herring Gulls

A seabird species that has received considerable attention for its breeding biology, population trends, relationship with people, and more recently biotransport of nutrient and chemical subsidies is the American herring gull (*Larus smithsoniansus* Coues). It is a habitat generalist; colonies may be located near the ocean but are also found around inland lakes across much of North America, from the Caribbean to the subarctic (Weseloh et al. 2020). Breeding colonies may consist of a few to thousands of breeding pairs. Moreover, herring gulls occupy high trophic positions in food webs (Weseloh et al. 2020), but are clearly dietary generalists that eat fish, molluscs, crustaceans, birds, small mammals, insects, worms, as well as scavenging carrion and multiple types of human industrial or residential waste products (the latter two notably in urban areas; Ewins et al. 1994; Weseloh et al. 2020). Herring gulls notoriously forage in landfills (e.g., Jardine et al. 2021; Lato et al. 2021), follow tractors in fields to catch invertebrates (e.g., Schwemmer et al. 2008), feed at fish plants and aquaculture sites (e.g., Gutowsky et al. 2021, 2023), swarm in flocks for bait or offal from fishing vessels (e.g., Skov and Durinck 2001), disturb and prey on nearby seabird colonies (e.g., Whittam and Leonard 1999; Stenhouse et al. 2000), and may even nest in parks and on rooftops in cities (Morris et al. 2003; Perlut et al. 2016). In the northeastern USA there are programs to control herring gull numbers (e.g., Donehower et al. 2007), and there have been calls for similar management options in some Canadian locations (e.g., Scopel and Diamond 2017) to conserve other seabird colonies. In maritime regions of eastern Canada and the northeastern USA, some airports have permits to operate lethal control programs for gulls, including herring gulls (Burger 1983, 1985; Seif et al. 2018).

12.2.2 Habitat Use by the Brier Island Gull Colony

The herring gull colony at Brier Island is the largest in Nova Scotia and perhaps the Canadian Maritimes, recently estimated at approximately 3000 nesting pairs (Geizer et al. 2021). Following drainage of the main bog (Big Meadow Bog) on the island in the 1950s, apparently to drain and dry soils for agriculture (LaRue 2016; Hill et al. 2019), gulls established a colony with nests on raised, drained areas within the vegetated part of the bog away from the treed areas along the bog margins; this main density of nest sites shifts somewhat among years (Mills 2022, B. Toms, Mersey Tobeatic Research Institute, pers. commun.). The location on Brier Island is very well-suited for gull nesting for three reasons: (a) it is separated from mainland Nova Scotia by strong ocean currents, which probably precludes arrivals of mammalian predators, although some coyotes (*Canis latrans* Say) have been observed on the island; (b) there is relatively little anthropogenic activity near the bog, so there is little disturbance to nesting birds; and (c) the island is surrounded by suitable foraging sites for herring gulls (Fig. 12.2), including fishing vessels in the Bay of Fundy, American mink (*Neovison vison* Schreber) farms < 40 km away, multiple fish plants < 20 km away, and aquaculture sites up and down the nearby coastline (Gutowsky et al. 2021). In the decades that researchers have been working at the site, no mass predation events have been observed at the colony, suggesting that if there are mammalian predators, their impact is minor. Thus, the presence of relatively safe nesting habitat and multiple, abundant food resources within their foraging range have resulted in a large gull colony at Brier Island.

To determine where gulls breeding on Brier Island went when they left the colony, and therefore dietary options and potential exposure to contaminants, Gutowsky et al. (2021) captured breeding herring gulls in 2014 and 2015 and attached GPS devices to the lower back of individuals (Fig. 12.3). These tracking units were solar-recharged (Fig. 12.4) and downloaded their stored data whenever they came within range of a

Fig. 12.2 Movements of herring gulls (*Larus smithsonianus*) breeding at Brier Island and tracked by GPS devices. Birds stopped to feed at mink farms, fish-processing plants, and aquaculture sites

receiver that was deployed centrally in the colony (Fig. 12.5). Scientists were able to follow movements of gulls with high accuracy (i.e., to within a few meters) during the breeding season at 15 min intervals. By overlaying locations from the tracking data on land cover maps, Gutowsky et al. (2021) linked gull movements to precise industrial sites or coastal habitats (Fig. 12.2; see also Shlepr et al. 2021). These data showed that gulls breeding on Brier Island spent much of their time at mink farms and fish plants during the breeding season (Gutowsky et al. 2021), although this changed at various temporal scales as activity levels at these industries changed, as did the energetic needs of the gulls through their annual cycle (Gutowsky et al. 2023). With the recent, large decline in the Nova Scotia mink industry (Campbell et al. 2022), these adaptable birds have likely reduced feeding at mink farms, and now rely proportionately more on other available resources, as shown by gulls at nearby colonies shifting diets through time (e.g., Ronconi et al. 2014). Additional tracking of individuals from nearby herring gull colonies found little overlap in foraging areas of birds nesting at different colonies (Gutowsky et al. 2023), although there was high diet overlap.

As noted, gulls nesting at Brier Island flew to mink farms in southwest Nova Scotia to scavenge, but it was unclear whether they were eating carcasses from the farms (which, according to regulations, were required to have been disposed of where avian scavengers could not access them; NSEFP 2023), or whether they were exploiting waste feed and refuse, as occurs at other agricultural sites. Geizer et al. (2021) deployed plastic platforms on posts in the Brier Island colony where gulls roosted, regurgitated and defecated; the birds left pellets or feces containing mink hair (Fig. 12.6). These findings confirmed not only that some of the main foraging

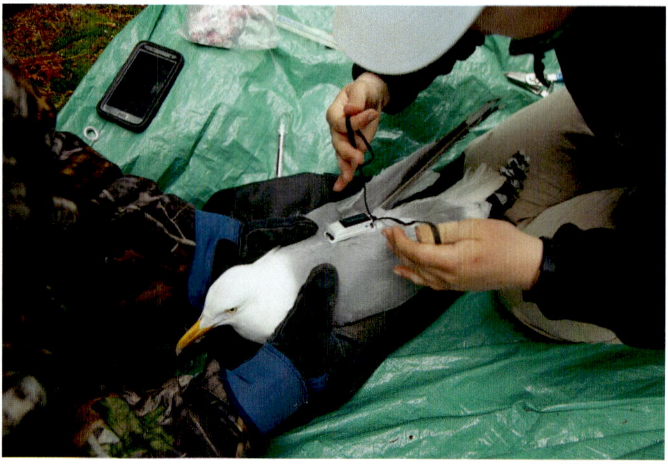

Fig. 12.3 A herring gull equipped with a solar-rechargeable GPS device. Photo by Julia Baak

Fig. 12.4 Herring gull in flight showing how GPS devices could recharge and transmit during activities. Photo by Julia Baak

locations for gulls nesting at Brier Island were mink farms, and that those gulls were scavenging mink carcasses.

By foraging on mink carcasses, mink feed, fish plant wastes (i.e., pieces of fish), and natural prey along coastlines, breeding herring gulls were feeding at high trophic levels (Gregory et al. 2022). Consequently, their feces not only had predictably high nutrients but also high mercury concentrations, acquired by ingesting mercury from

Fig. 12.5 A UHF receiver deployed near the colony. When gulls equipped with devices came within range, stored GPS locations with date and time would download to this receiving base station. Photo by Julia Baak

predominantly marine-derived prey from the Gulf of Maine/Bay of Fundy which have documented elevated mercury (English et al. 2015; Bradford et al. 2023).

Outside the breeding season, the same tracked gulls migrated south to winter from Cape Cod to Delaware Bay, except for one individual that remained resident in Nova Scotia spending most of its winter at the northern end of St. Mary's Bay (Anderson et al. 2019). The tracked gulls typically selected marine habitats close to shore, with a few seemingly specializing on urban habitats (Anderson et al. 2019). However, interactions with industry or diet during the non-breeding period are not yet known. Regardless, the potential for significant biotransport between wintering and breeding sites seems low, given the volume being transported back after migration would be very small compared to what is being deposited daily during breeding.

12.2.3 *Ornithogenic Nutrient and Contaminant Deposition in Big Meadow Bog*

What does all this deposited fecal matter do to the Big Meadow Bog ecosystem on Brier Island? There was an expectation of higher nutrient inputs (particularly nitrogen and phosphorus [notably orthophosphate, hereafter "reactive phosphorus"]) and higher concentrations of bioaccumulated contaminants such as mercury, given

Fig. 12.6 Plastic platforms **a** were erected in Big Meadow Bog, and gulls quickly used these to roost, as well as to defecate and regurgitate **b** Among the regurgitated or defecated materials, Geizer et al. (2021) could find many mink hairs, which confirmed that gulls were feeding on mink carcasses at nearby mink farms and transporting and releasing these materials back at their colony

the extensive evidence of this occurring near other bird colonies (e.g., Ellis et al. 2006; Otero et al. 2018). Geizer et al. (2021) quantified nutrient and chemical contaminant deposition on Brier Island based on concentrations of these in feces, the colony size, and the estimated amount of feces deposited per day. This formula, along with a count of ~ 6000 gulls (3000 pairs) nesting at the island was used to estimate the annual contributions to Big Meadow Bog (Table 12.1). For plastic debris, there is little published information on amounts of plastic in gull feces; most measures of plastic in gulls comes from boluses (Fig. 12.6; see also Stewart et al. 2020) or dissections of gizzards (e.g., Seif et al. 2018). Thus, we provide here a very coarse estimate of possible plastic debris deposition. We considered two scenarios: (a) that every gull at the colony regurgitated one bolus per breeding season (likely a gross underestimate), and (b) that every gull at the colony regurgitated ten boluses per season. Caldwell et al. (2020) showed that 68% of herring gull boluses in the Gulf of Maine contained plastic particles, and Almeida et al. (2023) found two pieces of plastic per yellow-headed gull (*Larus michahellis* Naumann) bolus, in 84% of the boluses collected. Note that these studies were not considering more ubiquitous microplastics < 1 mm in size (e.g., Provencher et al. 2018). We therefore used a value of one piece of plastic per bolus in our estimate. Second, herring gulls tend to retain less plastic in their gizzards (50–77% frequency of occurrence; Seif et al. 2018; Jardine et al. 2021) than other seabirds (e.g., northern fulmars *Fulmarus glacialis* L.; ~ 80–90%

frequency of occurrence; Provencher et al. 2017), in part because gulls can regurgitate these materials, so we are unsure how much plastic ingested by gulls is moved into their intestine. However, we have measures of plastic pieces in Arctic-feeding fulmar fecal precursors (mean of ~ 1 per single sample) and thick-billed murre (*Uria lomvia* L.) precursors (zero per sample; Bourdages et al. 2021; estimated at 1 g each). Therefore, as a first estimate of plastic deposition in feces, we assumed that 50% (a conservative estimate of frequency of occurrence) of the gulls each day were defecating out plastic debris, and we assumed that each fecal event produced 1 g dry mass of feces containing one piece of plastic. Using these values, we estimate that gulls deliver 1253 kg of reactive phosphorus to the bog each year (roughly akin to 7000 bags of 10:10:10 [nitrogen:phosphorus:potassium] fertilizer being added), along with 4.1 g of total mercury, 2.0 g of methylmercury, and just under 18 million pieces of plastic (almost all microplastics).

At Big Meadow Bog, Kickbush et al. (2018) measured considerable spatial variation in reactive phosphorus concentrations in water across the bog, but noted higher concentrations near the gull breeding areas. However, these high reactive phosphorus areas did not match areas in the bog that had greater total mercury concentrations in water, seemingly counterintuitive if gulls were excreting excess phosphorus and mercury in the same locations. Indeed, Kickbush et al. (2018) speculated that the patterns of high levels of total mercury in water may be more related to bog topography and drainage, or even runoff from the nearby runway (see also Ning et al. 2021). However, those regions in the bog that had high concentrations of phosphorus in water also had a strong spatial correspondence to regions of the bog that had higher concentrations of methylmercury in the bog water. Kickbush et al. (2018)

Table 12.1 Calculated estimates of nutrient and contaminant deposition in Big Meadow Bog, Brier Island, Nova Scotia, by breeding herring gulls

Parameter	Mean concentration in feces	Daily excretion rate (g/d)	Colony size (birds)	Time at colony (d)	Total
Phosphate	35 mg/g	39	6000	153	1253 kg
Total mercury	114 ng/g	39	6000	153	4.1 g
Methylmercury	55 ng/g	39	6000	153	2.0 g
Plastic debris—feces	1 piece/g	39	6000 * 0.5	153	17.9×10^6 pieces
Plastic debris—1 bolus/season	1 piece/bolus		6000		6000 pieces
Plastic debris—10 boluses/season	1 piece/bolus		6000		60,000 pieces

Estimates for THg, MeHg, and P are based on values in Geizer et al. (2021). Estimates for plastic deposition are explained in the text attributed this correlation to stimulation of microbial activity in locations in the bog with reactive phosphorus subsidies, which may have facilitated methylation of Hg.

Although water concentrations of nutrients and key contaminants were elevated compared to many other studies (Kickbush et al. 2018; Fig. 12.7), the anticipated, corresponding pattern in surface soils was not observed. McIntyre et al. (2022) sampled soils in Big Meadow Bog and a nearby Reference Bog (Fig. 12.8), and they predicted that relatively elevated mercury in soils would occur near the gull nesting colony. In fact, mean total mercury in soils was greater in the Reference Bog, and average soil concentrations of total mercury (~ 100–120 ng g^{-1}) were not particularly different than mercury concentrations from soils around the world (20–200 ng g^{-1}; De la Peña Lastra 2021), whether seabird-influenced or not. McIntyre et al. (2022) speculated that this unanticipated result may be attributable to the nature of the water table in Big Meadow Bog. Unlike other nesting colonies in southern Nova Scotia, or in Iceland (Mallory et al. 2023), in many parts of the bog (but not everywhere) the water is near the soil surface and soils may remain waterlogged most of the year. This may mean that mercury is more likely to remain attached to organic matter and flow with water through the bog (Ning et al. 2021), rather than become part of soil processes. Similarly, Mallory et al. (2023) found no pattern in mercury concentration with increasing depth in the soils, contrary to their prediction that concentrations would be greater on surface soils near sites of fresh deposits, which they attributed to water movement through soils.

12.3 Conclusion

Although gulls have for decades been transporting to, and depositing in, nutrients at Big Meadow Bog (~ 1200 kg/year), there is little obvious evidence at a bog-wide scale of "excessive" deleterious effects on local habitat: no tree die-offs or feces-encrusted ground as observed around cormorant colonies (e.g., Klimaszyk and Rzymski 2016), although there are some places of markedly altered vegetation species composition in nesting areas (e.g., ruderals, invasives) near some places of long-term, high density gull nesting (e.g., Baumberger et al. 2012; Hill et al. 2019). Despite the abundant nutrients, there is no obvious evidence of the bog being more lush than elsewhere due to nutrient additions (e.g., González-Bergonzoni et al. 2017). Nonetheless, we suspect that the bog is more productive than it would be without the gulls, and work is underway to test this hypothesis. A challenge to evaluating the effects of gulls, however, is that, to date, we presume that ditching and altered water regimes (Hill et al. 2019) have likely had a much greater influence on vegetation than ornithogenic subsidies. Nonetheless, wetlands are often production sites for methylmercury in watersheds, although they can serve as sources or sinks for total and methylmercury to downstream areas, depending on hydrology (St. Louis et al. 1996; Galloway and Branfireun 2004). We also predict that concentrations of ornithogenic subsidies of other contaminants, such as organochlorines that are in the mink and their feed (Gregory et al. 2022), are elevated in the bog. Moreover, recent documentation of

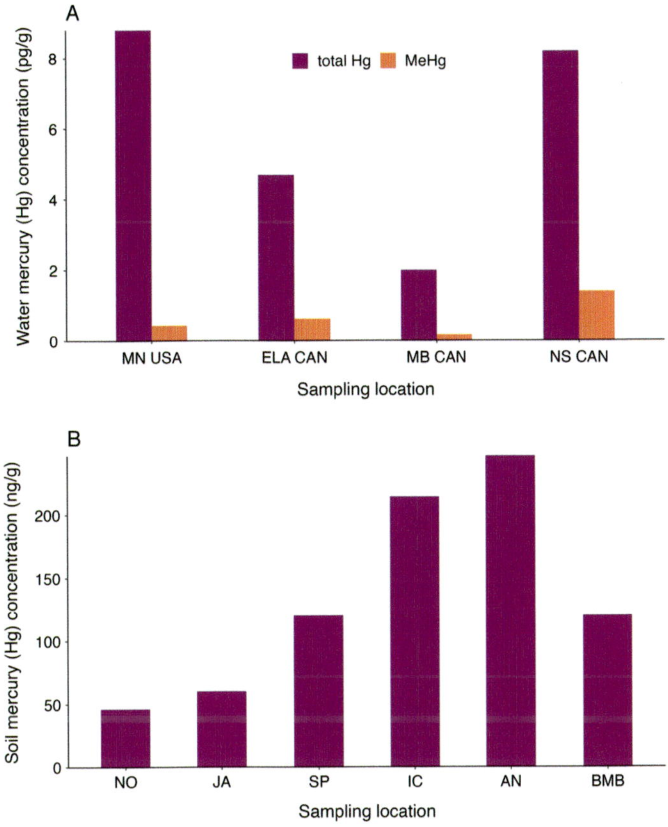

Fig. 12.7 Big Meadow Bog has relatively high methylmercury in water **a** but not obviously high total mercury in soil, **b** compared to some other sites. Locations of sampling for water comparisons were: MN USA (Minnesota, USA); ELA CAN (Experimental Lakes Area, Canada); MB CAN (Manitoba, Canada); and NS CAN (Nova Scotia, Canada) (from Kickbush et al. 2018). Locations of sampling for soils were: NO (Norway); JA (Japan); SP (Spain); IC (Iceland); AN (Antarctica); and BMB (Big Meadow Bog, Nova Scotia, Canada) (from McIntyre et al. 2022)

global transport of plastics and microplastics by seabirds (e.g., Provencher et al. 2017; Hammer et al. 2016), including by gulls (e.g., Jardine et al. 2021), deposition of these particulate contaminants at their colonies (e.g., Bourdages et al. 2021), and potential of these plastics to alter health of biota that consume them (Charlton-Howard et al. 2023; Fackelmann et al. 2023), should be a priority in future research. Collectively, the extent to which all these contaminants have become elevated in the terrestrial and aquatic biota of the bog (e.g., Malcata Martins et al. 2022) is not yet known, but that may in fact be the more prominent, deleterious effect of biotransport by herring gulls at Brier Island.

Fig. 12.8 Extracting a soil core with a metal corer. Cores were separated by a rinsed spoon and placed by centimeter depth into individual, sterile, marked plastic bags, and transported back to Acadia University for analysis

Acknowledgements We thank the many students and research assistants who contributed to the science base on which this paper was built, and in particular Haley Geizer, Jocelyn Kickbush, Carolyn Mallory, Jessie McIntyre, Rob Ronconi, Karel Allard, and Kate Shlepr. Financial support for this work came from Environment and Climate Change Canada, Mitacs, and the Natural Sciences and Engineering Research Council. Thanks to the referees for helpful comments on this manuscript.

References

Adame MF, Fry B, Gamboa JN, Herrera-Silveira JA (2015) Nutrient subsidies delivered by seabirds to mangrove islands. Mar Ecol Prog Ser 525:15–24. https://doi.org/10.3354/meps11197

Almeida FN, Leray C, Boutry J, Ter Halle A, Vittecoq M, Provencher JF, McCoy KD (2023) Changes in plastic ingestion by yellow-legged gulls (*Larus michahellis*) over the breeding season. Mar Pollut Bull 187:114483. https://doi.org/10.1016/j.marpolbul.2022.114483

Anderson CM, Gilchrist HG, Ronconi RA, Shlepr KR, Clark DE, Weseloh DC, Robertson GJ, Mallory ML (2019) Winter home range and habitat selection differs among breeding populations of herring gulls in eastern North America. Mov Ecol 7:8. https://doi.org/10.1186/s40462-019-0152-x

Bauer S, Hoye B (2014) Migratory animals couple biodiversity and ecosystem functioning worldwide. Science 344:1242552. https://doi.org/10.1126/science.1242552

Baumberger T, Affre L, Torre F, Vidal E, Dumas PJ, Tatoni T (2012) Plant community changes as ecological indicator of seabird colonies' impacts on Mediterranean island. Ecol Indicat 15:76–84. https://doi.org/10.1016/j.ecolind.2011.09.009

Belant JL (1997) Gulls in urban environments: landscape-level management to reduce conflict. Landscape Urban Plan 38:245–258. https://doi.org/10.1016/s0169-2046(97)00037-6

Blais JM, Kimpe LE, McMahon D, Keatley BE, Mallory ML, Douglas MS, Smol JP (2005) Arctic seabirds transport marine-derived contaminants. Science 309:445–445. https://doi.org/10.1126/science.1112658

Blais JM, Macdonald RW, Mackay D, Webster E, Harvey C, Smol JP (2007) Biologically mediated transport of contaminants to aquatic systems. Environ Sci Technol 41:1075–1084. https://doi.org/10.1021/es061314a

Blight LK, Bertram DF, Kroc E (2019) Evaluating UAV-based techniques to census an urban-nesting gull population on Canada's Pacific coast. J Unman Vehicle Syst 7:312–324. https://doi.org/10.1139/juvs-2019-0005

Boklund A, Hammer AS, Quaade ML, Rasmussen TB, Lohse L, Strandbygaard B, Jørgensen CS, Olesen AS, Hjerpe FB, Petersen HH, Jensen TK, Mortensen S, Calvo-Artavia FF, Lefevre SK, Nielsen SS, Halasa T, Belsham GJ, Bøtner A (2021) SARS-CoV-2 in Danish mink farms: course of the epidemic and a descriptive analysis of the outbreaks in 2020. Animals 11:164. https://doi.org/10.3390/ani11010164

Bourdages MP, Provencher JF, Baak JE, Mallory ML, Vermaire JC (2021) Breeding seabirds as vectors of microplastics from sea to land: evidence from colonies in Arctic Canada. Sci Total Environ 764:142808. https://doi.org/10.1016/j.scitotenv.2020.142808

Bradford MA, Mallory ML, O'Driscoll NJ (2023) Mercury bioaccumulation and speciation in coastal invertebrates: Implications for trophic magnification in a marine food web. Mar Pollut Bull 188:114647. https://doi.org/10.1016/j.marpolbul.2023.114647

Brimble SK, Blais JM, Kimpe LE, Mallory ML, Keatley BE, Douglas MS, Smol JP (2009a) Bioenrichment of trace elements in a series of ponds near a northern fulmar (*Fulmarus glacialis*) colony at Cape Vera, Devon Island. Can J Fish Aquat Sci 66:949–958. https://doi.org/10.1139/f09-053

Brimble SK, Foster KL, Mallory ML, Macdonald RW, Smol JP, Blais JM (2009b) High Arctic ponds receiving biotransported nutrients from a nearby seabird colony are also subject to potentially toxic loadings of arsenic, cadmium, and zinc. Environ Toxicol Chem 28:2426–2433. https://doi.org/10.1897/09-235.1

Burger J (1983) Bird control at airports. Environ Conserv 10:115–124. https://doi.org/10.1017/s0376892900012200

Burger J (1985) Factors affecting bird strikes on aircraft at a coastal airport. Biol Conserv 33:1–28. https://doi.org/10.1016/0006-3207(85)90002-3

Caliendo V, Lewis NS, Pohlmann A, Baillie SR, Banyard AC, Beer M, Brown IH, Fouchier RAM, Hansen RDE, Lameris TK, Lang AS, Laurendeau S, Lung O, Robertson G, van der Jeugd H, Alkie TN, Thorup K, van Toor ML, Waldenström J, Yason C, Kuiken T, Berhane Y (2022) Transatlantic spread of highly pathogenic avian influenza H5N1 by wild birds from Europe to North America in 2021. Sci Rep 12:11729. https://doi.org/10.1101/2022.01.13.476155

Caldwell A, Seavey J, Craig E (2020) Foraging strategy impacts plastic ingestion risk in seabirds. Limnol Oceanogr Lett 5:163–168. https://doi.org/10.1002/lol2.10126

Campbell J, Libera N, Smol JP, Kurek J (2022) Historical impacts of mink fur farming on chironomid assemblages from shallow lakes in Nova Scotia, Canada. Lake Reserv Manage 38:80–94. https://doi.org/10.1080/10402381.2021.2018631

Charlton-Howard HS, Bond AL, Rivers-Auty J, Lavers JL (2023) 'Plasticosis': Characterising macro-and microplastic-associated fibrosis in seabird tissues. J Hazard Mater 450:131090. https://doi.org/10.1016/j.jhazmat.2023.131090

Choy ES, Gauthier M, Mallory ML, Smol JP, Douglas MS, Lean D, Blais JM (2010a) An isotopic investigation of mercury accumulation in terrestrial food webs adjacent to an Arctic seabird colony. Sci Total Environ 408:1858–1867. https://doi.org/10.1016/j.scitotenv.2010.01.014

Choy ES, Kimpe LE, Mallory ML, Smol JP, Blais JM (2010b) Contamination of an arctic terrestrial food web with marine-derived persistent organic pollutants transported by breeding seabirds. Environ Pollut 158:3431–3438. https://doi.org/10.1016/j.envpol.2010.07.014
Clyde N, Hargan KE, Forbes MR, Iverson SA, Blais JM, Smol JP, Bump JK, Gilchrist HG (2021) Seaduck engineers in the arctic archipelago: nesting eiders deliver marine nutrients and transform the chemistry of island soils, plants, and ponds. Oecologia 195:1041–1052. https://doi.org/10.1007/s00442-021-04889-9
Coulson J (2002) Colonial breeding in seabirds. In: Biology of marine birds. CRC Press, New York, pp 87–113. https://doi.org/10.1201/9781420036305-7
Croxall JP, Butchart SH, Lascelles B, Stattersfield AJ, Sullivan B, Symes A, Taylor P (2012) Seabird conservation status, threats and priority actions: a global assessment. Bird Conserv Internat 22:1–34. https://doi.org/10.1017/s0959270912000020
De La Peña-Lastra S (2021) Seabird droppings: effects on a global and local level. Sci Total Environ 754:142148. https://doi.org/10.1016/j.scitotenv.2020.142148
De La Peña-Lastra S, Pérez-Alberti A, Otero XL (2021) Seabird colonies as the main source of nutrients for the coastal ecosystems in the Atlantic Islands of Galicia National Park (NW Spain). Chemosphere 275:130077. https://doi.org/10.1016/j.chemosphere.2021.130077
Dias MP, Martin R, Pearmain EJ, Burfield IJ, Small C, Phillips RA, Yates O, Lascelles B, Borboroglu PG, Croxall JP (2019) Threats to seabirds: a global assessment. Biol Conserv 237:525–537. https://doi.org/10.1016/j.biocon.2019.06.033
Donehower CE, Bird DM, Hall CS, Kress SW (2007) Effects of gull predation and predator control on tern nesting success at Eastern Egg Rock, Maine. Waterbirds 30:29–39. https://doi.org/10.1675/1524-4695(2007)030[0029:eogpap]2.0.co;2
Duda MP, Glew JR, Michelutti N, Robertson GJ, Montevecchi WA, Kissinger JA, Eickmeyer DC, Blais JM, Smol JP (2020) Long-term changes in terrestrial vegetation linked to shifts in a colonial seabird population. Ecosystems 23:1643–1656. https://doi.org/10.1007/s10021-020-00494-8
Ellis JC, Fariña JM, Witman JD (2006) Nutrient transfer from sea to land: the case of gulls and cormorants in the Gulf of Maine. J Anim Ecol 75:565–574. https://doi.org/10.1111/j.1365-2656.2006.01077.x
English MD, Robertson GJ, Mallory ML (2015) Trace element and stable isotope analysis of fourteen species of marine invertebrates from the Bay of Fundy, Canada. Mar Pollut Bull 101:466–472. https://doi.org/10.1016/j.marpolbul.2015.09.046
Evenset A, Carroll J, Christensen GN, Kallenborn R, Gregor D, Gabrielsen GW (2007) Seabird guano is an efficient conveyer of persistent organic pollutants (POPs) to Arctic lake ecosystems. Environ Sci Technol 41:1173–1179. https://doi.org/10.1021/es0621142
Ewins PJ, Weseloh DV, Groom JH, Dobos RZ, Mineau P (1994) The diet of herring gulls (*Larus argentatus*) during winter and early spring on the lower Great Lakes. Hydrobiol 279:39–55. https://doi.org/10.1007/978-94-011-1128-7_4
Fackelmann G, Pham CK, Rodríguez Y, Mallory ML, Provencher JF, Baak JE, Sommer S (2023) Current levels of microplastic pollution impact wild seabird gut microbiomes. Nat Ecol Evol 7:698–706. https://doi.org/10.1038/s41559-023-02013-z
Galloway ME, Branfireun BA (2004) Mercury dynamics of a temperate forested wetland. Sci Total Environ 325:239–254. https://doi.org/10.1016/j.scitotenv.2003.11.010
Gaston AJ (2004) Seabirds: a natural history. T. & A.D. Poyser. 222 p. https://doi.org/10.5040/9781472596840
Gaston AJ, Hipfner JM (2020). Thick-billed Murre (*Uria lomvia*), version 1.0. In: Billerman SM (ed) Birds of the world. Cornell Lab of Ornithology, Ithaca, NY USA. https://doi.org/10.2173/bow.thbmur.01
Geizer HD, Klapstein SJ, Mallory ML, O'Driscoll NJ (2021) Total mercury, methylmercury, phosphate, and sulfate inputs to a bog ecosystem from herring gull (*Larus smithsoniansus*) guano. Ecotoxicol Environ Safe 226:112845. https://doi.org/10.1016/j.ecoenv.2021.112845
Gillham ME (1961) Alteration of the breeding habitat by seabirds and seals in western Australia. J Ecol 49:289–300. https://doi.org/10.2307/2257263

González-Bergonzoni I, Johansen KL, Mosbech A, Landkildehus F, Jeppesen E, Davidson TA (2017) Small birds, big effects: the little auk (*Alle alle*) transforms high arctic ecosystems. Proc Roy Soc B 284:20162572. https://doi.org/10.1098/rspb.2016.2572

Grant ML, Lavers JL, Hutton I, Bond AL (2021) Seabird breeding islands as sinks for marine plastic debris. Environ Pollut 276:116734. https://doi.org/10.1016/j.envpol.2021.116734

Grant ML, Bond AL, Lavers JL (2022) The influence of seabirds on their breeding, roosting and nesting grounds: a systematic review and meta-analysis. J Anim Ecol 91:1266–1289. https://doi.org/10.1111/1365-2656.13699

Gregory BRB, Kissinger JA, Clarkson C, Kimpe LE, Eickmeyer DC, Kurek J, Smol JP, Blais JM (2022) Are fur farms a potential source of persistent organic pollutants or mercury to nearby freshwater ecosystems? Sci Total Environ 833:155100. https://doi.org/10.1016/j.scitotenv.2022.155100

Guerra AS, Bui A, Klope M, Orr DA, Shaffer SA, Young HS (2022) Leaving more than footprints: anthropogenic nutrient subsidies to a protected area. Ecosphere 13:e4371. https://doi.org/10.1002/ecs2.4371

Gutowsky SE, Studholme KR, Ronconi RA, Allard KA, Shlepr K, Diamond AW, McIntyre J, Craik SR, Mallory ML (2021) The influence of multiple industries on the behaviour of breeding gulls from four colonies across the eastern Gulf of Maine, Canada. Wildl Biol 2021:1–13. https://doi.org/10.2981/wlb.00804

Gutowsky SE, Baak JE, Craik SR, Mallory ML, Knutson N, d'Entremont AA, Allard KA (2023) Seasonal and circadian patterns of herring gull (*Larus smithsoniansus*) movements reveal temporal shifts in industry and coastal island interaction. Ecol Solut Evid 4:e12274. https://doi.org/10.1002/2688-8319.12274

Hammer S, Nager RG, Johnson PCD, Furness RW, Provencher JF (2016) Plastic debris in great skua (*Stercorarius skua*) pellets corresponds to seabird prey species. Mar Pollut Bull 103:206–210. https://doi.org/10.1016/j.marpolbul.2015.12.018

Hargan KE, Michelutti N, Coleman K, Grooms C, Blais JM, Kimpe LE, Gilchrist G, Mallory M, Smol JP (2017) Cliff-nesting seabirds influence production and sediment chemistry of lakes situated above their colony. Sci Total Environ 576:85–98. https://doi.org/10.1016/j.scitotenv.2016.10.024

Heatwole H (1971) Marine dependent terrestrial biotic communities on some cays in the Coral Sea. Ecology 52:363–366. https://doi.org/10.2307/1934598

Hill NM, Chinner K, Toms B, Garbary DJ (2019) Recovery succession of native plant communities at Big Meadow Bog after disturbance and eutrophication from herring gulls. Wetlands 39:533–548. https://doi.org/10.1007/s13157-018-1099-1

Jardine AM, Provencher JF, Pratte I, Holland ER, Baak JE, Robertson GJ, Mallory ML (2021) Annual plastic ingestion and isotopic niche patterns of two sympatric gull species at Newfoundland. Canada. Mar Pollut Bull 173:112991. https://doi.org/10.1016/j.marpolbul.2021.112991

Kharitonov SP, Siegel-Causey D (1988) Colony formation in seabirds. Curr Ornithol 5:223–272. https://doi.org/10.1007/978-1-4615-6787-5_5

Kickbush JC, Mallory ML, Murimboh JD, Rand J, Klapstein SJ, Loder AL, Hill NM, O'Driscoll NJ (2018) The influence of avian biovectors on mercury speciation in a bog ecosystem. Sci Total Environ 637:264–273. https://doi.org/10.1016/j.scitotenv.2018.04.224

Klimaszyk P, Rzymski P (2016) The complexity of ecological impacts induced by great cormorants. Hydrobiol 771:13–30. https://doi.org/10.1007/s10750-015-2618-1

LaRue D (2016) Natural regeneration of the globally rare *Geum peckii* (Rosaceae) on Brier Island, Nova Scotia. Rhodora 118:310–314. https://doi.org/10.3119/15-39

Lato KA, Madigan DJ, Veit RR, Thorne LH (2021) Closely related gull species show contrasting foraging strategies in an urban environment. Sci Rep 11:23619. https://doi.org/10.1038/s41598-021-02821-y

Malcata Martins B, O'Driscoll NJ, Mallory ML, Canário J (2021) A review of freshwater invertebrates as biomonitors of methylmercury: the importance of more complete physical and chemical reporting. Bull Environ Contam Toxicol 107:801–808. https://doi.org/10.1007/s00128-021-03274-9

Mallory ML, Mahon L, Tomlik MD, White C, Milton GR, Spooner I (2015) Colonial marine birds influence island soil chemistry through biotransport of trace elements. Water Air Soil Pollut 226:1–8. https://doi.org/10.1007/s11270-015-2314-9

Mallory ML, Petersen A, Thorstensen S, Spooner I, O'Driscoll NJ, Baak JE, McIntyre JA (2023) Mercury in soils of seabird nesting islands in West Iceland. Arctic 76:48–59. https://doi.org/10.14430/arctic76992

McIntyre JA, O'Driscoll NJ, Spooner I, Robertson GJ, Smol JP, Mallory ML (2022) Scavenging gulls are biovectors of mercury from industrial wastes in Nova Scotia. Canada. Chemosphere 304:135279. https://doi.org/10.1016/j.chemosphere.2022.135279

Michelutti N, Keatley BE, Brimble S, Blais JM, Liu H, Douglas MS, Mallory ML, Macdonald RW, Smol JP (2009a) Seabird-driven shifts in Arctic pond ecosystems. Proc Roy Soc B 276:591–596. https://doi.org/10.1098/rspb.2008.1103

Michelutti N, Liu H, Smol JP, Kimpe LE, Keatley BE, Mallory M, Macdonald RW, Douglas MS, Blais JM (2009b) Accelerated delivery of polychlorinated biphenyls (PCBs) in recent sediments near a large seabird colony in Arctic Canada. Environ Pollut 157:2769–2775. https://doi.org/10.1016/j.envpol.2009.04.025

Michelutti N, Blais JM, Mallory ML, Brash J, Thienpont J, Kimpe LE, Douglas MS, Smol JP (2010) Trophic position influences the efficacy of seabirds as metal biovectors. Proc Natl Acad Sci 107:10543–10548. https://doi.org/10.1073/pnas.1001333107

Mills EL (2022) The birds of Brier Island, Nova Scotia. Proc Nova Scotia Inst Sci 52:1–168. https://doi.org/10.15273/pnsis.v52i1.11371

Mitsch WJ, Gosselink JG, Zhang L, Anderson CJ (2009) Wetland ecosystems. Wiley, 304 p

Morris RD, Weseloh DV, Shutt JL (2003) Distribution and abundance of nesting pairs of Herring Gulls (*Larus argentatus*) on the North American Great Lakes, 1976 to 2000. J Great Lakes Res 29:400–426. https://doi.org/10.1016/s0380-1330(03)70447-x

Ning Y, Liu J, Huang X, Wang P, Yang S, Bi X, Yang X, Mo J (2021) Geochemical cycle of mercury associated with wet deposition and inflows in a subalpine wetland. Ecotoxicol Environ Safe 208:111507. https://doi.org/10.1016/j.ecoenv.2020.111507

NSEFP (2023) On-farm livestock mortality management. Nova Scotia Environmental Farm Plan, Nova Scotia Federation of Agriculture, East Mountain, Nova Scotia, Canada. https://www.nsefp.ca/wp-content/uploads/2018/08/On-farm-Livestock-Mortality-Management-NSDA.pdf

Otero XL, De La Peña-Lastra S, Pérez-Alberti A, Ferreira TO, Huerta-Diaz MA (2018) Seabird colonies as important global drivers in the nitrogen and phosphorus cycles. Nat Commun 9:246–247. https://doi.org/10.1038/s41467-017-02446-8

Perlut NG, Bonter DN, Ellis JC, Friar MS (2016) Roof-top nesting in a declining population of Herring Gulls (*Larus argentatus*) in Portland, Maine, USA. Waterbirds 39:68–73. https://doi.org/10.1675/063.039.sp113

Polis GA, Anderson WB, Holt RD (1997) Toward an integration of landscape and food web ecology: the dynamics of spatially subsidized food webs. Ann Rev Ecol System 28:289–316. https://doi.org/10.1146/annurev.ecolsys.28.1.289

Pollet IL, Ronconi RA, Jonsen ID, Leonard ML, Taylor PD, Shutler D (2014) Foraging movements of Leach's storm-petrels *Oceanodroma leucorhoa* during incubation. J Avian Biol 45:305–314

Provencher JF, Bond AL, Avery-Gomm S, Borrelle SB, Rebolledo ELB, Hammer S, Kühn S, Lavers JL, Mallory ML, Trevail A, van Franeker JA (2017) Quantifying ingested debris in marine megafauna: a review and recommendations for standardization. Anal Methods 9:1454–1469. https://doi.org/10.1039/c6ay02419j

Provencher JF, Vermaire JC, Avery-Gomm S, Braune BM, Mallory ML (2018) Garbage in guano? Microplastic debris found in faecal precursors of seabirds known to ingest plastics. Sci Total Environ 644:1477–1484. https://doi.org/10.1016/j.scitotenv.2018.07.101

Ronconi RA, Steenweg RJ, Taylor PD, Mallory ML (2014) Gull diets reveal dietary partitioning, influences of isotopic signatures on body condition, and ecosystem changes at a remote colony. Mar Ecol Prog Ser 514:247–261. https://doi.org/10.3354/meps10980

Sánchez-Piñero F, Polis GA (2000) Bottom-up dynamics of allochthonous input: direct and indirect effects of seabirds on islands. Ecology 81:3117–3132. https://doi.org/10.2307/177405

Santamans AC, Boluda R, Picazo A, Gil C, Ramos-Miras J, Tejedo P, Pertierra LR, Benayas J, Camacho A (2017) Soil features in rookeries of Antarctic penguins reveal sea to land biotransport of chemical pollutants. PLoS ONE 12:e0181901. https://doi.org/10.1371/journal.pone.0181901

Santana-Sagredo F, Schulting RJ, Méndez-Quiros P, Vidal-Elgueta A, Uribe M, Loyola R, Maturana-Fernández A, Díaz FP, Latorre C, McRostie VB, Santoro CM (2021) 'White gold' guano fertilizer drove agricultural intensification in the Atacama Desert from AD 1000. Nat Plants 7:152–158. https://doi.org/10.1038/s41477-020-00835-4

Schwemmer P, Garthe S, Mundry R (2008) Area utilization of gulls in a coastal farmland landscape: habitat mosaic supports niche segregation of opportunistic species. Landscape Ecol 23:355–367. https://doi.org/10.1007/s10980-008-9194-y

Scopel LC, Diamond AW (2017) The case for lethal control of gulls on seabird colonies. J Wildl Manage 81:572–580. https://doi.org/10.1002/jwmg.21233

Seif S, Provencher JF, Avery-Gomm S, Daoust PY, Mallory ML, Smith PA (2018) Plastic and non-plastic debris ingestion in three gull species feeding in an urban landfill environment. Arch Environ Contam Toxicol 74:349–360. https://doi.org/10.1007/s00244-017-0492-8

Shlepr KR, Ronconi RA, Hayden B, Allard KA, Diamond AW (2021) Estimating the relative use of anthropogenic resources by Herring Gull (*Larus argentatus*) in the Bay of Fundy, Canada. Avian Conserv Ecol 16:1. https://doi.org/10.5751/ace-01739-160102

Shoji A, Elliott KH, Aris-Brosou S, Mizukawa H, Nakayama SMM, Ikenaka Y, Ishizuka M, Kuwae T, Watanabe K, Escoruela Gonzalez J, Watanuki Y (2019) Biotransport of metallic trace elements from marine to terrestrial ecosystems by seabirds. Environ Toxicol Chem 38:106–114. https://doi.org/10.1002/etc.4286

Signa G, Tramati CD, Vizzini S (2013) Contamination by trace metals and their trophic transfer to the biota in a Mediterranean coastal system affected by gull guano. Mar Ecol Prog Ser 479:13–24. https://doi.org/10.3354/meps10210

Simpson LT, Canty SWJ, Cissell JR, Steinberg MK, Cherry JA, Feller IC (2021) Bird rookery nutrient over-enrichment as a potential accelerant of mangrove cay decline in Belize. Oecologia 197:771–784. https://doi.org/10.1007/s00442-021-05056-w

Sklepkovych BO, Montevecchi WA (1989) The world's largest known nesting colony of leach's storm-petrels on Baccalieu Island, Newfoundland. Am Bird 43:38–42

Skov H, Durinck J (2001) Seabird attraction to fishing vessels is a local process. Mar Ecol Prog Ser 214:289–298

Sobey DG, Kenworthy JB (1979) The relationship between herring gulls and the vegetation of their breeding colonies. J Ecol 67:469–496. https://doi.org/10.2307/2259108

Spelt A, Williamson C, Shamoun-Baranes J, Shepard E, Rock P, Windsor S (2019) Habitat use of urban-nesting lesser black-backed gulls during the breeding season. Sci Rep 9:10527. https://doi.org/10.1038/s41598-019-46890-6

St. Louis VL, Rudd JW, Kelly CA, Beaty KG, Flett RJ, Roulet NT (1996) Production and loss of methylmercury and loss of total mercury from boreal forest catchments containing different types of wetlands. Environ Sci Technol 30:2719–2729. https://doi.org/10.1021/es950856h

Stenhouse IJ, Robertson GJ, Montevecchi WA (2000) Herring gull *Larus argentatus* predation on Leach's storm-petrels *Oceanodroma leucorhoa* breeding on Great Island, Newfoundland. Atlantic Seabirds 2:35–44

Stewart LG, Lavers JL, Grant ML, Puskic PS, Bond AL (2020) Seasonal ingestion of anthropogenic debris in an urban population of gulls. Mar Pollut Bull 160:111549. https://doi.org/10.1016/j.marpolbul.2020.111549

Thomas GJ (1972) A review of gull damage and management methods at nature reserves. Biol Conserv 4:117–127. https://doi.org/10.1016/0006-3207(72)90012-2

Weseloh DV, Hebert CE, Mallory ML, Poole AF, Ellis JC, Pyle P, Patten MA (2020) Herring Gull (*Larus argentatus*), version 1.0. In: Billerman SM (ed) Birds of the world. Cornell Lab of Ornithology, Ithaca, NY, USA. https://doi.org/10.2173/bow.hergul.01

Whittam RM, Leonard ML (1999) Predation and breeding success in roseate terns (*Sterna dougallii*). Can J Zool 77:851–856. https://doi.org/10.1139/z99-047

Winter TC (1999) Relation of streams, lakes, and wetlands to groundwater flow systems. Hydrogeol J 7:28–45. https://doi.org/10.1007/s100400050178

Wright ME, Solo-Gabriele HM, Elmir S, Fleming LE (2009) Microbial load from animal feces at a recreational beach. Mar Pollut Bull 58:1649–1656. https://doi.org/10.1016/j.marpolbul.2009.07.003

Young HS, McCauley DJ, Dunbar RB, Dirzo R (2010) Plants cause ecosystem nutrient depletion via the interruption of bird-derived spatial subsidies. Proc Natl Acad Sci 107:2072–2077. https://doi.org/10.1073/pnas.0914169107

Chapter 13
A Multi-Year Study of Methylmercury and Nutrient Export in Surface Water from Big Meadow Bog, Brier Island, Nova Scotia

Nelson J. O'Driscoll, Jocelyn C. Kickbush, Holly E. Turner, Sara Klapstein, Mia Doncaster, Kelly Stevens, Rachel Clarke, Molly Bradford, Brianna Bowes, Josie Rogers, N. Kirk Hillier, and Mark L. Mallory

Abstract Mercury (Hg) is a naturally-occurring, toxic contaminant that accumulates in ecosystems, potentially having deleterious effects on the health of wildlife and habitat quality. Significant advances have been made in understanding Hg cycling in freshwater and marine ecosystems, but the processes controlling Hg accumulation in bogs and the confounding effects of both nutrient subsidies and increased flooding are not well known. Previous research by our group has shown that feces from avian biovectors are likely increasing certain trace metals and nutrients in shallow groundwater within the northern portion of recently restored Big Meadow Bog (BMB). There was significant variation in Hg, trace metals, and orthophosphate (PO_4^{3-}) concentrations, with higher concentrations of these elements located in the highest gull nesting density areas. Since then, our analyses have shown that gull guano is very high in water soluble PO_4^{3-} and is most likely the nutrient source to shallow groundwater. Total mercury (THg) and methylmercury (MeHg) concentrations were found to be low in guano but indicated a shift in food sources over a season. In this chapter we examine data from the bog's northern outflow which shows high concentrations of MeHg in the water from the fall of 2018 to summer 2019 (immediately post-restoration), and declining concentrations thereafter. Seasonal cycles in nutrients (PO_4^{3-}, total organic carbon (TOC)) were observed from 2019 to 2022 with decreasing summer maximums through time. The MeHg concentration in 2019 approached the Canadian Water Quality Guideline for the protection of low trophic level freshwater aquatic life (4 ng/L MeHg). Our data are consistent with various

N. J. O'Driscoll (✉) · J. C. Kickbush · H. E. Turner · S. Klapstein · M. Doncaster · K. Stevens · R. Clarke · M. Bradford · B. Bowes · J. Rogers
Department of Earth and Environmental Science, Acadia University, 32 University Avenue, Wolfville, Nova Scotia B4P 2R6, Canada
e-mail: nelson.odriscoll@acadiau.ca

N. K. Hillier · M. L. Mallory
Department of Biology, Acadia University, Wolfville, Nova Scotia, Canada

N. M. Hill et al. (eds.), *Wetland Restoration for Endangered Species Recovery*, Environmental Contamination Remediation and Management,
https://doi.org/10.1007/978-3-031-71344-6_13

studies examining terrestrial freshwater catchment flooding that indicate a short-term increase in the export of MeHg followed by a gradual decline over a decade in otherwise undisturbed catchments.

Keywords Methylmercury · Nutrients · Phosphate · Wetland restoration · Temporal analyses · Avian biovectors

13.1 Introduction

Mercury (Hg) is a naturally-occurring, ubiquitous heavy metal present at trace concentrations in most freshwater ecosystems. While there are natural sources of atmospheric Hg, such as volcanic activity and emissions from geological deposits, anthropogenic activities like fossil fuel combustion and gold mining have increased the atmospheric load of Hg tenfold over the past 100 years (Gustin et al. 2008; UNEP 2013). This problem is further exacerbated by the ability of Hg to remain in the atmosphere for 1–2 years, and to be transported long distances from the emission source (Gonzalez-Raymat et al. 2017; Chouvelon et al. 2018; O'Driscoll et al. 2005a; Zhang and Wong 2007). Mercury may enter ecosystems through atmospheric deposition, where this toxic contaminant can accumulate and negatively affect wildlife health and habitat quality (Malcata-Martins et al. 2021). However, in some situations, much greater amounts of Hg may be concentrated and moved from environmental reservoirs (e.g., marine food chains) and concentrated at receiving ecosystems through biotransport by a vector, such as birds (e.g., Blais et al. 2007). Significant advances have been made in understanding Hg cycling in freshwater and marine ecosystems (ECCC 2016), but the processes controlling Hg accumulation in freshwater bogs and the confounding effects of both nutrient subsidies and increased flooding are not well quantified.

Big Meadow Bog (BMB) (Brier Island, N.S.) underwent water table restoration, which was completed in the summer of 2018 (funded by the National Wetland Conservation Fund). The restoration was aimed at conserving the endangered Eastern Mountain Avens (*Geum peckii*) plant species. The bog is also host to thousands of American herring gulls (*Larus smithsoniansus*) clustered in a small area in the northern portion of the bog and feed at nearby American mink (*Neovison vison*) farming and aquaculture sites (McIntyre et al. 2022). Kickbush et al. (2018) found significant spatial variation in Hg, trace metals, and orthophosphate (PO_4^{3-}) concentrations in Big Meadow Bog's shallow groundwater, with higher concentrations of these elements located in areas with the highest densities of nesting gulls. Further research by Geizer et al. (2021) showed that gull feces contain high levels of water soluble PO_4^{3-} and approximately 1191 kg of PO_4^{3-} is deposited annually via feces to the bog, likely making it the largest nutrient source to shallow groundwater. Geizer et al. (2021) also showed that total mercury (THg) and methylmercury (MeHg) were low in feces, but THg in feces declined through the summer months while the percentage of methylmercury (%MeHg) significantly increased (ranging from 12 to 100% of THg), suggesting a switch in diet through the summer. This biovector subsidy of PO_4^{3-} is likely affecting many aspects of plant ecology and Hg speciation at this site.

13.1.1 Mercury Speciation in Aquatic Systems

The speciation of Hg controls its fate and distribution in ecosystems. Primary mercury species include elemental [Hg(0)], divalent [Hg(II)], and MeHg (O'Connor et al. 2019; Ma et al. 2019; O'Driscoll et al. 2005a). Due to its volatility, Hg(0) is the main species found in the atmosphere. Once in the atmosphere, Hg(0) is transformed via oxidation into the more soluble Hg(II), which is deposited via wet and dry deposition into terrestrial catchments and surface waters where Hg(II) binds with organic and inorganic ligands (O'Driscoll et al. 2005a). The methylation process of Hg(II) often occurs in anoxic and acidic conditions that encourage the presence of methylating bacteria (predominantly sulfate- and iron-reducing bacteria (SRB; IRB); Vaidya et al. 2000; Gochfeld 2003; Wiener 2013; Figueiredo et al. 2018). Methylmercury is the most toxic, bioavailable, and bioaccumulative form of Hg to organisms, hence posing the greatest health risks (Bravo et al. 2018).

13.1.2 Effects of Increased Water Table on Microbial Production of MeHg

When terrestrial catchments are flooded, the submergence of organic soils and vegetation leads to an increase in dissolved organic matter (DOM) in the water, a decrease in oxygen concentration, and increased bacterial metabolism. Sulfate- and iron-reducing bacterial activity (SRB; IRB) are affected by changes in temperature, DOM, acidity, nutrients, oxygen content, and the presence of a terminal electron acceptor (Munthe et al. 2007; Graham et al. 2012; Lehnherr 2014). Therefore, the conditions resulting from an increased water table may also largely control the microbial production of MeHg.

SRB use sulfate as a terminal electron acceptor in the oxidation of organic matter and are also largely responsible for the methylation of Hg in aquatic ecosystems. Mercury complexation with DOM can be a limiting factor for bacterial uptake of Hg due to the large size of DOM molecules, which generally cannot cross cell membranes (Ullrich et al. 2001; Barkay et al. 1997; Aiken et al. 2003; Höll et al. 2009). However, at low pH Hg(II) becomes more available to methylating bacteria as functional binding sites on DOM become protonated and less Hg-DOM binding occurs (Barkay et al. 1997). In environments with limited sulfate, microbes may use DOM as an energy source, therefore an increase in DOM would stimulate microbial growth which enhances methylation rates (Bravo et al. 2018; Chiasson-Gould 2014).

DOM can also enhance Hg methylation rates under sulfidic conditions by inhibiting the precipitation of Hg-sulfide complexes, thereby keeping Hg(II) in a soluble DOM complex and potentially available for methylation (Luo et al. 2017). The influence of DOM on MeHg production is complex: simultaneously decreasing and increasing the bioavailability of Hg(II) in many processes that are not well constrained.

Another influence in the production and bioavailability of MeHg in aquatic systems is pH. While the specific mechanisms by which pH influences Hg methylation are not known, it is theorized that acidity increases the decomposition of organic matter which in turn stimulates the activity of Hg-methylating SRB (Winfrey 1990; Kelly et al. 2003; Wiener et al. 2006). This is key in peat-dominated wetlands such as Big Meadow Bog, as these are naturally acidic due to the prevalence of sphagnum moss that relies on cation exchange to obtain nutrients from the surrounding environment, and in doing so releases hydrogen ions, that lower the pH of the water (in some cases from a pH between 6.5 and 7.5 to a pH between 3.5 and 4.5, depending on the dominance of sphagnum moss; Urban et al. 1995). Additionally, wetlands have high DOM compared to other aquatic systems such as lakes. Since DOM is a weak organic acid, its presence is another factor contributing to lower pH (O'Driscoll et al. 2005b; Gavin et al. 2018; Hansen et al. 2016).

13.1.3 Biovector Inputs to Ecosystems

While atmospheric circulation and deposition of Hg are useful to explain and predict the movement of contaminants on a global scale, biological vectors of contaminant transport have subverted this model, often concentrating contaminants in sensitive ecosystems (Fig. 13.1). Biovector transport refers to any contaminant that is carried within an organism (biovector) that then moves through the food web, within an ecosystem, or between ecosystems (Blais et al. 2007; Kallenborn and Blais 2015). While the study of biovectors is relatively novel, their influences in a variety of biologically sensitive ecosystems are compelling evidence of their importance on regional and global scales (Bauer and Hoye 2014). Nutrient enrichment has been observed in a number of avian biovector impacted sites. For example, Cape Vera (Nunavut, Canada) is the nesting location of a seabird colony numbering > 20,000 northern fulmars (*Fulmarus glacialis*), which input nutrients to nearby freshwater ponds via guano (Foster et al. 2011; Michelutti et al. 2011; Stewart et al. 2013). The hypereutrophic ponds situated directly below the nesting cliffs exhibited total nitrogen (TN) and total phosphorus (TP) concentrations (filtered) as high as 1590 mg/L and 52 mg/L, respectively. The effect of nutrient inputs on the Hg cycle in aquatic ecosystems is still an area requiring more study.

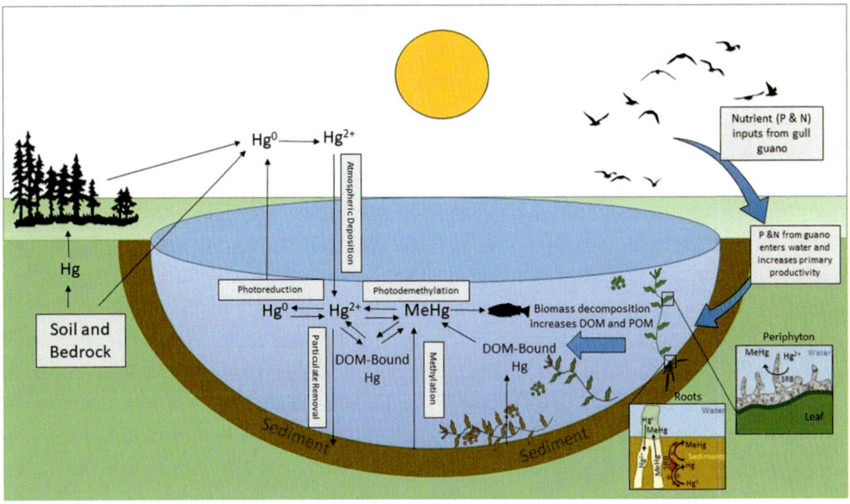

Fig. 13.1 Conceptual model of the Hg cycle in Big Meadow Bog, Brier Island, and the relationship between primary producers, dissolved organic matter (DOM), and Hg speciation, and the subsequent microenvironments for Hg speciation created by the roots of macrophytes, and the periphyton located on the surface of macrophytes

13.1.4 Nutrient Effects on Methylmercury Production in Surface Water

Biovector inputs can result in increased phosphorus and nitrogen, and thereby cause an increase in productivity and growth within an ecosystem (Brimble et al. 2009; Signa et al. 2012). With increased productivity and growth comes increased biomass of macrophytes and algae. This in turn can influence the fate of Hg via a number of different processes such as sediment trapping and a reduction of water flow, modification of sediment and abiotic physiochemical parameters, increasing populations of associated periphyton communities, and increasing dissolved and particulate organic matter (DOM and POM; Cosio et al. 2014).

The presence of macrophytes can reduce turbulence by up to 80% at the sediment–water interface and reduce water flow rates by a factor of 10, which consequently increases deposition of sediment (Vereecken et al. 2006; Kleeberg et al. 2010). This increased sedimentation could result in a higher rate of deposition of sediment-bound Hg which may become methylated under the anoxic conditions being created by the sedimentation and slower water flow (Bravo et al. 2011; Cosio et al. 2014).

Notably, while Hg may become methylated by these conditions, sedimentation still may have an overall effect as a Hg sink. This was demonstrated by the MESOSIM experiment which analyzed newly deposited Hg in boreal lake mesocosms, where Orihel et al. (2008) determined that while newly deposited Hg was reactive and accumulated in sediments throughout the two-year study, the availability of MeHg to the food web declined within the first year. However, this project did not consider MeHg concentrations in sediment from an area of high biomass density with a slow water flow rate and high sediment deposition rate, versus MeHg concentrations in sediments from an area of low biomass density (and the consequential increase in water flow rate and decrease in sediment deposition). Windham-Myers et al. (2014) examined effects of biomass density on MeHg concentrations in sediment and, while not incorporating sediment deposition rates, reported that vegetation increased MeHg concentrations in sediments by a median of 280% (versus a non-vegetated environment). However, this was attributed to physiochemical changes induced by vegetation rather than physical changes from increased biomass (Windham-Myers et al. 2014).

Sediment physiochemical parameters can be influenced by the roots of certain macrophytes, which can influence the abundance and structure of microbial communities, and hence Hg bioavailability and speciation (Windham-Myers et al. 2014). For example, Canário et al. (2007) found that sediments of estuaries colonized by *Sarcocornia fruticosa*, *Halimione portulacoides*, and *Spartina maritima* were 70 times higher in Hg than the maximum values measured in non-vegetated estuarine sediments. This finding was attributed to increases in labile carbon exudation from macrophyte roots that stimulated SRB activity. Regier et al. (2012) found that *Elodea nuttallii* stimulated two bacterial operational taxonomic units which were phylogenetically related to SRB in the rhizosphere, and therefore created a microenvironment favorable for Hg methylation. However, it only resulted in 2.4-fold increase of MeHg from 0.9 to 2.2% of total Hg in pore water. O'Driscoll et al. (2011) observed coastal wetland macrophytes creating microenvironments around their roots with redox conditions that allowed MeHg to bind to organic matter. This resulted in the uptake of MeHg into root material with concentrations of MeHg in roots ranging from 0 to 957 pg/g (O'Driscoll et al. 2011).

Decaying macrophytes release Hg, DOM, and POM to the water column and organic solids are integrated into the sediment (Cosio et al. 2014). Decomposition of organic matter largely controls the biological and chemical oxygen demand in aquatic systems. If the rate of organic decomposition is faster than the rate of aeration, anoxic conditions result which stimulate Hg-methylating microbial activity (Gochfeld 2003; Rabalais et al. 2010). Anoxia can also lead to the formation of hypoxic conditions in sediments and the water column given that there are slow water flow rates, minimal sediment–water exchange and ventilation, stratification, and increased DOM (Rabalais et al. 2010); all of these processes can be influenced by macrophyte biomass.

Macrophytes are also associated with periphyton: communities of microorganisms including microalgae, fungi, and prokaryotes attached to dead and living substrates that can be considered a class of mini-ecosystems with Hg methylation potentials. SRB have been found with periphyton likely due to the anoxic microenvironments created by periphyton, and therefore expose periphyton to MeHg (Correia et al. 2012; Gentès et al. 2017). Studies have observed increases in MeHg concentrations in periphyton located on floating macrophyte roots that are similar to, or greater than, those found in sediments and the water column (Fig. 13.1; Guimarães et al. 2000;

Mauro et al. 2002). Therefore, as biomass increases with nutrient input, so will the contribution to Hg methylation by periphyton.

Phytoplankton (microalgae) are another type of biota, influenced by DOM and nutrients that play a role in Hg speciation and the introduction of Hg to the food web. Phytoplankton bioconcentrate Hg from the water via MeHg accumulation in the cytoplasm and Hg(II) binding to the cell membrane (Mason et al. 1996; Pickhardt and Fisher 2007). Excessive nutrient inputs can lead to algal blooms which then cause hypoxia, resulting in MeHg accumulation (McCord et al. 2016). However, Hg binding sites on phytoplankton can be outcompeted by Hg-DOM binding, lowering the bioavailability of Hg(II) and MeHg at the base of the food web (Gorski et al. 2008). Hg-DOM binding is dependent on the structural composition of DOM which is determined by the organic source material as well as changes through time due to biological, chemical, and photochemical processes (Ravichandran 2004).

13.1.5 Research Gaps and Study Rationale

The influence of both flooding and nutrient inputs from avian biovectors on the aquatic speciation of Hg occurs through a complex array of secondary mechanisms that likely increases the production of MeHg. However, there are very few studies examining Hg speciation over multi-year time periods in flooded and nutrient-impacted sites. This work provides an overview of the monitoring of Hg speciation, DOM, and nutrient analyses in the primary outflow of Big Meadow Bog, post-water table restoration, from 2018 to 2023.

13.2 Methods

13.2.1 Water Sampling

COVID-19 restrictions on travel during the height of the pandemic meant that sampling was impossible during some time periods. From 2018 to 2023 surface water was collected from the north outflow of BMB. While most sampling was completed between April/May and November of each year, the sampling period varied among some years depending on ice cover in the bog. In 2021 and 2022 samples were collected from April to December while in 2023 samples were collected from January–June. Surface water (depth < 0.2 m) was collected from the north outflow of BMB using two pre-rinsed polyethylene bottles, then refrigerated in a cooler while undergoing transport. Once at the lab, water samples were frozen at -20 °C until processing and analysis.

Water samples were thawed then vacuum filtered using a 0.45 μm Super hydrophilic polyethersulfone membrane filter (Pall Corporation). All components of the vacuum filtration apparatus were triple-rinsed with Milli-Q water between samples and before use. Subsamples were taken from the filtered water for MeHg, THg, DOC (MeHg and THg subsamples were taken in triplicate). Subsamples were stored in 50 mL Falcon tubes and preserved according to their subsequent analyses: MeHg subsamples with 0.5% hydrochloric acid (HCl), THg subsamples with 0.5% bromine monochloride (BrCl), and cation subsamples with 1% nitric acid (HNO_3). Samples were then refrigerated with the exception of the anion and DOC subsamples, which were instead frozen at − 20 °C for preservation until analysis.

13.2.2 PO_4^{3-}, NO_3^-, SO_4^{2-}, and DOC Analysis

Phosphate (PO_4^{3-}), total phosphorus, nitrate (NO_3^-), and sulfate (SO_4^{2-}) concentrations in water were analyzed on a HACH DR3900 spectrophotometer (HACH, Loveland, Colorado) using their corresponding HACH pre-made reagent test kits. PO_4^{3-} and total phosphorus concentrations in water were analyzed using a HACH TNTplus 843 reagent test kit following the TNTplus method 10209/10210 (HACH 2022). Check standards (3 mg L^{-1} total phosphorus as PO_4^{3-}) were run alongside samples to ensure accuracy, and a portion of the samples were run in triplicate to ensure precision. NO_3^- concentrations were measured using a HACH TNTplus 835 reagent test kit following the TNTplus method 10206 while SO_4^{2-} concentrations were measured using a HACH TNTplus 865 reagent test kit following the TNTplus Method 10227 (HACH 2019a, b).

Dissolved organic carbon (DOC) concentrations in water samples from 2018 to 2021 were measured using a Shimadzu TOC-VCPH/TOC-CPN Total Organic Carbon Analyzer with an ASI-V autosampler (Shimadzu Scientific Instruments, Canby, Oregon) following the methods outline in Haverstock et al. (2012). DOC measurements were calculated from the difference in total carbon and inorganic carbon using thermal decomposition and wet oxidation with diffuse infrared detection. Analytical replicates are run internally in the TOC analyzer to calculate a coefficient of variation (< 2.0%) using the best three of five replicates. Water samples for DOC analysis from 2021 to 2023 were analyzed by the Clean Water Lab at Dalhousie University using a Shimadzu TOC-V CPH analyzer with a Shimadzu ASI-V autosampler (Shimadzu Scientific Instruments, Canby, Oregon). All analyses unless otherwise stated were performed in the Mercury Lab (www.mercurylab.acadiau.ca). Data were analyzed and plotted for significant trends using MS Excel and Sigma Plot (Table 13.1).

13.2.3 MeHg and THg Analysis

MeHg concentrations were analyzed in water samples by gas chromatography-atomic fluorescence spectroscopy (GC-AFS) with direct ethylation following U.S. EPA

Table 13.1 Summary of quality assurance and quality control measures in nutrient and organic carbon analysis

Variable	Reporting range (mg L^{-1})	Batch sampling dates	Method detection limit (mg L^{-1})	Ongoing check standard recoveries [SD]	Mean %RSD of triplicates [SD]
Phosphate (PO_4^{3-})	0.15–4.50	2018–2021	0.017 (n = 6)	95.1% ± 1.5% (n = 6)	7.3% ± 8.2% (n = 15)
		2022–2023	–	–	17.8% ± 23.0% (n = 2)
Total phosphorus (TP)	0.05–1.5	2018–2021	0.03 (n = 6)	95.1% ± 1.5% (n = 6)	3.4% ± 4.0% (n = 15)
		2022–2023	–	–	3.8% ± 0.7% (n = 2)
Nitrate (NO_3^-)	0.23–13.50	2018–2023	–	–	–
Sulfate (SO_4^{2-})	150–900	2018–2023	–	–	–
Dissolved organic carbon (DOC)	–	2018–2021	0.011 (n = 20)	89.9% ± 12.2% (n = 4)	–
		2021–2023	0.250 mg/L*	–	–

Method detection limits (MDL) were calculated as 3*x* the standard deviation of a repeated blank
*Reportable detection limit, calculated as 3–5*x* the method detection limit

method 1630 (U.S. EPA 1998) using a Brooks Rand Model III MERX analyzer (Brooks Rand Instruments, Seattle, Washington). Most samples were subsampled in triplicate then analyzed with triplicate spiked recoveries (50 pg). THg concentrations were measured in triplicate water subsamples using a Tekran Model 2600 CVAFS Mercury Analysis System (Tekran, Toronto, Ontario) following U.S. EPA method 1631 (U.S. EPA 2002). Samples were measured in four batches and MeHg and THg concentrations were method blank corrected after analysis (Table 13.2).

13.3 Results and Discussion

13.3.1 Phosphorous and Dissolved Organic Carbon

Total phosphorus (TP) ranged from < 0.5 to 2 mg/L and a large percentage of the TP (mean % = 71.4, SD 17.1% overall years) was present as orthophosphate (Fig. 13.2). There are several forms of P found in bird guano, but orthophosphate (PO_4^{3-}) most readily solubilizes into surface and groundwaters (Otero et al. 2018). While peat bogs often have large amounts of N and P in organic matter, the majority of this is

Table 13.2 Summary of quality assurance and quality control measures in methylmercury and total mercury analyses

Variable	Batch sampling dates	Method detection limit (ng L^{-1})	Mean % recovery of ongoing check standards [SD]	Mean % RSD of triplicates [SD]
Methylmercury (MeHg)	September 2018–July 2019	0.0098 (n = 20)	102.3% ± 1.5% (n = 7)	–
	July 2019–June 2021	0.021 (n = 11)	104.6% ± 4.4% (n = 5)	2.5% ± 2.2% (n = 15)
	June 2021–June 2022	0.089 (n = 12)	91.8% ± 6.1% (n = 5)	6.6% ± 3.8% (n = 10)
	June 2022–June 2023	0.025 (n = 29)	100.3% ± 5.3% (n = 10)	10.9% ± 5.8% (n = 14)
Total mercury (THg)	September 2018–October 2020	0.28 (n = 12)	90.2% ± 7.6% (n = 8)	8.9% ± 5.4% (n = 20)
	October 2020–June 2021	0.20 (n = 4)	116.4% ± 5.2% (n = 4)	3.1% ± 2.3% (n = 4)
	June 2021–June 2022	0.23 (n = 6)	105.3% ± 3.0% (n = 5)	3.5% ± 3.2% (n = 12)
	July 2022–June 2023	0.185 (n = 3)	102.6% ± 0.8% (n = 5)	2.4% ± 1.3% (n = 14)

Method detection limits were calculated as $3x$ the standard deviation of a repeated blank

unavailable for primary productivity (Mitsch and Gosselink 2000). Available forms of P such as PO_4^{3-} are normally limiting nutrients in bogs that are present at very low concentrations (Bedford et al 1999); PO_4^{3-} addition in small quantities has also been suggested to be useful post-restoration to stimulate plant reestablishment (Sottocornola et al. 2007). The total phosphorous (TP) levels observed in Bog Meadow Bog are all well above the Candian Water Quality Guideline definition of highly eutrophic (> 0.2 mg/L TP; CCME 2004).

DOC concentrations in the outflow also peaked in 2018 at 56 mg/L and varied seasonally with highest concentrations occurring in July and August (Fig. 13.3; Table 13.3). The mean DOC over the years 2018–2023 was 24.11 mg/L (SD = 11.09), with an observed minimum concentration of 3.18 mg/L and maximum of 56.06 mg/L (Table 13.3). The dramatic range in DOC concentration may be due to increased productivity and decomposition during the summer with nutrient inputs from guano combined with variations in discharge associated with precipitation events or meltwater outflow in spring (Fig. 13.1).

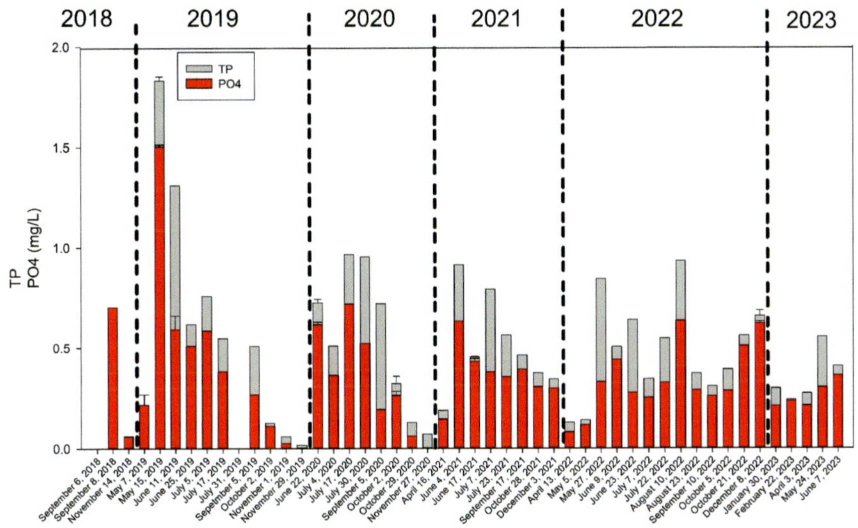

Fig. 13.2 TP/PO_4^{3-} concentration (triplicate mean with SD) for filtered (0.45 μm) water samples from the Northern outflow of Big Meadow Bog (Brier Island, N.S.) from 2018 to 2023

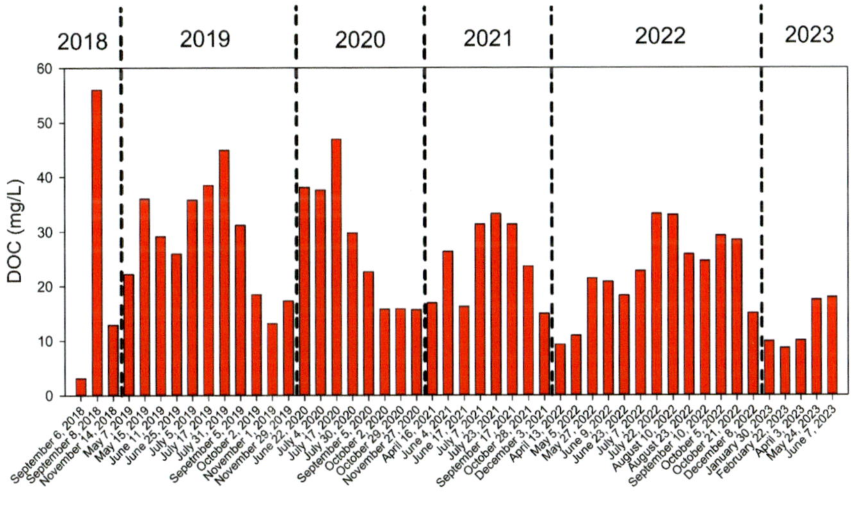

Fig. 13.3 Dissolved organic carbon (DOC) for filtered (0.45 μm) water samples from the Northern outflow of Big Meadow Bog (Brier Island, N.S.) from 2018 to 2023

Table 13.3 Summary of descriptive statistics for water quality parameters ($n = 48$ for each variable; 25 and 75% refer to percentile of the dataset)

Variable	Units	Max.	Min.	Mean	Std. Dev.	Median	25%	75%
Phosphate (PO_4^{3-})	(mg/L)	1.50	− 0.01	0.36	0.26	0.30	0.21	0.51
Total phosphorous (TP)	(mg/L)	1.83	− 0.17	0.48	0.37	0.46	0.23	0.68
Nitrate (NO_3^{2-})	(mg/L)	3.18	− 0.09	1.05	0.61	1.02	0.55	1.41
Sulfate (SO_4^-)	(mg/L)	511.00	57.70	83.36	73.26	61.05	59.30	65.15
Dissolved organic carbon (DOC)	(mg/L)	56.06	3.18	24.11	11.09	22.65	15.74	31.34
Total mercury (THg)	(ng/L)	8.10	0.33	3.64	1.46	3.44	2.62	4.34
Methyl mercury (MeHg)	(ng/L)	3.03	0.05	0.80	0.68	0.63	0.31	1.04
% MeHg	%	65.67	2.19	21.83	14.60	18.88	9.55	26.23

13.3.2 Mercury Speciation

The northern outflow Hg concentration data had an annual seasonal trend with summertime maximums and winter minimums. A decreasing trend in MeHg concentration maximums can be seen between 2018 and 2023. The highest MeHg concentrations and %MeHg occurred during the summer of 2019 and generally decreased between 2020 and 2023 (Fig. 13.4). Other research examining terrestrial freshwater catchment flooding indicates that export of MeHg increases 3–6 times over a 3–9-year time span in otherwise undisturbed catchments (Hsu-Kim et al. 2018). Kelly et al. (1997) found that MeHg export increased 39 fold two years post-flooding of an Experimental Lakes Area (ELA) reservoir in the boreal forest of Ontario. They observed a peak %MeHg of ~ 30% with concentrations reaching greater than 2 ng/L. At the BMB site, the percentage of THg present as MeHg increased to > 60% in July of 2019 and reduced to < 40% in the following years (with some annual minimums < 10%), suggesting either a decrease in Hg methylation or a dilution of MeHg with changes in outflow volume (Fig. 13.4).

Wang et al. (2020) studied THg in restored versus natural freshwater wetlands and found similar %MeHg values in surface and groundwaters (between 38 and 44%). They concluded that the surface water concentrations of THg and MeHg were not significantly different between restored and natural wetlands. However, the mean THg concentrations in groundwater at the restored wetlands were significantly higher than the natural wetlands ($p = 0.003$). They attributed the elevated THg input to the shallower soil depth, given that the depth of groundwater gathered for restored wetlands was less than that of the natural wetlands. Similar results to our study

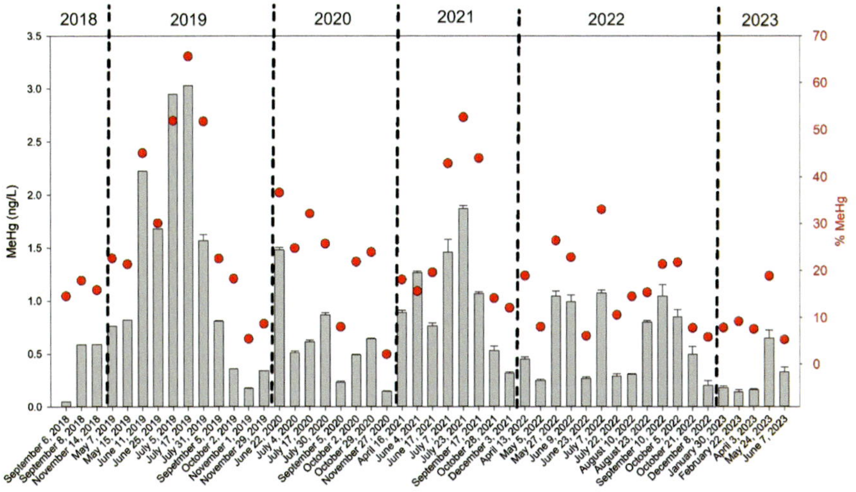

Fig. 13.4 Methylmercury concentration (triplicate mean with SD) and %THg present as MeHg (%MeHg) for filtered (0.45 μm) water samples from the northern outflow of Big Meadow Bog (Brier Island, N.S.) from 2018 to 2023

were reported by Eckley et al. (2015) who found higher concentrations of Hg in outflows from the Cottage Grove reservoir in Oregon during the summer with a peak of %MeHg (38%) in late summer as compared to < 5% MeHg during the colder seasons.

13.3.3 Conclusions

The water analyses of the northern BMB outflow indicated that there is an annual seasonal trend in MeHg, DOC, and PO_4^{3-} concentrations with highest values observed between September of 2018 and September of 2019 the first year after restoration. We hypothesize that initial high MeHg concentrations in the first year after water table restoration may be due to the decrease in oxygen in water-logged surface soil. In addition, increased nutrient release from previously dry soil and newly deposited biovector inputs of nutrients and Hg likely exacerbated the increased production of MeHg by both sulfate- and iron-reducing bacteria. Increased MeHg production and food web MeHg exposure has been well-documented in the literature in other flooded terrestrial ecosystems such as hydroelectric and recreational reservoirs (Calder et al. 2016; Eklöf et al. 2021). From our data, the MeHg export appears to be slowly returning to a stable state of lower levels of annual release. The nutrient export is also decreasing; however, the concentrations are still far higher than would be normal in an undisturbed bog system, likely due to the ongoing inputs of avian feces. We predict that the continued, elevated phosphate concentrations will likely contribute to enhanced primary productivity and vegetation changes at this site into the future, unless gull inputs are removed from the site (i.e., the colony is abandoned or moved).

Acknowledgements We acknowledge and appreciate the funding received for this work from the NS Habitat Fund (contributions from hunters and trappers). Sampling and some preparations were also assisted by summer technician Allyssa Gomez.

References

Aiken G, Haitzer M, Ryan JN, Nagy K (2003) Interactions between dissolved organic matter and mercury in the Florida Everglades. J De Physique IV (Proc) 107:29–32. https://doi.org/10.1051/jp4:20030235

Barkay T, Gillman M, Turner RR (1997) Effects of dissolved organic carbon and salinity on bioavailability of mercury. Appl Environ Microbiol 63(11):4267–4271. https://doi.org/10.1128/aem.63.11.4267-4271

Bauer S, Hoye BJ (2014) Migratory animals couple biodiversity and ecosystem functioning worldwide. Science 344. https://doi.org/10.1126/science.1242552

Blais JM, Macdonald RW, Mackay D, Webster E, Harvey C, Smol JP (2007) Biologically mediated transport of contaminants to aquatic systems. Environ Sci Technol 41:1075–1084. https://doi.org/10.1021/es061314a

Bravo AG, Bouchet S, Amouroux D, Poté J, Dominik J (2011) Distribution of mercury and organic matter in particle-size classes in sediments contaminated by a waste water treatment plant: Vidy Bay, Lake Geneva, Switzerland. J Environ Monit 13:974–982. https://doi.org/10.1039/C0EM00534G

Bravo AG, Peura S, Buck M, Ahmed O, Mateos-Rivera A, Herrero Ortega S, Schaefer JK, Bouchet S, Tolu J, Björn E, Bertilsson S (2018) Methanogens and iron-reducing bacteria: the overlooked members of mercury-methylating microbial communities in boreal lakes. Appl Environ Microbiol 84:e01774-e1818. https://doi.org/10.1128/AEM.01774-18

Brimble SK, Blais JM, Kimpe LE, Mallory ML, Keatley BE, Douglas MSV, Smol JP (2009) Bioenrichment of trace elements in a series of ponds near a northern fulmar (*Fulmarus glacialis*) colony at Cape Vera, Devon Island. Can J Fish Aquat Sci 66:949–958. https://doi.org/10.1139/F09-053

Bedford BL, Walbridge MR, Aldous A (1999) Patterns in nutrient availability and plant diversity of temperate North American wetlands. Ecology 80:2151–2169. https://doi.org/10.1890/0012-9658(1999)080[2151:PINAAP]2.0.CO;2

Calder RSD, Schartup AT, Li M, Valberg AP, Balcom PH, Sunderland EM (2016) Future impacts of hydroelectric power development on methylmercury exposures of Canadian indigenous communities. Environ Sci Technol 50:13115–13122. https://doi.org/10.1021/acs.est.6b04447

Canadian Council of Ministers of the Environment (CCME) (2004) Canadian water quality guidelines for the protection of aquatic life: phosphorus: Canadian guidance framework for the management of freshwater systems. In: Canadian environmental quality guidelines. Canadian Council of Ministers of the Environment, Winnipeg

Canário J, Caetano M, Vale C, Cesário R (2007) Evidence for elevated production of methylmercury in salt marshes. Environ Sci Technol 41:7376–7382. https://doi.org/10.1021/es071078j

Chiasson-Gould SA, Blais JM, Poulain AJ (2014) Dissolved organic matter kinetically controls mercury bioavailability to bacteria. Environ Sci Technol 48:3153–3161. https://doi.org/10.1021/es4038484

Chouvelon T, Cresson P, Bouchoucha M, Brach-Papa C, Bustamante P, Crochet S, Marco-Miralles F, Thomas B, Knoery J (2018) Oligotrophy as a major driver of mercury bioaccumulation in medium-to high-trophic level consumers: a marine ecosystem-comparative study. Environ Pollut 233:844–854. https://doi.org/10.1016/j.envpol.2017.11.015

Correia RS, Miranda MR, Guimarães JRD (2012) Mercury methylation and the microbial consortium in periphyton of tropical macrophytes: effect of different inhibitors. Environ Res 112:86–91. https://doi.org/10.1016/j.envres.2011.11.002

Cosio C, Flück R, Regier N, Slaveykova VI (2014) Effects of macrophytes on the fate of mercury in aquatic systems. Environ Toxicol Chem 33:1225–1237. https://doi.org/10.1002/etc.2499

Eckley CS, Luxton TP, McKernan JL, Goetz J, Goulet J (2015) Influence of reservoir water level fluctuations on sediment methylmercury concentrations downstream of the historical Black Butte mercury mine, OR. Appl Geochem 61:284–293. https://doi.org/10.1016/j.apgeochem.2015.06.011

Eklöf K, Drohan P, Needoba J, Landefeld S, Peterson TD, Hu H, Iavorivska L, Boyer EW (2021) Methylmercury in lake bed soils during re-flooding of an Appalachian reservoir in the Northeastern USA. Environmental Research Communications 3:085004. https://doi.org/10.1088/2515-7620/ac1d83

Environment and Climate Change Canada (ECCC) (2016) Canadian mercury science assessment: report. Cat. En84-130/3-2016E-PDF. ISBN: 978-0-660-04499-6

Figueiredo N, Serralheiro ML, Canário J, Duarte AC, Hintelmann H, Carvalho C (2018) Evidence of mercury methylation and demethylation by the estuarine microbial communities obtained in stable Hg isotope studies. Int J Environ Res Public Health 15:2141. https://doi.org/10.3390/ijerph15102141

Foster KL, Kimpe LE, Brimble SK, Liu H, Mallory ML, Smol JP, Macdonald RW, Blais JM (2011) Effects of seabird vectors on the fate, partitioning, and signatures of contaminants in a high arctic ecosystem. Environ Sci Technol 45:10053–10060. https://doi.org/10.1021/es202754h

Gavin AL, Nelson SJ, Klemmer AJ, Fernandez IJ, Strock KE, McDowell WH (2018) Acidification and climate linkages to increased dissolved organic carbon in high-elevation lakes. Water Resour Res 54:5376–5393. https://doi.org/10.1029/2017WR020963

Geizer H, Klapstein S, Mallory M, O'Driscoll NJ (2021) Total mercury, methylmercury, phosphate, and sulphate inputs to a bog ecosystem from herring gull (*Larus smithsoniansus*) guano. Ecotoxicol Environ Saf 226:112845. https://doi.org/10.1016/j.ecoenv.2021.112845

Gentès S, Taupiac J, Colin Y, André J, Guyoneaud R (2017) Bacterial periphytic communities related to mercury methylation within aquatic plant roots from a temperate freshwater lake (South-Western France). Environ Sci Pollut Res 24:19223–19233. https://doi.org/10.1007/s11356-017-9597-x

Gochfeld M (2003) Cases of mercury exposure, bioavailability, and absorption. Ecotoxicol Environ Saf 56:174–179. https://doi.org/10.1016/S0147-6513(03)00060-5

Gonzalez-Raymat H, Liu G, Liriano C, Li Y, Yin Y, Shi J, Jiang G, Cai Y (2017) Elemental mercury: Its unique properties affect its behavior and fate in the environment. Environ Pollut 229:69–86. https://doi.org/10.1016/j.envpol.2017.04.101

Gorski PR, Armstrong DE, Hurley JP, Krabbenhoft DP (2008) Influence of natural dissolved organic carbon on the bioavailability of mercury to a freshwater alga. Environ Pollut 154:116–123. https://doi.org/10.1016/j.envpol.2007.12.004

Graham AM, Aiken GR, Gilmour CC (2012) Dissolved organic matter enhances microbial mercury methylation under sulfidic conditions. Environ Sci Technol 46:2715–2723. https://doi.org/10.1021/es203658f

Guimarães JR, Davée M, Meili LD, Hylander E, de Castro e Silva M, Roulet JB, Narvaez M, Alves de Lemos R (2000) Mercury net methylation in five tropical flood plain regions of Brazil: high in the root zone of floating macrophyte mats but low in surface sediments and flooded soils. Sci Total Environ 261:99–107. https://doi.org/10.1016/S0048-9697(00)00628-8

Gustin MS, Lindberg SE, Weisberg PJ (2008) An update on the natural sources and sinks of atmospheric mercury. Appl Geochem 23:482–493. https://doi.org/10.1016/j.apgeochem.2007.12.010

HACH (2019a) Short working procedure: TNTplus 835 Nitrate, Low Range, Method 10206. HACH Company

HACH (2019b) Short working procedure: TNTplus 865 Sulfate, Method 10227. HACH Company

HACH (2022) Short working procedure: TNTplus 843 phosphorus total and reactive (Ortho), low range, method 10209/10210. HACH Company

Hansen AT, Dolph CL, Finlay JC (2016) Do wetlands enhance downstream denitrification in agricultural landscapes? Ecosphere 7:e01516. https://doi.org/10.1002/ecs2.1516

Haverstock S, Sizmur T, Murimboh J, O'Driscoll NJ (2012) Modeling the photo-oxidation of dissolved organic matter by ultraviolet radiation in freshwater lakes: Implications for mercury bioavailability. Chemosphere 88:1220–1226. https://doi.org/10.1016/j.chemosphere.2012.03.073

Höll BS, Fiedler S, Jungkunst HF, Kalbitz K, Freibauer A, Drösler M, Stahr K (2009) Characteristics of dissolved organic matter following 20 years of peatland restoration. Sci Total Environ 408:78–83. https://doi.org/10.1016/j.scitotenv.2009.08.046

Hsu-Kim H, Eckley CS, Achá D, Feng X, Gilmour CC, Jonsson S, Mitchell CPJ (2018) Challenges and opportunities for managing aquatic mercury pollution in altered landscapes. Ambio 47:141–169. https://doi.org/10.1007/s13280-017-1006-7

Kallenborn R, Blais J (2015) Tracking contaminant transport from biovectors. In: Blais JM, Robson M, Smol JP (eds) Environmental contaminants: using natural archives to track sources and long-term trends of pollution. Springer, pp 461–498

Kelly CA, Rudd JWM, Bodaly RA, Roulet NP, St. Louis VL, Heyes A, Moore TR, Schiff S, Aravena R, Scott KJ, Dyck B, Harris R, Warner B, Edwards G (1997) Increases in fluxes of greenhouse gases and methyl mercury following flooding of an experimental reservoir. Environ Sci Technol 31:1334–1344. https://doi.org/10.1021/es9604931

Kelly CA, Rudd JWM, Holoka MH (2003) Effect of pH on mercury uptake by an aquatic bacterium: Implications for Hg cycling. Environ Sci Technol 37:2941–2946. https://doi.org/10.1021/es0263660

Kickbush J, Mallory M, Murimboh J, Rand J, Klapstein S, Loder A, Hill N, O'Driscoll NJ (2018) Influence of avian biovectors on mercury speciation in a bog ecosystem. Sci Total Environ 637:264–273. https://doi.org/10.1016/j.scitotenv.2018.04.224

Kleeberg A, Köhler J, Sukhodolova T, Sukhodolov A (2010) Effects of aquatic macrophytes on organic matter deposition, resuspension and phosphorus entrainment in a lowland river. Freshw Biol 55:326–345. https://doi.org/10.1111/j.1365-2427.2009.02277.x

Lehnherr I (2014) Methylmercury biogeochemistry: a review with special reference to Arctic aquatic ecosystems. Environ Rev 22:229–243. https://doi.org/10.1139/er-2013-0059

Luo HW, Yin X, Jubb AM, Chen H, Lu X, Zhang W, Lin H, Yu H-Q, Liang L, Sheng G-P, Gu B (2017) Photochemical reactions between mercury (Hg) and dissolved organic matter decrease Hg bioavailability and methylation. Environ Pollut 220:1359–1365. https://doi.org/10.1016/j.envpol.2016.10.099

Ma M, Du H, Wang D (2019) Mercury methylation by anaerobic microorganisms: a review. Crit Rev Environ Sci Technol 49:1893–1936. https://doi.org/10.1080/10643389.2019.1594517

Malcata-Martins B, O'Driscoll NJ, Mallory ML, Canário J (2021) A review of freshwater invertebrates as biomonitors of methylmercury: the importance of more complete physical and chemical reporting. Bull Environ Contam Toxicol 107:801–808. https://doi.org/10.1007/s00128-021-03274-9

Mason RP, Reinfelder JR, Morel FM (1996) Uptake, toxicity, and trophic transfer of mercury in a coastal diatom. Environ Sci Technol 30:1835–1845. https://doi.org/10.1021/es950373d

Mauro J, Guimarães J, Hintelmann H, Watras C, Haack E, Coelho-Souza S (2002) Mercury methylation in macrophytes, periphyton, and water—comparative studies with stable and radio-mercury additions. Anal Bioanal Chem 374:983–989. https://doi.org/10.1007/s00216-002-1534-1

McCord SA, Beutel MW, Dent SR, Schladow SG (2016) Evaluation of mercury cycling and hypolimnetic oxygenation in mercury-impacted seasonally stratified reservoirs in the Guadalupe River Watershed, California. Water Resour Res 52:7726–7743. https://doi.org/10.1002/2016WR019061
McIntyre JA, O'Driscoll NJ, Spooner I, Robertson G, Smol JP, Mallory ML (2022) Scavenging gulls are biovectors of mercury from industrial wastes in Nova Scotia. Canada. Chemosphere 304:135279. https://doi.org/10.1016/j.chemosphere.2022.135279
Michelutti N, Mallory ML, Blais JM, Douglas MSV, Smol JP (2011) Chironomid assemblages from seabird-affected high arctic ponds. Polar Biol 34:799–812. https://doi.org/10.1007/s00300-010-0934-5
Mitsch WJ, Gosselink JG (eds) (2000) Wetlands, 3rd edn. Wiley, New York
Munthe J, Bodaly RA, Branfireun BA, Driscoll CT, Gilmour CC, Harris R, Horvat M, Lucotte M, Malm O (2007) Recovery of mercury-contaminated fisheries. AMBIO 36:33–44. https://doi.org/10.1579/0044-7447(2007)36[33:romf]2.0.co;2
O'Connor D, Hou D, Ok YS, Mulder J, Duan L, Wu Q, Wang S, Tack F, Rinklebe J (2019) Mercury speciation, transformation, and transportation in soils, atmospheric flux, and implications for risk management: a critical review. Environ Int 126:747–761. https://doi.org/10.1016/j.envint.2019.03.019
O'Driscoll N, Canário J, Crowell N, Webster T (2011) Mercury speciation and distribution in coastal wetlands and tidal mudflats: Relationships with sulphur speciation and organic carbon Water. Air Soil Pollution 220:313–326. https://doi.org/10.1007/s11270-011-0756-2
O'Driscoll NJ, Rencz AN, Lean DRS (2005a) The biogeochemistry and fate of mercury in the environment. In: Sigel A, Sigel H, Sigel RKO (eds) Metal ions in biological systems, 43, 221–238. https://doi.org/10.1201/9780824751999
O'Driscoll NJ, Rencz AN, Lean DRS (eds) (2005b) Mercury cycling in a wetland-dominated ecosystem: a multidisciplinary study. Society of Environmental Toxicology and Chemistry, Pensacola, Florida, USA
Orihel D, Paterson M, Blanchfield P, Bodaly RA, Gilmour CC, Hintelmann H (2008) Temporal changes in the distribution, methylation, and bioaccumulation of newly deposited mercury in an aquatic ecosystem. Environ Pollut 154:77–88. https://doi.org/10.1016/j.envpol.2007.12.030
Otero XL, De La Peña-Lastra S, Pérez-Alberti A, Ferreira TO, Huerta-Diaz MA (2018) Seabird colonies as important global drivers in the nitrogen and phosphorus cycles. Nat Commun 9(246). https://doi.org/10.1038/s41467-017-02446-8
Pickhardt PC, Fisher NS (2007) Accumulation of inorganic and methylmercury by freshwater phytoplankton in two contrasting water bodies. Sci Total Environ 41:125–131
Rabalais NN, Díaz RJ, Levin LA, Turner RE, Gilbert D, Zhang J (2010) Dynamics and distribution of natural and human-caused hypoxia. Biogeosciences 7:585–619. https://doi.org/10.5194/bg-7-585-2010
Ravichandran M (2004) Interactions between mercury and dissolved organic matter—a review. Chemosphere 55:319–331. https://doi.org/10.1016/j.chemosphere.2003.11.011
Regier N, Frey B, Converse B, Roden E, Grosse-Honebrink A, Bravo AG, Cosio C (2012) Effect of *Elodea nuttallii* roots on bacterial communities and MMHg proportion in a Hg polluted sediment. PLoS ONE 7(9). https://doi.org/10.1371/journal.pone.0045565
Signa G, Mazzola A, Vizzini S (2012) Effects of a small seagull colony on trophic status and primary production in a mediterranean coastal system (Marinello Ponds, Italy). Estuar Coast Shelf Sci 111:27–34. https://doi.org/10.1016/j.ecss.2012.06.008
Sottocornola M, Boudreau S, Rochfort L (2007) Peat bog restoration: effect of phosphorous on plant re-establishment. Ecol Eng 31:29–40. https://doi.org/10.1016/j.ecoleng.2007.05.001
Stewart E, Michelutti N, Blais J, Mallory M, Douglas M, Smol J (2013) Contrasting the effects of climatic, nutrient, and oxygen dynamics on subfossil chironomid assemblages: a paleolimnological experiment from eutrophic high arctic ponds. J Paleolimnol 49:205–219. https://doi.org/10.1007/s10933-012-9658-6

U.S. EPA (1998) Method 1630: methyl mercury in water by distillation, aqueous ethylation, purge and trap, and cold vapor atomic fluorescence spectrometry. U.S. Environmental Protection Agency, Office of Water, Washinton, D.C.
U.S. EPA (2002) Method 1631, revision E: mercury in water by oxidation, purge and trap, and cold vapor atomic fluorescence spectrometry. United States Environmental Protection Agency, Office of Water, Washington, D.C.
Ullrich SM, Tanton TW, Abdrashitova SA (2001) Mercury in the aquatic environment: a review of factors affecting methylation. Crit Rev Environ Sci Technol 31:241–293. https://doi.org/10.1080/20016491089226
UNEP (2013) Minamata convention on mercury. Global Treaty. United Nation Environment Programme, Geneva, Switzerland. http://www.mercuryconvention.org/Portals/11/documents/Booklets/Minamata%20Convention%20on%20Mercury_booklet_English.pdf
Urban NR, Verry ES, Eisenreich SJ (1995) Retention and mobility of cations in a small peatland: trends and mechanisms. Water Air Soil Pollut 79:201–224. https://doi.org/10.1007/BF01100438
Vaidya OC, Howell GD, Leger DA (2000) Evaluation of the distribution of mercury in lakes in Nova Scotia and Newfoundland (Canada). Water Air Soil Pollut 117:353–369. https://doi.org/10.1023/A:1005190429095
Vereecken H, Baetens J, Viaene P, Mostaert F, Meire P (2006) Ecological management of aquatic plants: effects in lowland streams. Hydrobiologia 570:205–210. https://doi.org/10.1007/s10750-006-0181-5
Wang T, Driscoll CT, Hwang K, Chandler D, Montesdeoca M (2020) Total and methylmercury concentrations in ground and surface waters in natural and restored freshwater wetlands in northern New York. Ecotoxicology 29:1602–1613. https://doi.org/10.1007/s10646-019-02155-6
Wiener JG (2013) Mercury exposed: advances in environmental analysis and ecotoxicology of a highly toxic metal. Environ Toxicol Chem 32:2175–2178. https://doi.org/10.1002/etc.2333
Wiener JG, Knights BC, Sandheinrich MB, Jeremiason JD, Brigham ME, Engstrom DR, Woodruff LG, Cannon WF, Balogh SJ (2006) Mercury in soils, lakes, and fish in Voyageurs National Park (Minnesota): importance of atmospheric deposition and ecosystem factors. Environ Sci Technol 40:6261–6268. https://doi.org/10.1021/es060822h
Windham-Myers L, Marvin-DiPasquale M, Stricker CA, Agee JL, Kieu LH, Kakouros E (2014) Mercury cycling in agricultural and managed wetlands of California, USA: experimental evidence of vegetation-driven changes in sediment biogeochemistry and methylmercury production. Sci Total Environ 484:300–307. https://doi.org/10.1016/j.scitotenv.2013.05.028
Winfrey MR, Rudd JWM (1990) Environmental factors affecting the formation of methylmercury in low pH lakes. Environ Toxicol Chem 9:853–869. https://doi.org/10.1002/etc.5620090705
Zhang L, Wong MH (2007) Environmental mercury contamination in China: sources and impacts. Environ Int 33:108–121. https://doi.org/10.1016/j.envint.2006.06.022

Chapter 14
Foliar Endophytic Fungi from *Geum peckii* in Canada

Sarah J. Adams, Brent M. Robicheau, Diane LaRue, Robin D. Browne, and Allison K. Walker

Abstract In Canada, the *Geum peckii* population has declined over the past forty years due in part to changing habitat conditions. We investigated the culturable foliar fungi present in *G. peckii* leaves at five locations with differing habitat conditions (water and nutrient levels, degree of shrubby plant ingrowth) within this plant species' Canadian range. Fungal identifications were made using *ITS* rDNA barcoding of axenic fungal cultures isolated from leaf tissue. Differences in foliar fungal communities among sites were documented, with a predominance of Gnomoniaceae (Class: Sordariomycetes, Phylum: Ascomycota). Habitats showing altered conditions showed lower endophytic diversities (10–16 species) compared to the reference habitat (27 species). Intriguingly, several fungi may represent previously unknown taxa. Our work represents a significant step towards understanding *G. peckii*'s mycobiome and provides relevant data to inform conservation of this rare and endangered plant.

Keywords Foliar fungal endophytes · Eastern Mountain Avens · Mycobiome · Plant conservation · Rosaceae

---

Content in this chapter has been reproduced with permissions from Adams SJ et al (2021) Plants 10(5):1026. See https://doi.org/10.3390/plants10051026 for original article.

S. J. Adams · A. K. Walker (✉)
Department of Biology, Acadia University, Wolfville, NS B4P 2R6, Canada
e-mail: allison.walker@acadiau.ca

S. J. Adams
e-mail: sarah.adams@acadiau.ca

B. M. Robicheau
Biology Department, Clark University, Worcester, MA, USA

D. LaRue
Mersey Tobeatic Research Institute, Mount Merritt, NS B0T 1B0, Canada

R. D. Browne
K.C. Irving Environmental Science Center, Acadia University, Wolfville, NS B4P 2R6, Canada

N. M. Hill et al. (eds.), *Wetland Restoration for Endangered Species Recovery*, Environmental Contamination Remediation and Management,
https://doi.org/10.1007/978-3-031-71344-6_14

14.1 Introduction

The majority of land plants interact with fungi either as endophytes, which live asymptomatically within plant tissues showing no signs of disease, or as arbuscular or ectomycorrhizae living in symbiosis with plant roots (Chen et al., 2018; Martin et al., 2016). Endophytes can be symbionts, latent decomposers that will eventually aid in the rapid breakdown of the plant upon its senescence, or latent pathogens, symptoms of which only appear when the host plant experiences certain types of physiological stress (Rodriguez et al., 2009; Yan et al., 2019). An individual endophyte (species) can play one or all three of these roles during its time within a host, or it may simply remain inert until it re-enters the environment upon the plant's death and decomposition (Rodriguez et al., 2009). Endophytic fungi can be beneficial in a variety of ways, some improve plant health and confer resilience to biotic and abiotic stressors, such as through the production of anti-herbivory or anti-microbial toxins, or by the acquisition and biosynthesis of compounds for nutrition (Rana et al., 2019; Yan et al., 2019; Chen et al., 2018). Through these multifarious roles and activities, the endophyte helps its host to adapt to changes in the surrounding habitat or environment (Yan et al., 2019). Since endophyte-plant interactions can represent an important facet of a plant's lifecycle—and perhaps even its overall success and survival—it is crucial to understand the endophytic fungi of rare and endangered plants as this knowledge can inform restoration goals (Gooden et al., 2020; Sikes et al., 2016). For example, exposure to select endophytes may aid in the growth and development of host tissue and pest protection (Sikes et al., 2016; Rodriguez et al., 2009). The foliar endophyte of spruce, *Phialocephala scopiformis*, produces rugulosin, which inhibits the feeding of the spruce budworm, *Choristoneura fumiferana*. Consequently, the Canadian forestry industry inoculates spruce seedlings with *P. scopiformis* to improve host health (Frasz et al., 2014).

Over the summer of 2015 and prior to restoration efforts, we examined *Geum peckii* plants from several locations within the Nova Scotian population (see sites in Fig. 14.1 insets I–A and I–B). It is known from 40 years of previous monitoring that *G. peckii* populations on Brier Island and later the Digby Neck have been observed to be in decline (COSEWIC 2010). Factors contributing to changing habitat conditions at some sites included: the former use of bog land as pasture, bog drainage for a failed agricultural enterprise, presence of a large Herring gull (*Larus argentatus*) colony, as well as old unused ATV trails (LaRue 2016a). Prior to our study, the endophytes of *G. peckii* had not been examined. Using several sites with differing water and nutrient levels, substrate disturbance (drying and exposure of mineral soil and bog peat), and degree of shading shrubby plant ingrowth into habitat surrounding *G. peckii*, we compared the foliar endophytic fungi present in *G. peckii*. Our two aims were: (a) to first characterise endophytic species present in *G. peckii*, and then (b) to investigate whether a relationship between immediate habitat conditions and endophyte assemblages is evident. Our approach considered two sampling times (June and July) and two nutrient media types (MEA and $^1/_2$ MS + A). Our dataset provides both valuable insight into the endophytes that can reside within *G. peckii* in

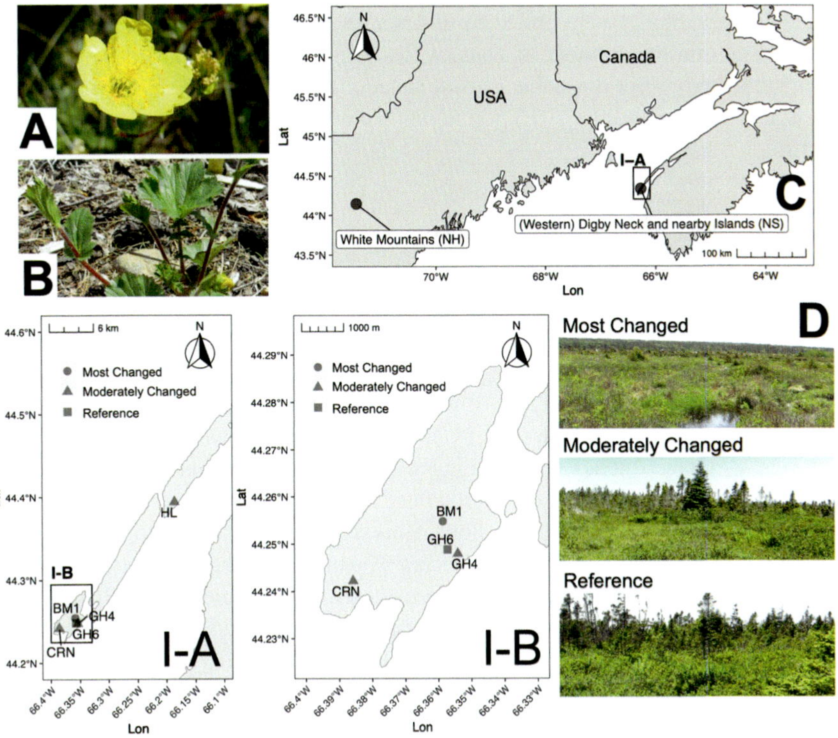

Fig. 14.1 Study species and study sites. **A** *Geum peckii* flower (average leaf diameter = 2.5–3.5 cm) and **B** leaves (average width = 5–10 cm). **C** Locations of the global populations of *Geum peckii*. Locations sampled herein from the Nova Scotian populations are shown in map inset (**I–A**) and inset (**I–B**) (HL = Western Digby Neck, NS; all other sites = Brier Island, NS). (**D**) Examples of foliage at sites across the gradient of observed changes to habitat conditions (moderately changed site shown is CRN)

connection with habitat differences, as well as a list of fungi that will be of interest for future restoration and conservation applications. From a broader perspective, our methodology may also be of interest to the general study of endangered plant mycobiota.

14.2 Materials and Methods

The full methods for this study can be found in Adams et al. (2021) at https://doi.org/https://doi.org/10.3390/plants10051026. Briefly, leaves were collected from four sites on Brier Island: Camp Road North (CRN), Big Meadow Bog 1 (BM1), Green Head 4 (GH4), and Green Head 6 (GH6). GH6 is the reference bog for this study. We also collected at one site on the Digby Neck at Harris Lake (HL) (see Fig. 14.1).

Geum peckii samples were collected under a scientific collection permit issued by the Nova Scotia Department of Natural Resources. The sites named and sampled during this study were the same as those listed in the 2010, COSEWIC Assessment and Status Report and the 2012 population survey completed by LaRue and Toms (2013).

14.2.1 Site Descriptions

The following site descriptions come from LaRue and Toms 2013. Also included are observations taken during sampling of the sites.

BM1 had the greatest number of observed changing habitat conditions. It was the most impacted of the BI sites by the lowering of the water table due to the historical drainage ditches and standing water was only present in one part of the site. This change allowed for the ingrowth of shrubby plants and trees, into the bog, and appeared to be transitioning the flora at this site from bog species towards invasive plant, and early successional native, species. The increased shrubs and trees, resulted in the establishment of a herring gull, *Larus argentatus*, colony on the Big Meadow bog which deposited additional nutrients on the site; primarily nitrogen, which in turn has increased phosphorus availably in the water and soils.

GH4, **CRN**, and **HL** were three sites (differentiated by water levels) that were considered to have fewer, but varying, observed changes to habitat conditions. **GH4** (Dry) was a non-bog site, situated on exposed mineral soil, beside an ATV trail and Gull Rock Road. It was the driest of the sampled sites, with no standing water and only a small seepage at one side of the site. The ingrowth of shrubby plants was ongoing. **CRN** (Moist) was a bog site, with bare soil, elevated above the water table. Spruce trees shaded the site and atypical bog plants, Soft Rush (*Juncus effuses*) and Common Horsetail (*Equisetum arvense*), were present and indicated possible mineral enrichment of the soil. Slender Sedge (*Carex lasiocarpa*) was also present indicating prolonged flooding, at this site, though this was not observed during our visits. ATV trails crossed the site and had disturbed soil. **HL** (Wet) was a fen site, whose water levels had been altered by the construction of a dam in the 1960s to form Harris Lake adjacent to the sampled site, which in turn inundated the fen. Holes in the dam had allowed the water levels to drop but standing water was present throughout this site. Changes to the water level have allowed for ingrowth of trees and shrubby plants though these plants were present on hummocks.

GH6 was considered the reference bog for our study, with no observable changes to habitat condition. Its water levels were considered minimally impacted by the drainage efforts in the adjacent Big Meadow Bog, with standing water present throughout the site. Shrubby plant growth was only present at the edge of this site. At the time of our study, this site had the highest number of observable *G. peckii* and they were the 'healthiest' in appearance (large plants with vibrant green, large leaves, no reddening of the leaves) of the sampled sites.

14.2.2 *Leaf Sampling*

For each leaf, 8–24 discs (1 cm wide) were taken and then sterilised using 30 s in 70% ethanol, immersion in 20% bleach for 7.5 min, a second 30 s rinse in 70% ethanol, and then a final 30 s rinse in autoclaved distilled H_2O (Schulz et al., 1993). Half of the total number of tissue discs from each leaf were plated on 2% Malt Extract Agar (2% MEA) and half were plated on $^1/_2$ Murashige and Skoog + Agar ($^1/_2$ MS + A; Murashige and Skoog 1962). Petri dishes were sealed with Parafilm and incubated at 25 °C. Plates were observed daily, and axenic fungi were isolated and transferred to 2% MEA. DNA was extracted from each axenic culture and the ITS ribosomal DNA (the accepted species-level barcode region for fungi; Schoch et al., 2012) was amplified using ITS4 and ITS5 primers (Raja et al., 2017; White et al., 1990). PCR products were sent for Sanger sequencing at the Génome Québec Innovation Centre (McGill University, Montreal, Canada). DNA sequences were compared to sequences in the standard and TYPE sequence nucleotide BLAST databases of GenBank (at NCBI (National Centre for Biotechnology Information)) using a $\geq$ 97% sequence identity threshold for making preliminary identifications. Maps and the MCA analysis were constructed in RStudio Version 1.3.1093 (RStudio Team 2020) using R Version 4.0.3 (R Core Team 2020). Map making required the following additional R packages: ggplot2 (Wickham 2016), sf (Pebesma 2018), tidyverse (Wickham et al., 2019), ggspatial (Dunnington 2023), and ggrepel (Slowikowski et al., 2020).

14.3 Results

Using ITS rDNA barcoding we identified 327 endophytic isolates from 844 leaf discs collected from five study sites across two months, representing 51 fungi (Table 14.1). Of these, 29 could only be identified to genus based only on the closest type material sequence matches in NCBI's GenBank database showing pair-wise identity scores of > 97% for the isolates in question (see Adams et al., 2024 Supplemental T1). Several fungi recovered may represent species that are unrepresented in the current GenBank database.

It is important to note that the sampling effort differed by month due to (i) the number of sites that could be sampled, as outlined by the scientific permit issued by the Nova Scotia Department of Natural Resources, and (ii) during June sampling the oldest + youngest leaves were collected from each plant to determine if there was a difference in their fungal community. Since no difference was observed between old and young leaves, this procedure was not done in July to avoid oversampling this endangered plant species; only youngest leaves were collected in July. Consequently, the total number of leaf discs collected from each site varied: BM1 = 250 (148 in June + 102 in July), GH4 = 112 (July only), CRN = 146 (116 in June + 74 in July), HL = 102 (July only), GH6 = 234 (130 in June + 104 in July). As a result of uneven sampling, the differing numbers of leaf discs per site hampers in depth

Table 14.1 Fungi identified according to site, collection month, and isolation media. Data used in MCA indicated by an asterisk (*)

	Taxa		Collection Site		# Sites		Month		Media	
	GH6 *	HL *	CRN	GH4	BM1 *		June	July *	MEA	MS + A
Dothideomycetes										
Alternaria sp.	●					1	●		●	●
Cladosporium sp.		●				1		●	●	
Didymocyrtis cladoniicola					●	1		●	●	
Neostagonospora elegiae				●		1		●	●	
Phaeosphaeria poagena					●	1		●		●
Ramularia sp.					●	1		●	●	
Stagonospora perfecta					●	1	●			●
Eurotiomycetes										
Penicillium sp.	●					1		●		●
Articulospora sp.	●					1	●			●
Coccomyces sp.	●					1		●		●
Leotiomycetes										
Godronia sp.	●		●			2	●	●	●	●
Helotiaceae sp. 1	●					1		●		●
Helotiaceae sp. 2	●					1		●	●	●
Lachnum virgineum			●			1		●		●
Mollisia melaleuca	●					1		●	●	●
Mollisia sp.	●	●	●		●	4	●	●	●	●
Phacidium sp.	●					1	●		●	
Phlyctema sp.	●					1		●	●	●
Phlyctema phoenicis	●					1	●		●	
Rhexocercosporidium sp.			●	●		2		●	●	●
Varicosporium elodeae					●	1		●		●
Sordariomycetes										
Apiognomonia hystrix					●	1		●	●	
Asteroma alneum	●	●	●		●	4		●	●	●
Colletotrichum sp. 1					●	1	●		●	
Colletotrichum sp. 2				●		1		●	●	
Cryptodiaporthe aubertii	●	●	●	●	●	5	●	●	●	●
Diaporthe sp. 1				●		1		●		●
Diaporthe sp. 2				●		1		●	●	●
Diaporthe sp. 3			●			1		●		●

(continued)

Table 14.1 (continued)

	Taxa		Collection Site		# Sites		Month		Media	
	GH6 *	HL *	CRN	GH4	BM1 *		June	July *	MEA	MS + A
Diaporthe sp. 4	●	●		●	●	4	●	●	●	●
Discula sp.	●	●				2		●	●	●
Fusarium sp.					●	1		●	●	●
Gaeumannomycella caricicola		●	●			2	●	●		●
Gnomoniaceae sp. 1		●				1		●		●
Gnomoniaceae sp. 2	●					1		●	●	
Gnomoniopsis idaeicola					●	1		●	●	●
Gnomoniopsis macounii		●				1		●		●
Gnomoniopsis occulta				●		1		●		●
Microacsospora sp.	●	●		●	●	4	●	●	●	●
Ophiognomonia acadiensis	●	●	●	●		4		●	●	●
Ophiognomonia alni-viridis				●		1		●		●
Ophiognomonia aff. *gardiennetii*	●					1		●	●	●
Ophiognomonia intermedia	●	●		●		3		●	●	●
Ophiognomonia ischnostyla	●					1		●		●
Ophiognomonia sp. 1	●	●		●	●	4		●	●	●
Ophiognomonia sp. 2	●					1		●	●	
Ophiognomonia sp. 3		●				1		●		●
Physalospora vaccinii		●				1		●		●
Plagiostoma lugubre	●	●	●	●	●	5	●	●	●	●
Plagiostoma sp.	●					1		●		●
Trichoderma sp.	●					1	●			●
Totals	**27**	**16**	**10**	**14**	**16**		**14**	**44**	**31**	**40**

comparisons of the fungal communities present at each location. Nevertheless, these data provide a snapshot of total culturable foliar fungal endophyte richness. Fungal isolation from the leaf discs varied by site 74% (CRN)–98% (GH4) and media type 62–96% (MEA) and 87–100% (1/2 MS + A). Though specifics on isolation success and total number of fungal isolates per disc were not recorded, the emergence of more than one endophyte was observed for the majority of discs.

The total number of fungi observed by site (Table 14.1) was: GH6 (reference bog) = 27, BM1 (most observed habitat condition changes) = 16, CRN (Moist) = 10, GH4 (Dry) = 14, and HL (Wet) = 16; the last three sites had varying habitat condition changes. Of the 51 taxa observed, only two species, *Cryptodiaporthe aubertii* and *Plagiostoma lugubre*, were found at all five sites (Table 14.1). Conversely, 38 individual fungi were only isolated from a single site each, four species were isolated from only two sites, one species was isolated from only three sites, and six species were isolated from only four sites (Table 14.1). Only six fungi were isolated during both months. We found a difference in the fungi recovered depending on isolation media used.

In an effort to distil site trends related to changes to habitat condition, multivariate analysis was conducted on a subset of our data. For this analysis only data for sites GH6, HL, and BM1 from July were compared, as these had approximately equivalent numbers of leaf discs (for GH6 $n = 104$, for HL $n = 102$, and for BM1 $n = 102$). The overlap in fungi for just these three sites is shown in the Venn diagram presented in Fig. 14.2a. Overall, seven fungi were shared by all three sites, three fungi were shared between GH6 (reference bog) and HL (wet, varying habitat changes), while no fungi were shared between both GH6 and BM1 (most habitat changes of sampled sites) and HL (wet, varying habitat changes) and BM1 (Fig. 14.2a) pairings. In addition to observing an overlap in species composition, a multiple correspondence analysis (MCA) was used to assess presence/absence trends between the observed fungal species and the media on which they were isolated. MCA results are shown in Fig. 14.2b. For the MCA, Dimension 2 (Dim2) captured 40.4% of the data and showed that the fungi present at HL differ from those of GH6 or BM1 (Fig. 14.2b). The remaining 59.6% of the dataset was captured by Dimension 1 (Dim1) of the MCA and showed a separation of the sites by the number of observed changes to habitat conditions (Fig. 14.2b). Fungi that were unique to GH6 were isolated on $^1/_2$ MS + A media or both media types, while fungi unique to site BM1 were isolated on 2% MEA or both media types. The seven fungi that were isolated from all sites (GH6, HL, & BM1) in July were found on both media types (Fig. 14.2).

14.4 Discussion

Our results uncovered for the first time a large fraction of the culturable fungi associated with *Geum peckii*. We observed 51 ascomycete fungi from the classes Sordariomycetes, Leotiomycetes, Dothideomycetes, and Eurotiomycetes. We observed foliar endophytic richness was higher at the reference bog site relative to the sites where there were observable changes to the habitat conditions. When attempting to correct for sampling bias, this trend was still upheld, suggesting that along the range of observable changes to the sampled sites habitat conditions, the endophytic richness changes and may act in response to habitat conditions present at each site. In the discussion that follows, we will review not only the trends in the fungi observed, but

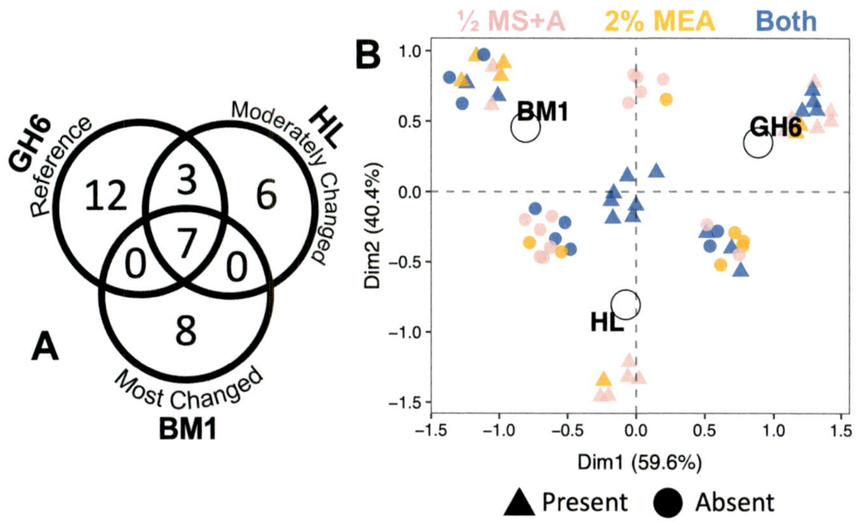

Fig. 14.2 Venn diagram **A** and MCA **B** comparing the fungal communities reported for BM1 (most observed changes to habitat conditions), HL (wet, moderately changed habitat conditions), and GH6 (reference bog) and the media type used for fungal isolations. Please note that only data from July sampling were used in Fig. 14.2

also past studies of Sordariomycetes, the taxonomic group most frequently recovered in our samples.

14.4.1 Trends in the Endophytic Fungi from Geum peckii

Previous work has shown that a plant species living across a landscape will share a common core fungal assemblage (Henning et al., 2021). With this in mind, one would expect that all sample sites would to some degree have similar fungal diversities that differ in species due to the limitations of spore dispersal and ecological niches present at each site, while some fungal species would be found regardless of site type (Henning et al., 2021; Seabloom et al., 2019). There were seven species that were observed at all three sites (i.e. for the sites that had similar sampling intensity). These fungi may be part of a fungal community that is 'common' to the bog ecosystems in Digby Neck (NS), including Brier Island. Endophyte communities have been found to have strong associations with their host plant species and across the geographic region wherein the host plant species is found (Darcy et al., 2020; DeMers and May 2020). This indicates that presence of the fungal endophytes observed during our study could be shaped by either: a) similar ecological niches being present at all sites, and/or b) these fungal species are associates of *G. peckii*. The notion of fungi being associated with the mycobiome of *G. peckii* is of some merit. For instance, *Cryptodiaporthe aubertii*, which was found at all five sites observed during our study, was also found in aseptically grown *G. peckii* transplants, which were out planted to a comparable bog habitat on Digby Neck at which *G. peckii* had not been observed previously (Fancy et al., 2020; Bradbury 2017 unpublished data).

Our study found that sites which had more observable changes to their habitat conditions had lower fungal richness than that of the reference site. Due to the changes in habitat conditions which have occurred, these sites display altered plant communities, plant health, and available ecological niches, compared to the reference site. These factors are known to impact fungal richness present within a habitat (Henning et al., 2021; Lumibao et al., 2019; Seabloom et al., 2019). Increased nitrogen availability has been found to decrease the plant diversity and in turn reduced the fungal endophyte richness present (Henning et al., 2021; Oono et al., 2020). Diverse plant communities have been shown to be better able to resist fungal colonisation by pathogens (Henning et al., 2021). Whereas plant communities undergoing stressors are less diverse and are more likely to host fast growing generalist fungi (Lumibao et al., 2019; Seabloom et al., 2019). The majority of the species observed during our study (38 of 51) were documented only once, which is a common finding of endophytic work from temperate climates investigating a single host species (Rodriguez et al., 2009). These fungi often have generalist ecologies and may represent the fastest growing species in culture (Rodriguez et al., 2009). Our small study size (844 leaf discs across five sites) further indicates that only a small portion of the foliar fungal community of *G. peckii* may have been observed during this study.

Our MCA analysis (Fig. 14.2b) also showed that all three of the similarly sampled sites cluster separately based on species composition, suggesting that geography, climate, and/or habitat status may be at play in shaping endophytic fungal assemblages. Site HL is located approximately 20 km northeast of the BM1 and GH6 sites (see Fig. 14.1 inset I–A). Although this geographical distance may be enough to alter the fungal assemblages (Lumibao et al., 2019), it is important to note that the degradation at BM1 is ongoing, whereas the ATV trail disturbance and active water level increase at HL ended during the early to mid 2000s, though the ingrowth of shrubby species is ongoing (LaRue 2016b). No species were shared between HL and BM1 nor GH6 and BM1, while three species were shared between HL and GH6. This would indicate that despite a greater geographic distance, the fungal community present at HL is more similar to GH6 then it is to BM1, yet HL still remains less diverse than GH6. These findings are likely due to the differences in plant communities present at the three sampled sites. The HL site hosts a wide variety of herbaceous and woody plants that were not present at BM1 or GH6 (Adams et al., 2024 Supplemental T2). This difference in plant community could account for the unique fungi at HL, while the recovered state of the HL bog could explain the species shared between HL and GH6.

Despite all of the fungi being new records for *G. peckii*, the majority of the fungi found are known plant pathogens or saprobes of either conifers, hardwoods, or herbaceous plant hosts (Crocker et al., 2016; Mirhosseini et al., 2015; Walker et al., 2013; Żukiewicz-Sobczak et al., 2012). The encroachment of woody shrubs and competition with other herbaceous plants contributes to the population loss of *G. peckii* and was observed to varying degrees at all sites sampled (Adams et al., 2024

Supplemental T2). Woody plant species (e.g. huckleberry (*Gaylussacia baccata*), Labrador tea (*Rhododendron groenlandicum*), lambkill (*Kalmia angustifolia*), bog laurel (*Kalmia polifolia),* sweet gale (*Myrica gale*), and saplings of black spruce (*Picea mariana*) and speckled alder (*Alnus incana* ssp. *rugosa*) were common to all or most of our sampled sites (Adams et al., 2024 Supplemental T2). The fungi observed in *G. peckii* are potentially pathogenic to these surrounding woody plant species indicating that *G. peckii* may serve as a reservoir of numerous latent fungal pathogens, while further indicating that the fungi may not be pathogenic for the duration of their life cycles or that they may only be pathogenic to specific species in this bog ecosystem (Vašutová et al., 2021; Karolina et al., 2019; Filippova and Thormann 2015; Rodriguez et al., 2009). Though pathogenicity of these fungal species was not documented for *G. peckii,* it should be considered that as new plant species establish in what was historically *G. peckii* habitat, new fungal species are arriving as well. These fungi may be antagonists of *G. peckii,* playing an unseen role in the decline of this species. Further investigation of the fungal assemblages of the ingrowing species in the *G. peckii* habitats is required to determine which fungi are associated with which plants and whether they are antagonistic to the continued survival of *G. peckii*. In 2017, major restoration efforts returned the level of the water table in Big Meadow Bog to its original level (East Coast Aquatics 2018). It would be of interest to repeat sampling at BM1 to see if the changes to the habitat conditions have had an impact on the fungal species richness and what species are now present.

Due to sampling differences between months, with more samples be taken in July compared to June, it was not possible to determine the effect of time on observed fungal species richness. This said, previous studies have found that the foliar fungal diversity of plants in temperate to boreal climates can change with season and appears to peak between July and August (Filippova and Thormann 2015; Albrectsen et al., 2010). This would suggest that in June, we likely sampled just prior to fungal peak, and in the future, it would be advantageous to sample over a longer time period to see if additional endophytic fungi might be uncovered. In the future we recommend a metagenomics approach to capture not only the culturable but also the unculturable fungal species richness present. However, a strength of our approach is the availability now of isolated fungi of interest for future research and use in plant propagation and restoration studies.

14.4.2 Known Ecologies of Sordariomycetes Recovered from Geum peckii

Of the 51 fungal taxa observed during our study, the majority (59%) were Sordariomycetes. Previous studies have found that foliar fungal endophytes are typically comprised of plant pathogens, endophytes, and latent saprobes from Sordariomycetes (most often isolated from woody material), as well as Dothideomycetes that are isolated from herbaceous material (Luo et al., 2018; Arnold 2007). Saprophytic fungi are responsible for the decomposition of plant litter, which rapidly accumulates in bog habitats in temperate climates, and is broken down and incorporated into the substrate (Luo et al., 2018; Lau et al., 2013; Raja et al., 2008). Sordariomycetes produce an array of enzymes capable of degrading cellulose, lignocellulose and hemicellulose present in woody plant cells (Bomble et al., 2017). Along Digby Neck, the bog soils are comprised of peat and mineral soil, and degradation of plant litter by microbes (including fungi) allows for the cycling of bioactive nutrients to occur (Luo et al., 2018; Lau et al., 2013; Raja et al., 2008). Of the Sordariomycetes observed, 16 of the species were from the fungal family Gnomoniaceae. Fungi from this family typically occur on or in plant leaves and have been documented globally (Walker et al., 2013; Walker et al., 2010).

Gnomoniopsis species have narrow host ranges and are often pathogens of *Rosaceae* (Walker et al., 2010). The species herein observed from this genus, *G. idaeicola*, *G. occulta*, and *G. macounii*, have been reported as only occurring within *Rubus* sp., *Potentilla* sp., and *Spirea* sp., respectively. We can now report that *G. peckii* is another *Rosaceae* host for these species. *Rubus* sp., *Potentilla* sp., and *Spirea* sp. were documented at most of our sites, though not always within the immediate vicinity of the *G. peckii* plants we sampled (Adams et al., 2024 Supplemental T2).

Similarly, four previously described *Ophiognomonia* species known from *Picea rubens* (*O. acadiensis*) (detected at all sites other than the highly impacted site in our study), or from *Betula* and *Alnus* sp. (*O. alni-viridis*, *O. intermedia*) (located at three of our sites) and *O. ischnostyla* were reported by our study (Becker et al., 2020; Melo et al., 2020; Johnston et al., 2019; Shestibratov et al., 2018; Kowalski et al., 2016; Tobias et al., 2016; Hüseyin and Selçuk 2014; Udayanga et al., 2014; Walker et al., 2014; Walker et al., 2013; Mejía et al., 2011; Walker et al., 2010; Polashock et al., 2009; Holm and Holm 1991). *Ophiognomonia* species were also found during our study. The finding of these species in *G. peckii* is unsurprising, as this plant is a member of the *Rosaceae* family, of which *Ophiognomonia* are known pathogens (Walker et al., 2013). To fully understand the ecology of this fungal genus expanded host sampling is needed to determine if the host range of this species is restricted to the *Rosaceae* (Walker et al., 2013, 2010). *Asteroma alneum* was found at all our sites and all habitat types except GH4 and is a known pathogen of alder (*Alnus* spp.) (Hüseyin and Selçuk 2014); *Alnus incana* ssp. *rugosa* was present at all our sites.

Of special note was the species *Cryptodiaporthe aubertii*, which was found as a *G. peckii* foliar endophyte at all sites. Intriguingly, this species has also been reported from the foliar tissues of *G. peckii* grown from seed in sterile tissue culture prior to ex situ planting in the Digby Neck bog habitat, where no *G. peckii* have previously been observed (Bradbury 2017 unpublished data). This fungus is a known endophyte and is possibly a latent saprophyte of Sweet Gale (*Myrica gale*), collected in bogs in Sweden and Russia (Tobias et al., 2016; Holm and Holm 1991). *Plagiostoma lugubre* was also found at all sites and is a known endophyte and latent saprophyte with a wide range of host plants throughout the northern hemisphere (Mejía et al., 2011). *Diaporthe* sp. 4 was observed at all sites except CRN, and members of this genus have broad plant associations in temperate climates, where they have been reported as endophytes and minor plant pathogens (Udayanga et al., 2014). Some species prefer damp habitats and are quick to colonise plant tissues, outcompeting other species (Becker et al., 2020; Kowalski et al., 2016).

Another commonly encountered endophyte was the hyphomycete *Microascospora* sp., found at four sites (BM1, HL, GH4, GH6). Fungi in this genus may be coprophilous, with some species known from rotting plant matter and soil (Melo et al., 2020).

Fourteen fungi from the classes Leotiomycetes and Eurotiomycetes were isolated during our study, 11 of which were only found at a single site. Leotiomycetes are often found as plant endophytes and pathogens, showing low host specificity, and often co-occurring between hosts (Johnston et al., 2019). In the case of the coastal bog environment inhabited by *G. peckii*, Leotiomycete species have previously been reported from *Rhododendron* sp., *Vaccinium* sp. (*Coleophoma* sp., *Godronia* sp., and *Phacidium* sp.) (Filippova and Thormann 2015; Polashock et al., 2009; Wu et al., 1996), *Chamaedaphne calyculata*, *Andromeda polifolia*, *Kalmia angustifolia*, *Alnus* sp., *Betula* sp. (*Godronia* sp.) (Filippova and Thormann 2015; Filippova 2012), and *Abies balsamea* (*Mollisia melaleuca*) (Tanney and Seifert 2020). The aquatic hyphomycetes *Varicosporium elodeae*, and *Articulospora* sp. have been isolated from the soil and roots of plants including *Picea glauca* (Walsh et al., 2018; Asemaninejad et al., 2017). These plant species were common to the sites that were sampled for this study. Eurotiomycetes are commonly saprobes (Visagie et al., 2014).

Understanding the fungal community associated with *G. peckii* will be key to ex situ efforts to propagate this species for in situ or potential assisted migration of the species. Our study provides preliminary fungal species which require further investigation and study to determine which roles they play within this endangered plant species. Previous work on root endophytes by Sikes et al. (2016) found that the fungal assemblage present in 'adult' plants did not always confer the same benefits to seedlings, as the fungi may interact in a pathogenic manner. Further investigation of both young and adult *G. peckii* is also required to elucidate the fungal richness and assemblages of each of these life stages.

14.5 Conclusions

In summary, we present the first preliminary survey of foliar endophytic fungi from the rare and endangered Eastern Mountain Avens, *Geum peckii*. We documented 51 ascomycete fungi from the classes Sordariomycetes, Leotiomycetes, Dothideomycetes, and Eurotiomycetes. Endophytic fungal species richness changed with observed changes to habitat condition, suggesting that this may influence the fungal assemblage present within leaves. A core fungal assemblage was documented for the first time in this plant host, from multiple sites throughout its Canadian range. Additionally, we present new host records for Rosaceae and make valuable linkages between host plants in the habitats and shared endophytes within *G. peckii* that warrant further ecological investigation. We have provided the first mycobiome assessment of *G. peckii*, an understanding that may inform the conservation and propagation of this endangered plant species.

14.6 Data Availability Statement

DNA sequence data are available at NCBI Genbank under accession numbers MW478644–MW478695. Additional data are available upon request from corresponding author.

14.7 Conflicts of interest

The authors declare no conflict of interest.

Acknowledgements Work was conducted under a Nova Scotia Department of Natural Resources Scientific Collection permit. Technical assistance from L. O'Halloran, K. Heo, H. Elliott, G. Proulx, A. Young, S. Winters, D. Kristie, L. Taul and Génome Québec Sequencing and Innovation Centre (McGill University) is gratefully acknowledged.

Funding This research was funded by the Mersey Tobeatic Research Institute, the Sage Environmental Fund, the Nova Scotia Department of Natural Resources Habitat Conservation Fund, a Nova Scotia Museum Research Grant, a Mycological Society of America Martin-Baker Research Award and NSERC Discovery Grant (No. NSERC—2017–04325) to A.K. Walker. An Acadia University Honours Student Research Award and an Arthur L. Irving Family Foundation scholarship to S.J. Adams further supported this study. B.M. Robicheau was supported by an NSERC Canadian Graduate Scholarship–Doctoral award (CGS-D).

References

Adams SJ, Robicheau BM, LaRue D et al (2021) Foliar endophytic fungi from the endangered Eastern Mountain Avens (Geum peckii, Rosaceae) in Canada. Plants 10:1026. https://doi.org/10.3390/plants10051026

Adams S, Walker A, Robicheau B, Browne R, LaRue D (2024) Foliar endophytic fungi from Geum peckii in Canada. Borealis (VI). https://doi.org/10.5683/SP3/O3TLTN

Albrectsen BR, Björkén L, Varad A et al (2010) (2010) Endophytic fungi in European aspen (*Populus tremula*) leaves: Diversity, detection, and a suggested correlation with herbivory resistance. Fungal Divers 41:17–28. https://doi.org/10.1007/s13225-009-0011-y

Arnold AE (2007) Understanding the diversity of foliar endophytic fungi: progress, challenges, and frontiers. Fungal Biol Rev 21:51–66. https://doi.org/10.1016/j.fbr.2007.05.003

Asemaninejad A, Thorn RG, Lindo Z (2017) Vertical distribution of fungi in hollows and hummocks of boreal peatlands. Fungal Ecol 27:59–68. https://doi.org/10.1016/j.funeco.2017.02.002

Becker R, Ulrich K, Behrendt U et al (2020) Analyzing ash leaf-colonizing fungal communities for their biological control of *Hymenoscyphus fraxineus*. Front Microbiol 11:590944. https://doi.org/10.3389/fmicb.2020.590944

Bomble YJ, Lin CY, Amore A et al (2017) Lignocellulose deconstruction in the biosphere. Curr Opin Chem Biol 41:61–70. https://doi.org/10.1016/j.cbpa.2017.10.013

Bradbury J (2017) Research topics report: foliar endophytes of tissue culture propagated *Geum peckii* from ex-situ field site. Unpublished data, Acadia University

Chen M, Arato M, Borghi L et al (2018) Beneficial services of arbuscular mycorrhizal fungi: from ecology to application. Front Plant Sci 9:1270. https://doi.org/10.3389/fpls.2018.01270

COSEWIC (2010) COSEWIC assessment and status report on the Eastern Mountain Avens *Geum peckii* in Canada. Committee on the Status of Endangered Wildlife in Canada. Ottawa. p 33 http://www.sararegistry.gc.ca/status/status_e.cfm

Crocker EV, Lanzafane JJ, Karp MA et al (2016) Overwintering seeds as reservoirs for seedling pathogens of wetland plant species. Ecosphere 7:01281. https://doi.org/10.1002/ecs2.1281

Darcy JL, Swift SOI, Cobian G et al (2020) Fungal communities living within leaves of native Hawaiian dicots are structured by landscape-scale variables as well as by host plants. Mol Ecol 29:3102–3115. https://doi.org/10.1111/mec.15544

DeMers M, May G (2020) Habitat-scale heterogeneity maintains fungal endophyte diversity in two native prairie legumes. Mycologia 113(1):20–32. https://doi.org/10.1080/00275514.2020.1813487

Dunnington D (2023) Spatial data framework for Ggplot2. https://paleolimbot.github.io/ggspatial/

East Coast Aquatics Inc (2018) Big meadow bog hydrology restoration: as built report. http://eastcoastaquatics.ca/projects.html

Fancy S, López-Gutiérrez JC, Walker AK et al (2020) Evaluating out-planting success and mycorrhizal status of endangered Geum peckii Pursh (Rosaceae), the Eastern Mountain Avens, in Nova Scotia. PNSIS 50:269. https://doi.org/10.15273/pnsis.v50i2.10002

Filippova NV, Thormann MN (2015) The fungal consortium of *Andromeda polifolia* in bog habitats. Mires Peat 16:29

Filippova NV (2012) Discomycetes from plant, leave and sphagnum litter in ombrotrophic bog (West Siberia). Environ Dyn Glob Clim Chang 3:1–20. https://doi.org/10.17816/edgcc311-20

Frasz SL, Walker AK, Nsiama TK et al (2014) Distribution of the foliar fungal endophyte *Phialocephala scopiformis* and its toxin in the crown of a mature white spruce tree as revealed by chemical and qPCR analyses. Can J for Res 44:1138–1143. https://doi.org/10.1139/cjfr-2014-0171

Gooden B, Thompson ER, French K (2020) Do native plant associations with arbuscular mycorrhizal fungi and dark septate endophytes differ between reconstructed and remnant coastal dunes? Plant Ecol 221:757–771. https://doi.org/10.1007/s11258-019-00959-4

Henning JA, Kinkel L, May G et al (2021) Plant diversity and litter accumulation mediate the loss of foliar endophyte fungal richness following nutrient addition. Ecology 102:03210. https://doi.org/10.1002/ecy.3210

Holm K, Holm L (1991) Ascomycetes on *Myrica gale* in Sweden. Nord J Bot 11:675–687. https://doi.org/10.1111/j.1756-1051.1991.tb01279.x

Hüseyin E, Selçuk F (2014) Coelomycetous fungi in several forest ecosystems of Black Sea provinces of Turkey. Agric for 60:19–32

Johnston PR, Quijada L, Smith CA et al (2019) A multigene phylogeny toward a new phylogenetic classification of Leotiomycetes. IMA Fungus 10:1–22. https://doi.org/10.1186/s43008-019-0002-x

Karolina G, Ewa W, Rafał S et al (2019) Endophytic fungi and latent pathogens in the sedge *Carex secalina* (Cyperaceae), a critically endangered species in Europe. Plant Prot Sci 55:102–108. https://doi.org/10.17221/120/2018-pps

Kowalski T, Kraj W, Bednarz B (2016) Fungi on stems and twigs in initial and advanced stages of dieback of European ash (*Fraxinus excelsior*) in Poland. Eur J for Res 135:565–579. https://doi.org/10.1007/s10342-016-0955-x

LaRue D (2016a) In situ and ex situ propagation of the globally rare *Geum peckii* (Rosaceae). Rhodora 118:315–319. https://doi.org/10.3119/15-38

LaRue D (2016b) Natural regeneration of the globally rare *Geum peckii* (Rosaceae) on Brier Island, Nova Scotia. Rhodora 118:310–314. https://doi.org/10.3119/15-39
LaRue D, Toms B (2013) Building for the future: establishing a foundation for the recovery of Eastern Mountain Avens: final report. HSP Project: 2012HSP6304. Mersey Tobeatic Research Institute for Environment Canada and the Eastern Mountain Avens Recovery Team. Mount Merrit, NS. p 104
Lau MK, Arnold AE, Johnson NC (2013) Factors influencing communities of foliar fungal endophytes in riparian woody plants. Fungal Ecol 6:365–378. https://doi.org/10.1016/j.funeco.2013.06.003
Lumibao CY, Condon BET, B, et al (2019) Site-specific responses of foliar fungal microbiomes to nutrient addition and herbivory at different spatial scales. Ecol Evol 9:12231–12244. https://doi.org/10.1002/ece3.5711
Luo B, Sun H, Zhang Y et al (2018) Habitat-specificity and diversity of culturable cold-adapted yeasts of a cold-based glacier in the Tianshan Mountains, northwestern China. Appl Microbiol Biotechnol 103:2311–2327. https://doi.org/10.1007/s00253-018-9512-5
Martin F, Kohler A, Murat C et al (2016) Unearthing the roots of ectomycorrhizal symbioses. Nat Rev Microbiol 14:760–773. https://doi.org/10.1038/nrmicro.2016.149
Mejía LC, Castlebury LA, Rossman AY et al (2011) A systematic account of the genus *Plagiostoma* (Gnomoniaceae, Diaporthales) based on morphology, host-associations, and a four-gene phylogeny. Stud Mycol 68:211–235. https://doi.org/10.3114/sim.2011.68.10
Melo RFR, Gondim NHDB, Santiago ALCMDA et al (2020) Co-prophilous fungi from Brazil: updated identification keys to all recorded species. Phytotaxa 436:104–124. https://doi.org/10.11646/phytotaxa.436.2.2
Mirhosseini H, Rahimian H, Babaeizad V et al (2015) Outbreak of leafspot on blackberry (Rubus fruticosus) caused by Gnomoniopsis sp.in Iran. N Dis Rep 31:9. https://doi.org/10.5197/j.2044-0588.2015.031.009
Murashige T, Skoog F (1962) A revised medium for rapid growth and bio assays with tobacco tissue cultures. Plant Phys 15:473–497. https://doi.org/10.1111/j.1399-3054.1962.tb08052.x
Oono R, Black D, Slessarev E et al (2020) Species diversity of fungal endophytes across a stress gradient for plants. New Phytol 228:210–225. https://doi.org/10.1111/nph.16709
Pebesma E (2018) Simple features for R: standardized support for spatial vector data. R J 10:439–446. https://doi.org/10.32614/rj-2018-009
Polashock JJ, Caruso FL, Oudemans PV et al (2009) The North American cranberry fruit rot fungal community: a systematic overview using morphological and phylogenetic affinities. Plant Pathol 58:1116–1127. https://doi.org/10.1111/j.1365-3059.2009.02120.x
R Core Team (2020) R: a language and environment for statistical computing. R foundation for statistical computing, Vienna
Raja HA, Schmit JP, Shearer CA et al (2008) Latitudinal, habitat and substrate distribution patterns of freshwater ascomycetes in the Florida peninsula. Biodivers Conserv 18:419–455. https://doi.org/10.1007/s10531-008-9500-7
Raja HA, Miller AN, Pearce CJ et al (2017) Fungal identification using molecular tools: a primer for the natural products research community. J Nat Prod 80:756–770. https://doi.org/10.1021/acs.jnatprod.6b01085
Rana KL, Kour D, Sheikh I et al (2019) Biodiversity of endophytic fungi from diverse niches and their bio-technological applications. In: Singh BP (ed) Advances in endophytic fungal research: Present status and future challenges. Springer International Publishing, Switzerland, pp 105–144
Rodriguez RJ, White JF Jr, Arnold AE et al (2009) Fungal endophytes: diversity and functional roles. Tan Rev N Phytol 182:314–330. https://doi.org/10.1111/j.1469-8137.2009.02773.x
RStudio Team (2020) RStudio: integrated development for R. RStudio, PBC, Boston
Schoch CL, Seifert KA, Huhndorf S et al (2012) Nuclear ribosomal internal transcribed spacer (ITS) region as a universal DNA barcode marker for Fungi. Proc Natl Acad Sci USA 109:6241–6246. https://doi.org/10.1073/pnas.1117018109

Schulz B, Wanke U, Draeger S et al (1993) Endophytes from herbaceous plants and shrubs: effectiveness of surface sterilization methods. Mycol Res 97:1447–1450. https://doi.org/10.1016/s0953-7562(09)80215-3

Seabloom EW, Condon B, Kinkel L et al (2019) Effects of nutrient supply, herbivory, and host community on fungal endophyte diversity. Ecology 100:e02758. https://doi.org/10.1002/ecy.2758

Shestibratov KA, Baranov OY, Subbotina NM et al (2018) Early detection and identification of the main fungal pathogens for resistance evaluation of new genotypes of forest trees. Forests 9:732. https://doi.org/10.3390/f9120732

Sikes BA, Hawkes CV, Fukami T (2016) Plant and root endophyte assembly history: interactive effects on native and exotic plants. Ecology 97:484–493. https://doi.org/10.1890/15-0635.1

Slowikowski K, Schep A, Hughes S et al (2020) Ggrepel: automatically position non-overlapping text labels with Ggplot2

Tanney J, Seifert K (2020) Mollisiaceae: An overlooked lineage of diverse endophytes. Stud Mycol 95:293–380. https://doi.org/10.1016/j.simyco.2020.02.005

Tobias A, Balashova N, Kiselev G (2016) The materials to study of Lebyazhy nature reserve microfungi (Lomonosov district of Leningrad region). Biol Comm 4:37–47. https://doi.org/10.21638/11701/spbu03.2016.403

Udayanga D, Castlebury LA, Rossman AY et al (2014) Insights into the genus *Diaporthe*: Phylogenetic species delimitation in the *D. eres* species complex. Fungal Divers 67:203–229. https://doi.org/10.1007/s13225-014-0297-2

Vašutová M, Jiroušek M, Hájek M (2021) High fungal substrate specificity limits the utility of environmental DNA to detect fungal diversity in bogs. Ecol Indic 121:107009. https://doi.org/10.1016/j.ecolind.2020.107009

Visagie C, Houbraken J, Frisvad J (2014) Identification and nomenclature of the genus *Penicillium*. Stud Mycol 78:343–371. https://doi.org/10.1016/j.simyco.2014.09.001

Walker DM, Castlebury LA, Rossman AY et al (2010) Systematics of genus *Gnomoniopsis* (Gnomoniaceae, Diaporthales) based on a three gene phylogeny, host associations and morphology. Mycology 102:1479–1496. https://doi.org/10.3852/10-002

Walker DM, Castlebury LA, Rossman AY et al (2013) Host conservatism or host specialization? Patterns of fungal diversification are influenced by host plant specificity in *Ophiognomonia* (Gnomoniaceae: Diaporthales). Biol J Linn Soc 111:1–16. https://doi.org/10.1111/bij.12189

Walker A, Hirooka Y, Walker D et al (2014) *Ophiognomonia acadiensis*. Fungal planet 274. Per-Soonia 32:290–291

Walsh E, Duan W, Mehdi M (2018) *Cadophora meredithiae* and *C. interclivum*, new species from roots of sedge and spruce in a western Canada subalpine forest. Mycology 110:201–214. https://doi.org/10.1080/00275514.2017.1406748

White TJ, Bruns TJ, Lee SJ et al (1990) Amplification and direct sequencing of fungal ribosomal RNA genes for phylogenetics. In: PCR: protocols and applications: a laboratory manual. San Diego: Academic Press. p 315–322

Wickham H (2016) Ggplot2: elegant graphics for data analysis. Springer, New York

Wickham H, Averick M, Bryan J et al (2019) Welcome to the Tidyverse. J Open Source Softw 4:1686

Wu W, Sutton BC, Gange AC (1996) *Coleophoma fusiformis* sp. nov. from leaves of Rhodo-dendron, with notes on the genus *Coleophoma*. Mycol Res 100:943–947. https://doi.org/10.1016/s0953-7562(96)80046-3

Yan L, Zhu J, Zhao X et al (2019) Beneficial effects of endophytic fungi colonization on plants. Appl Microbiol Biotechnol 103:3327–3340. https://doi.org/10.1007/s00253-019-09713-2

Żukiewicz-Sobczak W, Cholewa G, Krasowska E et al (2012) Pathogenic fungi in the work environment of organic and conventional farmers. Adv Dermatol Allergol 4:252–262. https://doi.org/10.5114/pdia.2012.30463

Chapter 15
Ex Situ Conservation of the Globally Imperiled Eastern Mountain Avens (*Geum Peckii*) in Nova Scotia, Canada

Robin D. Browne, Sarah Hines, Alain Belliveau, Sarah J. Adams, Juan C. López-Gutiérrez, Katie King, Rebecca S. Dodge, and Allison K. Walker

Abstract Seed banks and botanical gardens worldwide seek to preserve plant germplasm ex situ as a complementary conservation tool in support of in situ habitat restoration efforts. Studies on population stability of Eastern Mountain Avens (*Geum peckii*), an endangered plant species found in Nova Scotia, identified a need for additional conservation strategies. Our work represents a significant advancement toward development of ex situ conservation methods for *G. peckii*. Ex situ conservation research on *G. peckii* began in 2012 at the K.C. Irving Environmental Science Centre's Harriet Irving Botanical Gardens and Acadia Seed Bank. Methods for seed collection, processing, storage, germination, propagation, and outplanting were developed for this species and implementation of ex situ conservation of *G. peckii* is now possible. This research marks the first time in Nova Scotia that an endangered plant species was successfully retrieved from seedbank storage, propagated by tissue culture and outplanted back into a natural habitat. This research also confirmed for the first time *G. peckii* grows in association with arbuscular mycorrhizal fungi, and *G. peckii* propagated in tissue culture can be effectively inoculated with these fungi and outplanted successfully.

Keywords Arbuscular mycorrhizal fungi · Ex situ conservation · *Geum peckii* · Seed bank

R. D. Browne · S. J. Adams · J. C. López-Gutiérrez · K. King · R. S. Dodge · A. K. Walker
Department of Biology, Acadia University, 33 Westwood Avenue, Wolfville, NS B4P 2R6, Canada

R. D. Browne · S. Hines (✉) · A. Belliveau · K. King
K.C. Irving Environmental Science Centre, Acadia University, 32 University Avenue, Wolfville, NS B4P 2R6, Canada
e-mail: sarah.hines@acadiau.ca

A. Belliveau
e-mail: alain.belliveau@acadiau.ca

N. M. Hill et al. (eds.), *Wetland Restoration for Endangered Species Recovery*, Environmental Contamination Remediation and Management,
https://doi.org/10.1007/978-3-031-71344-6_15

15.1 Introduction

15.1.1 The Value of Ex Situ Conservation

Native plants around the world face threats from habitat loss and degradation, population fragmentation, invasive alien species, and climate change. It is imperative to adopt appropriate conservation strategies to protect native biodiversity and maintain the natural resources that sustain life on earth. Conservation measures can be divided into in situ measures, which occur within native habitat, and ex situ measures, which occur outside native habitat (Convention on Biological Diversity 2012). Ex situ approaches are often complementary to in situ work by providing material for safeguarding, studying, and restoring the species while minimizing stress on the extant populations in their natural habitats. Environmental policy is increasingly recognizing the value of ex situ conservation. Conservation initiatives supported by Canada, such as the Global Strategy for Plant Conservation, recognize the importance of integrating ex situ methods with in situ conservation (Convention on Biological Diversity 2012; Galbraith, 2010; New York Declaration 2011) and promote the development and application of ex situ conservation.

Seed banks serve as an ex situ strategy to preserve genetic diversity among and within species in the long term. There are two established seed banks in the Maritime Provinces: The National Tree Seed Centre in Fredericton, New Brunswick, and the Acadia Seed Bank (ASB) at the K.C. Irving Environmental Science Centre (KCIESC), Acadia University, Wolfville, Nova Scotia. The National Tree Seed Centre holds collections of many woody species from across Canada, though material is generally not available for restoration activities. The KCIESC and ASB are dedicated to conservation of Wapna'ki plants through a seed bank, a tissue culture lab, botanical gardens and a herbarium. Institutions worldwide such as the Millennium Seed Bank of the Royal Botanic Gardens in Britain play a major role in promoting ex situ conservation strategies globally (Hird and Kramer 2013).

To ensure optimal ex situ conservation outcomes, seed banks use different propagation methods to continually assess stored material for its viability and usability for research. A common and inexpensive method is to sow seeds or other plant material (e.g., cuttings) in soil or wet paper towel and document germination rates. Over the past two decades, tissue culture (or in vitro) propagation has also been used to an increasing extent as an additional strategy (Maunder et al. 2004; Bunn et al. 2007; Pence 2011; Kulak et al. 2022; Radomir et al. 2023).

The use of tissue culture techniques has pros and cons for ex situ conservation work. Tissue culture can be advantageous by yielding higher seed germination rates for some species, multiplying specimens rapidly using only a small amount of material, and by providing the opportunity for cryopreservation at ultra-low temperatures (− 150 to − 192 °C) which can significantly extend seed and vegetative tissue life (Sharma et al. 2015). Some disadvantages include the higher cost and time requirements and the aseptic environment required (i.e., lack of beneficial microbial species) (Kumar and Rao, 2012). Although tissue culture is a clonal propagation technique, genetic diversity for restoration efforts can be initiated and sustained through judicious selection of multiple genetic sources. Given the benefits for some species' research and conservation scenarios, tissue culture and cryopreservation have become key elements of ex situ conservation programs in other parts of the world, such as Royal Botanic Gardens in Britain and the Center for Plant Conservation in the United States (Galbraith 2010; Paunescu 2009; Hird and Kramer 2013).

15.1.2 A Globally Imperiled Species: Geum Peckii *(Eastern Mountain Avens)*

Nova Scotia is part of the ecologically unique Wapna'ki / Acadian Forest Region (herein referred to as Wapna'ki) and is home to many plant species that are rare or at risk of extinction, yet not represented in ex situ collections. One example is the *Geum peckii* (Eastern Mountain Avens), a yellow-flowered herbaceous species of the rose family that globally only grows in two small areas within Wapna'ki. These are Brier Island (BI) and Digby Neck in Digby County, Nova Scotia, and the White Mountains of New Hampshire, United States. Generally, *G. peckii* grow in lower acidity peaty fen or bog habitat.

G. peckii is considered a provincially and federally endangered species, and is globally imperiled (COSEWIC 2010), making it a priority species for conservation efforts. Habitat degradation, declining populations, and changing climate are all significant threats and have prompted initiatives for ongoing in situ conservation and recovery (Book chapters 4, 5, 10). Beginning in 2012, the KCIESC and ASB began ex situ research on *G. peckii* while concurrently developing capacity for seed banking and tissue culture technology. For over ten years, this work evolved and adapted to, helping ensure the long-term persistence of this species while minimizing the impact on wild populations (LaRue 2016b).

15.2 Trials, Trials, and More Trials: Summary of *G. Peckii* Ex Situ Research from 2012-Present

Research began in 2012 when the Mersey Tobeatic Research Institute (MTRI) contacted Acadia University to engage in collaborative ex situ research involving *G. peckii*. These studies began with Diane LaRue (Research Associate, MTRI) in partnership with the KCIESC and ASB which were keen to expand their ex situ conservation program for the preservation of native plant species within Wapna'ki. In addition to seed-related studies, rhizome growth and multiplication were also investigated to examine this clonal strategy for sustaining *G. peckii* populations. The *G. peckii* ex situ research program then expanded to include tissue culture propagation, fungal relationship investigation, outplanting, and field studies.

Based on varying collaborations and research opportunities, the timeline of *G. peckii* ex situ conservation research activity at KCIESC-HIBG can be divided into three phases: Phase 1 (2012–2014), Phase 2 (2015–2019) and Phase 3 (2020 to present). Each phase of the work was completed in concordance with permits obtained from Nova Scotia Department of Natural Resources and Renewables. Across all three phases, *G. peckii* ex situ conservation research activities at KCIESC and ASB from 2012 to 2024 have explored the following research themes in Fig. 15.1.

Fig. 15.1 A summary of *G. peckii* ex situ conservation research themes from 2012-present. **A** collecting seed from Brier Island populations **B** Processed seeds of *G. peckii* **C** *G. peckii* growing in tissue culture on a biochar media **D** working in the biosafety cabinet under aseptic conditions for tissue culture research **E** *G. peckii* outplanted into bins in the greenhouse **F** a *G. peckii* two months after outplanting on Long Island **G** Outplanting *G. peckii* into the 2 × 2 m grid on Long Island. Photo credit A: Alain Belliveau, B: Robin Browne, C: David Kristie, D through G: Sarah Hines

15.2.1 Phase 1: Preliminary Germination Trials (2012–2014)

Phase 1 was completed primarily through a research partnership with MTRI, with initial efforts directed toward the addition of *G. peckii* into the ASB and the development of methods for germination and propagation testing. It supported in situ conservation efforts by providing additional information concerning the sexual and asexual reproductive strategies of *G. peckii* and helped increase the capacity of the KCIESC and ASB. Preliminary trials involved the investigation of seed germination, propagation via seedlings and rhizomes, and subsequent outplanting of propagules. Grow-out studies (i.e., outdoor growth of specimens to maturity) were also undertaken with plants from germination and propagation trials. The research included investigations of tissue culture-based propagation and outplanting methods. As Phase 1 progressed, it became clear that a more rigorous approach was needed to further advance the objective of successful ex situ conservation for this species. Subsequently, the results of these efforts laid the foundation for several more research projects in Phase 2.

15.2.1.1 Seed Banking

The first high-priority goal was the establishment of a genetically diverse ex situ collection representing multiple Brier Island sites and the development of effective short- and long-term storage protocols. The objectives for seed banking were threefold: (1) work with the permit-issuing government to ethically collect *G. peckii* seeds to minimize the impact on wild populations, and (2) determine the viability of seeds collected, and (3) determine the number of collected seeds required for adequate long-term storage and use for further research opportunities.

In 2012, 2013, and 2014, the KCIESC, ASB and partnering organizations conducted three seed collection events on Brier Island. The first two were completed by MTRI from three BI sites (GH4, CRN, GR1; see Chapter 14 for map) in late summer 2012 and from four BI sites (GH4, GR1-5, GR1-1, CRN; see Chapter 14; in late summer 2013. Seeds were delivered to the ASB, held in 4 °C storage, and used for the preliminary germination and propagation trials. Seeds from these collections were not processed for long-term storage due to insufficient amounts; however, the logistics of processing *G. peckii* seeds for long-term storage (i.e., after-ripening, cleaning, count-estimating, desiccation) were explored in anticipation of longer-term low-temperature storage in the future. In the fall of 2014, a third and larger BI seed collection was made with support from the Nova Scotia Habitat Conservation Fund, a provincially administered program that in 2014 supported the growth of the ASB and ex situ conservation of rare native species of Wapna'ki. Following protocols adapted from Kew Gardens Millennium Seed Bank, these seeds were cleaned of non-seed debris, dried to less than 15% moisture content in desiccation chambers, and stored in a freezer at − 18 °C as sample ASB000018 (Table 15.1). A subsample was also held at 4 °C for continued germination testing and propagation trials.

Table 15.1 Complete long-term holdings of *Geum peckii* seed with the Acadia Seed Bank, K.C. Irving Environmental Science Centre and Harriet Irving Botanical Gardens (additional information available in the Seed Bank Registry with KCIESC-HIBG)

Accession #	Location	Date of collection	Estimated # seeds*
SB000018	BI (BM) 44.25431; –66.3592	16/10/2014	5,000
SB000052	Experimental garden KCIESC-HIBG	01/09/2017	< 500
SB000053	Experimental garden KCIESC-HIBG	01/09/2017	< 500
SB000076	Experimental garden KCIESC-HIBG	07/10/2017	< 500
SB000081	Experimental garden KCIESC-HIBG	01/10/2017	< 500
SB000101	Experimental garden KCIESC-HIBG	28/08/2018	5,000
SB000103	Experimental garden KCIESC-HIBG	29/08/2018	500
SB000148	BI (BM) 44.250662; –66.363377	19/10/ 2018	2,000
SB000149	BI (BM) 44.246241; –66.361028	19/10/2018	< 500
SB000150	BI (BM) 44.252736; –66.355364	19/10/2018	2,000
SB000169	Experimental garden KCIESC-HIBG	15/09/2019	< 500
SB000170	Experimental garden KCIESC-HIBG	15/09/2019	< 500
SB000171	Experimental garden KCIESC-HIBG	15/09/2019	27,960
SB000193	Experimental garden KCIESC-HIBG	19/08/2020	< 500
SB000281	Experimental garden KCIESC-HIBG	23/08/2021 to 10/ 09/2021	1,140
SB000336	BI 44.2406397; –66.3849284	06/10/2021	9,150
SB000337	BI 44.2458176; –66.3582371	06/10/2021	1,720
SB000338	BI 44.2440066; –66.3594492	06/10/2021	4,130
SB000339	BI 44.2334553; –66.3684129	06/10/2021	5,190
SB000340	BI 44.2483006; –66.3574251	06/10/2021	15,330

(continued)

Table 15.1 (continued)

Accession #	Location	Date of collection	Estimated # seeds*
SB000341	BI 44.245481; –66.36825	06/10/2021	4,420
SB000342	BI 44.2442522; –66.368277	06/10/2021	4,980
SB000343	BI 44.2435756; –66.3644818	06/10/2021	250
SB000344	BI 44.2427979; –66.3653169	06/10/2021	1,830
SB000359	BI 44.241542; –66.3627683	06/10/2021	1,910
SB000360	BI 44.2420828; –66.3623396	06/10/2021	6,750

15.2.1.2 Seed Germination and Growth Trials

Exploratory Ex Situ Trial

An initial series of ex situ germination and seedling growth trials (Trial 1) was undertaken in 2012 under controlled greenhouse and laboratory conditions to explore the basic needs of *G. peckii* seeds for successful germination. Based on knowledge of *G. peckii* habitat, seed ecology knowledge of cousin species, and general knowledge of seed ecology for the region, seeds were cold stratified on moist peat for 60 days before the germination trials started, and then sown into similar soil in pots or placed in Petri plates with moistened filter paper. Many seeds were likely not viable, and the germination rates and the timing of germination varied among different sites. Germination rates were lower than expected in both pots and plates, based on previous unpublished findings (LaRue, personal communication). Based on observations, seed size was identified as a potential factor affecting germination rates and further explored later in phase 1. Many of the seeds placed on Petri plates were quickly overgrown with microbial contamination, thereby limiting germination and young seedling survival. Viable seedlings from the pot and plate germination tests were transplanted and monitored under greenhouse conditions (typically 20 °C, with ambient light, and seasonally fluctuating humidity levels). These seedlings grew slowly, and a high proportion died during this greenhouse outplanting trial. This low survival rate limited ex situ research possibilities, and all remaining viable plants were destroyed in 2013 in compliance with the provincially issued permit. Based on these exploratory 2012 trials, the following trials were set up in subsequent years to further the development of effective germination protocols.

Table 15.2 A summary of the six germination trials of *G. peckii* seeds conducted during phase 1 research, from 2012 to 2014

Trial #	Year	Germination conditions	Germination rates	Notes
1	2012	Ex situ: Cold stratified on moist peat for 60 days, then sown into similar soil in pots or placed in Petri plates with moistened filter paper	Generally low	Petri dish contamination was an issue; slow growth and high mortality rate of outplanted seedlings; seed size identified as a potential factor affecting germination rate
2	2013	In situ: Native soils in bins	11–22%	Resulting seedlings exhibited slow growth from 2013 to 2015
3	2013	Ex situ: Direct sowing into native BI mineral or peat soils, or on moistened filter paper in Petri plates; some seeds cold stratified, some not	24% among cold-stratified seeds; 3% among non-stratified seeds	Germination timing ranged from 9 to 14 days; 10% of seeds produced seedlings, then 35% of these died after outplanting; surviving plants healthy for two additional growing seasons outside; all plants deceased by 2021
4	2013	Ex situ: Moistened filter paper Petri plates, and either placed directly under warm lighting with or subjected to cold-moist stratification at 4 °C first then moved to the warm lit conditions	11.7–53.3% for stratified seeds; 7.5–64% among non-stratified seeds	Germination asynchronous; effects of cold stratification initially unclear, but later deemed positive; > 100 seedlings produced and outplanted; all plants deceased by 2021
5	2013	Ex situ: Like Trial 4 but with agar-only plates	30–45%	Most consistent germination rates to date
6	2014	Ex situ: Same as Trial 4, testing the rate of different-sizes seeds	Data unavailable, but likely similar to Trial 4 germination rates	Selection of larger seeds was deemed advantageous for germination trials intended for propagation

In Situ Germination Trial in Native Soils

In 2013, seeds collected were used for in situ germination trials (Trial 2) to observe the germination of *G. peckii* seeds on BI. Immediately after seeds were collected, they were sown into bins with native soil (i.e., a mix of mostly wet, peaty soil influenced by nearby low-acid basalt-derived mineral soil) and left at two BI sites in September 2013 (LaRue 2016b). The seed germination rates were monitored in 2014 and found to be 11–22% (Table 15.2). Thirty-one seedlings resulted from these in situ trials and the seedlings remained small during the evaluation period up to the end of 2015. These results confirmed the viability of in situ germination, with results of viability and the initial slow growth pattern similar to the earlier ex situ trials.

Ex Situ Germination Trials to Test Different Substrate and Stratification Types

After insight gained from Trial 1, two more germination trials (Trial 3 and Trial 4) were initiated in 2013 (Table 15.2). Larger seeds were selected for both trials, and for Trial 3 seeds were divided equally between two germination methods: direct sowing into native BI soils, and placing seeds on moistened filter paper in Petri plates (LaRue 2016b). In a previous side project, it was found larger seeds (>2 mm length and 0.75mg fresh weight) had a 3X higher gemination rate than small seeds (<2 mm, 0.45 mg). For both substrates, seeds were further split into cold-stratified and non-stratified groups (LaRue 2016b). The soil pots were placed in greenhouse conditions while the plates were kept at 20 °C with a light regime of 16 h light and 8 h dark (LaRue 2016b). The overall germination rate was found to be 24% among the cold-stratified groups, compared with 3% among the non-stratified groups. Germination timing ranged from 9–14 days, with the non-stratified groups being slower than the cold-stratified groups. The overall viable seedling production was 10% (LaRue 2014). To allow for unobstructed root and plant growth, seedlings from the pots and plates were transferred to 4-inch liner pots with ProMix LP-15 soil and monitored under greenhouse conditions. This population of seedlings initially had strong survival, but 35% eventually succumbed to transplant shock or other causes of mortality. Most of the seedlings that did not survive originally germinated in BI mineral soil. This suggested that BI mineral soil substrate may be less suitable for seedling survival. The surviving plants produced large rosettes with ramet proliferation over two growth seasons under greenhouse and outdoor garden conditions.

For Trial 4, seeds from the same BI sites were placed on moistened filter paper in 100 × 15 mm Petri plates, and either placed directly under lighting with a 12-h photoperiod and a temperature range of 21–25 °C or subjected to cold-moist stratification in darkness at 4 °C for 6 weeks then moved to the warm lit conditions. Germination rates for cold-stratified groups ranged from 11.7 to 53.3%, and for non-stratified groups ranged from 7.5 to 64%. Notably, germination was asynchronous, occurring over a four-month period. Stratification treatments had no clear beneficial effect on the test groups. However, stratification treatments were continued in subsequent germination trials, and they indicated a positive effect on timing of germination. Interestingly, germination rates varied significantly by the site of collection. Over 100 viable seedlings from the plates were outplanted into small pots filled with ProMix LP-15 soil and monitored under greenhouse conditions. This group was added to the population of seedlings from Trial 3, for inclusion in subsequent investigations. From these trials, seed dormancy, site specific germination variables, and the potential for seed priming (i.e., pre-trial treatments) arose as potential areas of future research.

Through 2012 and 2013 there was a trend of excessive microbial contamination and drying of filter paper during the petri plant germination trials. Many of the seedlings harvested from the filter paper plates were quite small with little to no epicotyl growth and few roots and did not survive subsequent outplanting.

Concurrently at the ASB, preliminary trials of Largeleaf Avens (*Geum macrophyllum*), a related non-endangered species, showed improved germination rates on agar medium (88.3%) compared to moistened filter paper (58.3%). Because of the utility of these results, Largeleaf Avens was used as a proxy species for protocol testing before moving to testing of *G. peckii* seeds. Some *G. peckii* seeds and germinants from 2012 and 2013 were experimentally transferred from the filter paper plates to plates containing solidified agar. After some testing, researchers determined that a mild surface sterilization treatment (10% bleach for 10 min) following a one-hour imbibition period under cold running tap water was an effective method for significantly reducing microbial contamination of agar-only plates. This seed pre-treatment is commonly used in tissue culture research, and ASB researchers began to explore other ways tissue culture science could improve *G. peckii* ex situ conservation research (see 15.25).

With refined methods and using mostly agar-only plates, several germination trials (Trial 5) were completed in 2014 soon after seed collection and processing. The germination rates from these trials were found to be 30–45%. This consistent and relatively high germination rate showcased the effectiveness of agar-only trials of pre-treated seeds and contributed significantly to ongoing efforts to test viability in relation to storage time which extended into Phase 2.

Grow-Out and Long-Term Survival of Plants

Post-trial plant mortality is a common occurrence after germination trials, therefore the ~ 150 viable *G. peckii* seedlings from 2013 germination trials were transplanted into larger pots and moved outdoors in the Harriet Irving Botanical Gardens (HIBG) experimental gardens in the spring of 2014. Seedlings were placed in overwintering pots buried under soil and mulch cover in shaded conditions. Overwintering survival, vegetative growth, and flowering were monitored over the following four years. None of the plants persisted beyond the spring of 2021. The decline occurred gradually, with incidences of leaf reddening observed among most plants. Sensitivity to full light and overwintering losses were noted. The population of seedlings from germination Trials 3 and 4 survived long enough to be used in growth studies in 2014 as well as subsequent Phase 2 research projects in 2015 and 2016. Second generation seed collections of surviving plants were completed in 2017, 2018, and 2019, and placed in the ASB for long-term storage at – 18 °C (Table 15.1). The knowledge gained during this outplanting research provided a roadmap for propagating ex situ *G. peckii* indefinitely or until the possible issue of genetic bottleneck limits viable seed production, providing an important conservation tool for *G. peckii* in Nova Scotia and globally.

15.2.1.3 Establishing Tissue Culture

Seedling Tissue Culture Experimentation

The low rate of germination and viability of non-sterile seeds in pots and on moistened filter paper in plates prompted research into the use of sterilized, agar filled plates. This opened the door to the potential for tissue culture propagation and subsequent multiplication of seedlings. This research was twofold: (1) attempt to generate seedling lines (i.e., replicates) of test material for use in ongoing greenhouse and field trials and (2) determine the feasibility of prolonged maintenance of seedling lines in aseptic vials for long-term banking of *G. peckii* plant material. Monitoring and learning from these tests led to more controlled germination testing and the development of in vitro tissue culture techniques for micropropagation of selected seedling lines (i.e., the replication of *G. peckii* from one in vitro specimen into one or several other in vitro replicates). This Phase 1 research into tissue culture propagation thus involved the four recognized stages of micropropagation: initiation, stabilization, multiplication, and outplanting (Bunn et al. 2007). These are described below in more detail.

Initiation

The use of an enriched nutrient medium was investigated as a method to enhance post-germination seedling development and to improve outplanting success. A basal nutrient medium (MS, obtained from Sigma-Aldrich) was tested as part of the initial micropropagation trials. Nutrient media made with MS are widely used in plant tissue culture, MS contains essential nutrients for plant growth (Murashige and Skoog 1962). Based on initial results and results from other species, our inclusion rate of MS was modified to half of the originally indicated amount (half rate is 2.2g/L), which proved beneficial in propagation of *G. peckii* and has since been used successfully with other native species. After seedlings were transferred from agar-only plates, they were observed to have more extensive shoot and root growth, compared to seedlings left on the agar-only medium.

Seed sterilization pre-treatments were added to agar medium to further mitigate microbial growth. Plant Preservative Mixture (PPM) (sourced from Plant Cell Technology), a commercially available product used to control microbial overgrowth in tissue, was found to help control microbial overgrowth during this critical early stage of seedling development. The concentration and exposure time of PPM was adjusted on several occasions to effectively control potential contamination carried over from germination trials while not adversely affecting subsequent shoot and root growth of seedlings.

Stabilization and Multiplication

Once seedlings were successfully initiated in culture, shoot growth and rooting were monitored across several transfers from old to fresh nutrient media. Seedlings with low shoot and/or root vigor or microbial contamination were eliminated, to establish a clean and uniform culture population. During the stabilization phase, rosette proliferation was noted among individual seedlings, which indicated the potential for division and multiplication. Several cultures were then divided successfully to increase the overall number of in vitro specimens. Researchers estimated a multiplication rate of 1.5 for every six weeks of tissue culture; however, specimen health decline was noted based on exudate accumulation in the medium, leaf browning and declines in shoot multiplication over time. These issues were addressed in Phase 2 studies.

Outplanting

Eventually, some seedlings were removed from culture, placed into small pots filled with ProMix LP-15 soil, and placed inside plastic bins to regulate humidity levels. These bins were moved to a greenhouse environment with shading for the first two weeks of transition. Outplanting success with culture-derived *G. peckii* plantlets was found to be less than 30%. Plantlets were prone to desiccation during the transition period and were slow to regrow if they survived. Further investigations were necessary in Phase 2 to develop a more effective outplanting procedure.

Overall, tissue culture methodology improved but outplanting losses were high among plantlets and this negated the primary advantages of tissue culture to generate test material for research and restoration purposes. Further improvements were necessary, the technical challenges described above were identified as priorities for Phase 2 research.

15.2.1.4 Rhizome Growth and Propagation Trials

Rhizome Greenhouse and Garden Trials

Rhizome-based propagation techniques were explored starting in 2013. Reproduction via seed by *G. peckii* can become infrequent and decline, particularly in disturbed environments with dense sphagnum moss and woody vegetation encroachment, in which case asexual reproduction by rhizome can offer a developmental process which enables *G. peckii* to persist in more competitive environments (LaRue 2016b). Potential advantages of clonal growth in situ include resource sharing, resource foraging, and capacity for plastic growth responses (i.e., more variable in time and space) to heterogenous environments (LaRue, 2014; Shurgurkar et al. 1996). Very little was previously known about rhizome-based propagation as a natural reproductive strategy of *G. peckii*, though various aspects had already been characterized in relation to competition for resources (LaRue 2014, 2016a; Chapter 4). Anecdotally, growth and spread via rhizomes was frequently observed in situ and this contributed significantly to the persistence of the species on BI, particularly in sites where flowering frequency was declining (LaRue 2016b). Therefore, a study of rhizome growth was useful for understanding in situ population dynamics globally and for the propagation of ex situ material for research purposes. It represented an alternate means to generate clonal lines of test plants for research and conservation purposes.

The first attempt to map rosettes on from rhizomes on BI began in May 2013 (LaRue 2014), but initial observations indicated that *G. peckii* rhizome networks were complex and difficult to document, especially in shrubbier areas. Due to this significant challenge, a small sample of rhizome material was collected later in 2013 and brought to the KCIESC with the idea that in simpler, ex situ conditions, rhizomes could be monitored more accurately, and some material could be used for ex situ propagation testing (LaRue 2014). Soils samples (a mineral soil and a peaty/sandy soil) were also collected and incorporated into ex situ propagation testing.

Some collected rhizome sections were carefully and selectively subdivided then planted into pots of either (1) BI mineral soil, (2) BI peaty/sandy soil, or (3) ProMix LP-15 soil. The rhizome sections were placed in greenhouse conditions and rhizome growth and rosette proliferation were monitored and documented. The viability and proliferation of new rosettes was soon confirmed, along with rhizome sections' ability to survive the winter. The highest rates of survival were among plants grown in BI peaty/sandy soil and the ProMix soil. As with Trial 3 seedlings produced from seeds, outplantings from rhizomes on BI mineral soil ultimately died. This question of soil type and outplanting success was later revisited in Phase 2 research.

Other rhizome sections were used to attempt the propagation of more *G. peckii* specimens from rhizome cuttings, which had likely not been done successfully before (LaRue 2014, 2016a). These attempts were successful with a multiplication rate of 1.42 in the first two months of experimentation. However, based on the observed frequency of rosette proliferation over time, the actual rate is undoubtedly higher for *G. peckii*. The results showed that rapid propagation of *G. peckii* is possible from rhizome sections and cuttings. Researchers now consider this propagation method suitable for research and restoration efforts.

The resulting rhizome-generated plants were then re-potted into ProMix-based soil (3:1 with BI peaty/sandy soil) and studied under greenhouse conditions (LaRue 2016a,b). Plant survival declined to 24% during the first year. Reddening leaves were prevalent and was attributed to high light intensity. A subset of plants was then placed outdoors under a layer of straw in the HIBG and monitored for survival. After observing a 100% survival rate among these plants, the rest of the rhizome-generated plants were also transferred to the outdoors. Each year, researchers continued to investigate the effects of varying light and moisture levels, often attempting to simulate in situ conditions in order to increase survival rates (LaRue 2016a,b). New rosettes continued to proliferate among the plants and flowering occurred over several growing seasons. Notably, several plants contributed to leaf tissue endophyte research by Adams (2016; Chapter 14). Furthermore, second generation seeds were collected from rhizome-generated plants in the fall of 2017, 2018, and 2019 and stored in the ASB. Despite appearing healthy for several years, the plants eventually exhibited decreasing rosette and flower production and none of the plants survived beyond the winter of 2021. Many of the plants were frost heaved during their last two years of life and greater winter protection of plants may be necessary for future outplantings. Although plants did survive eight years, more research will be required to maintain ex situ plant populations in the longer term.

Tissue Culture Propagation from Rhizome Explants

To test the potential for tissue culture propagation of rhizome sections, several 2cm long sections were removed from the trials described above and subjected to sterilization following the protocol in Fig. 15.2 (with rhizome tissue instead of seeds). Sterilized tissue was placed on MS nutrient medium as described in Sect. 15.15 and monitored for propagation success. No specimens survived this preliminary testing. Exudate leakage into the culture medium and extensive necrosis of the rhizome portions were observed, indicating a need for modifications to the sterilization procedure and/or culture conditions. High rates of microbial contamination, which tended to overgrow the rhizome material, were also observed. Further tissue culture testing of rhizome sections was not undertaken; however, the potential for mass clonal propagation merits further research consideration. Successful results from soil-borne explants for other species studied at the KCIESC (e.g., *Sporobolus pumilus* and *Lilium canadense*) suggest that appropriate sterilization and tissue culture propagation methods could be developed for *G. peckii* rhizomes and that this option could contribute meaningfully to ex situ conservation of *G. peckii.*

15.2.2 Phase 2: Refining Propagation and Outplanting Techniques (2015–2019)

Phase 2 involved additional development and refining of tissue culture methodologies and investigating factors affecting outplanting success of tissue culture-derived propagules. These efforts culminated in the first field trial of *G. peckii* plants originating from seeds retrieved from long-term storage at the ASB, the in-depth study of AMF interactions with *G. peckii*, and testing of the survivorship of young plants in a wild setting near BI.

15.2.2.1 Refining Seed Germination Protocol

Additional germination trials of *G. peckii* and the proxy species *G. macrohyllum* (Largeleaf Avens) seeds were completed to improve the protocol and to provide

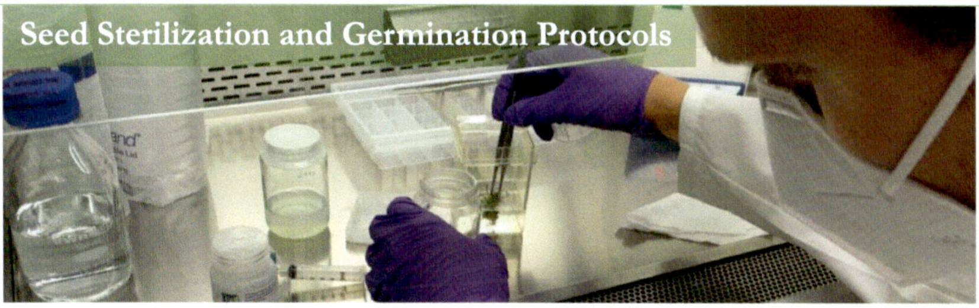

Count out seeds (typically 60-120 for a standard germination test) from working samples at 4°C, or storage at -18°C. Place seeds into metal tea infuser, wrap edge of infuser with Parafilm to prevent seeds escaping. Soak seed for one hour under cold running tap. Place infuser in a beaker and cover with 70 % isopropyl alcohol for 30 sec, pour off alcohol and rinse 2 x with RO water. Add 5.0 g/L NaDCC to cover infuser for 10 minutes, pour off NaDCC and rinse 2 x with RO water. Move infuser to sterile container under aseptic conditions (ideally inside a HEPA controlled flow bench). Place infuser in 20 % v/v NaOCL (bleach) for 15 minutes with two drops of Tween 80 surfactant, pour off liquid and rinse 4 x with sterile (autoclaved) reverse osmosis water. Aseptically open container and transfer seeds individually to agar Petri plates (100 x 15 mm) with 10 mL/L PPM; seal with Parafilm; place 10 seeds on agar solidified surface of each plate.

A standard germination test of G. peckii includes non-stratified and stratified groups: place half of the plates under 60-80 µmol LED lights with a 12 hour photoperiod at 21-25°C continuous temperature, and half in complete darkness at 4°C. After 3 days, aseptically transfer all seeds from PPM plates to agar only plates, seal with Parafilm. Monitor germination in non stratified group as measured by radicle emergence, and cotyledon and epicotyl development at 3, 7, 10, 14 days and weekly thereafter. After 6 weeks at 4°C, move the stratified group into same conditions as the non stratified group and continue monitoring. Germination trials are typically terminated after 4 months of evaluation; germination trials for the purpose of tissue culture micro-propagation are terminated sooner, moving germinants into tissue culture media at 28 days.

Fig. 15.2 Seed sterilization and germination testing protocol for *G. peckii*. A photo shows a researcher working aseptically in a HEPA controlled flow biosafety cabinet. Photo credit: Sarah Hines

seedlings for tissue culture trials. Researchers examined the effects of various exposure times and concentrations of sterilization reagents, the inclusion of PPM in the germination medium, the inclusion of various nutrients in media, and the seed plating density (Table 15.2). After much trial and error, a combination approach of sterilization reagents, seed soak times in these reagents, placement of seeds onto PPM containing media before transfer to agar-only media, and limiting the number of seeds in a plate resulted in the best observed germination rates (39%). The resultant seed sterilization and germination protocol is detailed in Fig. 15.2.

15.2.2.2 Developing Tissue Culture Protocol for Micropropagation

With an effective germination protocol established, focus shifted to test various culture medium components to support longer-term seedling growth and development to achieve a more successful transition from germination to tissue culture based micropropagation. The use of activated charcoal in the growth media was found to produce germination rates of 49.3%, compared with 39.0% for the agar-only group. The seedlings from the activated charcoal were larger and more robust than the agar-only group and improved the transition to tissue culture and outplanting. Tissue culture studies continued to test media additives such as activated biochar, activated

After 28 days, seedlings can be transferred from germination Petri plates to nutrient media as needed for micro-propagation. Primary nutrient medium is comprised of ½ x MS basal components including 2.2 g/L MS macro / micro-nutrients, myo-inositol, thiamine, adenine sulphate according to Murashige and Skoog 1962 plus 20 g/L sucrose and 7.0 g/L agar. Adjust pH to 5.8 with NaOH. Autoclave medium for 15 minutes (15 psi, 121 °C) and pour into 95 x 25 mm glass vials under aseptic conditions; slant vials at 45 degrees during agar hardening.

Initiation stage media (first 4-6 weeks), when seedlings are moved from agar only germination plates, add 2.0 mL/L filter sterilized PPM; post autoclaving. For media to be used during the stabilization stage (6-8 weeks), the addition of 250 mg/L activated biochar (or activated charcoal) improves vigor of cultures – can be added to media used after initiation stage – autoclave with regular basal components.

Seedlings are placed on the surface of the culture medium (slanted to facilitate the transfer and prevent condensation from pooling around the seedling); roots will penetrate the medium. Each culture vial is labelled with a seedling line code for traceability to the seed they originated from. Cultures are monitored and all contaminated and low vigor (shoot and/or root) cultures are terminated. As rosettes multiply from original seedlings, cultures can be regularly divided and transferred to fresh medium, typically in vials or larger vessels (jars or Magenta containers) as the plants grow.

Fig. 15.3 Tissue culture initiation and multiplication protocol for *G. peckii*. Photo credit: Robin Browne

charcoal, and growth regulators including Benzyladenine (a.k.a., BA) and Indole acetic acid (a.k.a., IAA). Notably, the addition of activated biochar was found to contribute to vigorous shoot cultures and continued multiplication of rosettes (Fancy 2017). Phase 2 tissue culture trials resulted in the protocol described in Fig. 15.3 for effective initiation and multiplication of *G. peckii*.

15.2.2.3 Developing Protocol for Outplanting from Tissue Culture

A challenge of using tissue culture to propagate *G. peckii* was the need to gradually transition delicate plantlets from the sheltered culture environment (i.e., low light, high humidity, supplied nutrients) to harsher greenhouse conditions (i.e., lower humidity, higher light intensity, nutrient limitations). During Phase 1 trials, seedling survival was initially > 75% but many experienced leaf reddening, necrosis and die back. Many of the 2013 plants did not survive overwintering. Phase 2 proxy trials with *G. macrophyllum* showed an 89.7% survival rate one year after outplanting, compared with 20% for *G. peckii*. This indicates a greater sensitivity of *G. peckii* to outplanting and a challenge for the long- term viability of plants propagated using tissue culture methods. Notably, similar difficulties with other legally at-risk species such as Rockrose (*Crocanthemum canadense*) and Thread-leaved Sundew (*Drosera filiformis*) with niche or specialized habitat requirements had been encountered at the KCIESC and ASB.

Modifications were made to the plantlet acclimatization conditions such as an extended period in culture prior to outplanting, longer periods in humidity-controlled bins, regular misting of the plantlets, gradual decrease of humidity, introduction of higher light intensities under greenhouse conditions, and subsequent shading to further protect plants. These measures were found to improve outplanting success and were incorporated into the outplanting protocol shown in Fig. 15.4.

It was noted that despite improved plant survival, plants moved from the greenhouse to the outdoor plots continued to experience leaf and petiole reddening, decreased flowering rates, overall decline in shoot vigor, and overwintering losses. These issues underscored the need for continued research to study factors affecting outplanting success and plant establishment, which led to the investigation of AMF inoculation (Fancy 2017; Fancy et al. 2020) and the inclusion of biochar in the growing substrate (Bradbury 2018).

Once plant cultures are established with a substantial root mass, outplanting to soil can occur. On a sterile work bench space prepare a moistened towel surface with forceps, scalpels and planting instruments, 2 large beakers with RO water, mister bottle, bins of moistened ProMix LP-15 soil or equivalent, labeling supplies, tape. Note: bin must have a clear lid and be perforated with a series of holes on lid, bottom and sides to allow air flow and/or drainage.

Assign a code to all plantlets to be outplanted and record to provide traceability back to the original seedling line. Remove each plantlet from the culture vessel and place immediately in a beaker of RO water. Tease off adhering agar from roots and trim any necrotic shoots or leaves. Make hole in soil and insert plantlet by gently pulling the roots into the hole and covering. Mist the plantlet once inserted in substrate and place label to with ID and out plant date, beside plantlet. Once a group has been outplanted into a bin (typically up to 30 per bin), mist the entire group, place lid on bin and cover holes on lid and sides of bin with tape. Place bin under growth conditions (60-80 µmol LED lights , 16 hour photoperiod, 21-25°C). Check bin daily, mist once per day – for two weeks. After one week, begin to remove portions of tape gradually each day. After two weeks, move bin to greenhouse under shaded conditions. Remove lid gradually over a one-week period, then leave lid off. Water as needed (every 2-3 days) under partially shaded conditions. Once plants are established, transfer to 4 – 5.5 inch liner pots for continued growth. Move pots to outdoor garden plots, or transfer to field conditions.

Fig. 15.4 Outplanting protocol for *G. peckii* to move plants from tissue culture to soil medium under greenhouse conditions. Photo credit: Robin Browne

15.2.2.4 Role of Arbuscular Mycorrhizal Fungi and Field Trial

Tissue culture conditions limit the microbial communities that support plant growth in typical soil environments, including mutualistic symbionts such as Arbuscular Mycorrhizal Fungi (AMF) (Kapoor et al. 2008). Therefore, when attempting to maximize outplanting success of tissue culture-derived plants, the lack of microbial partners associated with the tissue culture material should always be considered (Dhiman et al. 2020; Mortier et al. 2023). Natural associations of AMF are known to protect plants from abiotic stresses (Smith and Read 2008). In fact, 80% of all land plant species form symbiosis with AMF (Smith and Read 2008) and this partnership allows plants to access more soil volume resulting in a gain of mineral nutrients, particularly inorganic phosphate. In exchange, the obligate fungal symbiont receives fixed carbon, sugar and lipids from the plant (Ferrol et al. 2019). High AMF colonization rates had been reported for Water Avens (*Geum rivale*) in the literature (Uibopuu et al. 2009), however, it wasn't known whether *G. peckii* formed AMF associations on BI. We also hypothesized the lack of microbial associations in tissue culture-derived plantlets could impact survival and establishment at outplanting (Fancy 2017). Researchers began AMF studies with two objectives: 1) to determine whether *G. peckii* formed AMF associations naturally, and, if the first objective was successful, 2) to develop an AMF inoculant for testing in outplanting trials. This research was supported by the establishment of a mycological research laboratory in the Biology Department of Acadia University by Dr. Allison Walker in collaboration with Dr. Juan Carlos López, starting in 2014.

Soon after research began, the ability of *G. peckii* to form AMF association was observed when microscopic AMF structures in the prepared roots of outplanted ex situ culture-propagated *G. peckii* were noted (LaRue 2016b; Fancy 2017; Fancy et al. 2020). AMF spores were also successfully isolated from BI soil (see Phase 1 section, LaRue 2016b). To our knowledge, this was the first report of AMF associations with *G. peckii* (Fancy 2017; Fancy et al. 2020). For outplanting trials, researchers considered two options for inoculating AMF into growth substrate: commercial MykePro WP powder containing AMF species *Glomus intraradices*, and native BI AMF inoculum created using a modified trap pot method (Schenck, 1982). To create the latter inoculum, researchers used a mixture of soil collected from the *G. peckii* root zone (which included AMF), sterilized sand, and Corn (*Zea mays*) as a host (Fancy 2017; Fancy et al. 2020). Corn was then grown in this soil mixture to promote the growth of AMF. Reestablishing microbial connections between micropropagated *G. peckii* plantlets and AMF were then included in the following trials.

In 2016, a permit was obtained to outplant a population of 30 individual *G. peckii* plants into a $2m^2$ grid in Balancing Rock Bog, Long Island (LI), Nova Scotia (Fig. 15.5), where no naturally occurring *G. peckii* plants are found. Although it is not a suitable long-term habitat for *G. peckii* due to its high acidity (low pH) soil and possibly its hydrology, this test site is only 20km from BI and has an adequately similar ecosystem structure to mimic some of the conditions present in suitable habitat on BI. This bog shares many similar plant species with the BI sites, although it does lack some key lower acidity indicating species such as Shrubby Cinquefoil

(*Dasiphora fruticosa*) and Livid Sedge (*Carex livida*). The LI bog is fed primarily by rainwater and is hydrologically stagnant, while most of the suitable BI habitat is fen and is influenced by underlying basalt bedrock and the lower acidity water seeping from it via ground or surface water flow. As a result, the LI and BI bogs have a pH of 3.9-4-3 and 4.0–5.6, respectively (Dodge 2019; Chapter 4), suggesting that the general availability of nutrients for plant roots at LI is insufficient for healthy vigorous growth and long-term survival. Additionally, water table levels in the LI bog may be significantly lower than BI bogs, possibly affecting habitat suitability for *G. peckii* (Fancy 2017, Chapter 4). However, the LI site was deemed suitable enough for experimental research and gaining valuable ecological insight.

In conjunction with the outplanting trial on LI, a replicate trial for another 30 individual *G. peckii* plants was conducted under greenhouse conditions at the KCIESC. This effort marked the first time in Nova Scotia that an endangered plant species had been successfully retrieved from ex situ seed bank storage, propagated by tissue culture, and outplanted back into a wild setting (Fancy 2017; Fancy et al. 2020).

While transitioning *G. peckii* from in vitro conditions to the greenhouse or LI bog, each third of plants underwent one the following three treatments: commercial mycorrhizal inoculant (MykePro WP from Premier Tech in Rivère-du-Loup, Quebec), native BI mycorrhizal inoculant developed from BI soil, or a sterile non-inoculated control (Fancy et al. 2020). Once acclimatized to the soil substrate, these

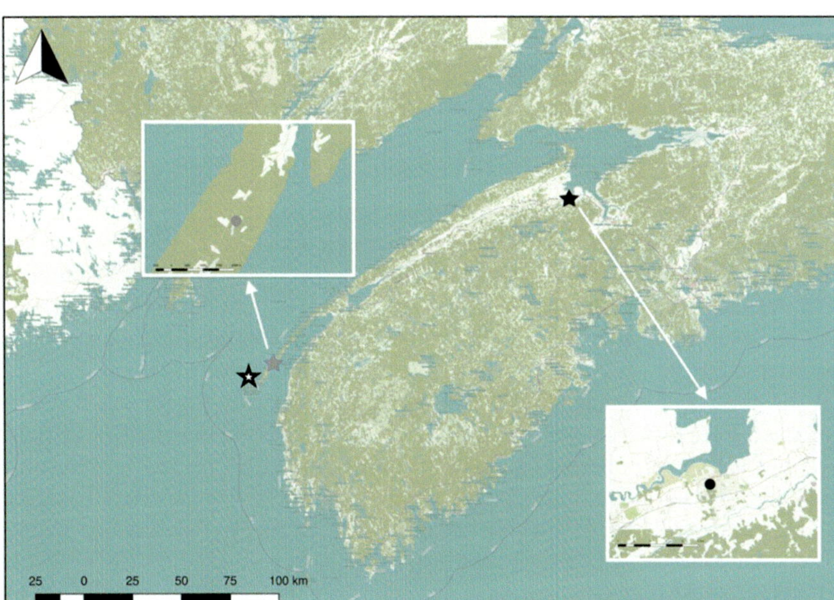

Fig. 15.5 Location of ex situ populations of *G. peckii* in Nova Scotia. Shown is Balancing Rock bog on LI (gray star), the KCIESC at Acadia University, Wolfville (black star), and BI (white-centered star). Figure adapted from Dodge 2019

At the field site, estimate number of plants and flowering stems. For G. peckii collections on BI, September or October seemed to be the ideal time. Select up to 20-30 or <50% of visible plants on site. Harvest up to 20% of flowers from an individual plant. Place mature flowers in paper bags labeled with site and collection date. After-ripen flowers in bags for at least 2 weeks prior to processing (keep in dry conditions at ~20 ℃. For processing, place flowers in glass Petri plates with filter paper, tease out individual seeds with forceps. Remove debris off seeds. Count out 100 seeds for mass determination (100 seed weight). Measure mass of the entire sample to estimate total number of seeds (total wt/100seed wt). Place the Petri plate with seeds into a humidity chamber set for <20%RH for 2-3 weeks. Measure seed moisture content with hygrometer (target is <15%). Pour all seeds into a sample envelope, enter sample data in seed bank accession system. Envelopes are placed in a fridge at 4 ℃ for 24 hours (note: working subsamples are sometimes retained at 4 ℃ for initial germination and propagation trials). After 24 hours move samples to -18 ℃ freezer for long term storage in the seed bank. Expected germination tests (per Figure 2) every 3-5 years for long term storage.

Fig. 15.6 Seed Bank Protocol for collecting and processing *G. peckii* seeds for long-term storage. Photo credit: Robin Browne

transitioning plants were split into two groups with 30 remaining in the greenhouse and 30 moved to the LI bog. Later during the same growing season, researchers noted that 29 of the LI bog specimens and all greenhouse specimens survived (Fancy et al. 2020). A small mass of roots was collected from five greenhouse plants to determine AMF presence using root ink staining and microscopy, but no AMF colonization was observed (Fancy 2020). LI bog plants were left undisturbed.

By the third field season, in 2018, 77% of the plants at the LI bog remained, two or three plants from each treatment group had died. At the end of summer 2018, 18 of the surviving plants were collected for analysis to determine AMF presence and potential correlation to plant growth rate (Dodge 2019). Plant roots were selectively stained for fungi, and AMF were observed using light microscopy. Colonization counts showed that 40–49% of *G. peckii* root length had been colonized by AMF. This is the first time an AMF in *G. peckii* roots was confirmed in an ex situ population. DNA barcoding techniques were unsuccessful in obtaining fungal taxonomic identities from the LI bog plant roots, but concurrent DNA extractions and colonization counts done on roots from ex situ plants in HIBG experimental beds found that *G. peckii* contained AMF in their roots, including three unique DNA barcodes representing three species from the fungal family Glomeraceae: *Glomus* sp. and 2 *Dominikia* spp. (Dodge 2019).

During these trials, additional soil and *G. peckii* root samples were collected from BI sites in the summer of 2017 to test for additional AMF species (Bradbury 2018). Two species of AMF (both Glomeraceae), *Archaeospora* sp. and *Acaulospora trappei*, were identified from the roots of Corn grown in BI soil (Bradbury 2018). Further research of the AMF associated with *G. peckii* would undoubtedly enhance our understanding of *G. peckii* soil biology and assist with in situ and ex situ conservation.

Overall, results from Phase 2 research led to breakthrough refinements of methodologies for seed sterilization, germination and micropropagation, and outplanting methodology, along with significant insights into AMF species, associations, and *G. peckii* soil biology. Researchers were now more confident in ensuring that seeds could be used after long-term storage, and that plant material could be transitioned out of tissue culture and reintroduced to the wild.

15.2.3 Phase 3: Monitoring Surviving Ex Situ Trial Plants for More Insight (2020–2024)

Starting in 2020, Phase 3 focused on the monitoring of surviving ex situ plants and the collection of more seeds for long-term storage. Mature potted plants from previous germination trials were monitored and maintained in the HIBG experimental gardens at KCIESC. Although they declined gradually over time, many specimens survived for eight years with little assistance. This suggests that the initial propagation and outplanting from Phase 1, which was not yet as refined as in Phase 2, were still adequate for producing some robust plants. The LI bog site had six surviving plants at the close of the project in 2018. Terminal leaflet width was found to be 20% narrower and survivorship was 10% lower on LI compared to *G. peckii* transplants between BI sites (Fancy 2017; Dodge 2019; Chapter 4). The site was surveyed again in July of 2021 when five plants were noted as alive (Belliveau, pers. comm. May 2024). Individual plants were relatively small, were not forming rhizomes, were not producing flowers or seeds, and were experiencing significant leaf reddening compared to naturally occurring and healthy BI plants. Poirier (2016) found that leaves and petioles of *G. peckii* may turn red as a stress response to waterlogging in wet bogs. Red leaves and petioles were also consistently observed in the drier HIBG soils, though not as frequently compared to LI plants. Another possible symptom of poor plant health was the frequency of insect herbivory of LI plants, which ranged from 48 to 60% and is likely higher than BI or HIBG specimens (Fancy 2017). Although terminal leaflet width did not vary in browsed individuals, suggesting a lack of change in plant vigor (Fancy 2017), a lack of anti-herbivore secondary metabolites in leaf tissue may indicate the opposite though this has yet to be researched. Although the specific metrics of suitable habitat are not completely understood and it is difficult to determine the exact cause of poor plant health, the survival of these outplanted individuals at LI and HIBG strongly suggests that propagation and outplanting methods from KCIESC studies, especially from Phase 2, represent an effective ex situ conservation tool in suitable habitat.

By 2020, researchers could successfully produce robust *G. peckii* plants from seeds stored for long periods of time. The research goal was to collect as many seeds as possible from as many sites as possible on BI. This would ensure that as much genetic diversity as possible was preserved at the ASB for future conservation efforts. In 2021 a targeted seed collection effort occurred (Table 15.1). All major sites on BI (i.e., sites capable of producing at least a few hundred seeds) were visited for seed collection. While some locations were not suitable for collections because either no *G. peckii* plants were found (due to degrading habitat) or because a collection would have posed a potential risk to the long-term persistence of that site in the future, 11 individual collections were made. This resulted in a collection of over 55,000 seeds. To ensure adequate viability, one small ad hoc germination trial was conducted on seeds from a variety of sites, and satisfactory germination was noted. The rest of the seeds were desiccated to approximately 15% relative humidity and stored at − 20 °C at the ASB.

15.3 Conclusion

15.3.1 Future Research Topics

Significant knowledge was gained from ex situ conservation research from 2012 to 2024 at the ASB and KCIESC, too much to fit into this book chapter. Future research could include developing metrics to accurately describe suitable habitat. Improved morphological data to effectively assess plant vigor, both in situ (e.g., quality of natural sites on BI) and ex situ (e.g., quality of propagated *G. peckii*), could help assess the quality of habitat. Analyzing physiological differences between in situ plants and ex situ plants could also shed light on plant vigor and reproductive abilities. For example, differences have been noted in plantlets from tissue culture compared to conventionally propagated material (Chandra et al. 2010). Differences in stomatal configuration and function, cuticular composition, composition and abundance of secondary metabolites, or root-shoot vascular connections might impact outplanting success and subsequent photosynthetic capacity during the acclimatization period, and such factors may influence carbohydrate allocation to roots and subsequent rhizome development. Further research on *G. peckii* AMF relationships would undoubtedly discover more species and their interactions and contributions to healthy *G. peckii.* To optimize propagation success, learning more about the timing of AMF connections with *G. peckii* could help with AMF applications during outplanting trials. Additionally, the possibility of long-term seed storage (e.g., decades or centuries) by cryopreservation could be further explored for this globally imperiled species. These are just some of the outstanding questions that future researchers can tackle.

15.3.2 The Future of G. Peckii *and the ASB*

G. peckii is a sentinel species for conservation in Canada and globally, and an integral part of BI's identity. Its loss would trigger irreversible, immeasurable changes, both ecologically and socially. Its preferred habitat is unique, and its protection helps collectively draw a line between an excessive appropriation of nature and a sustainable relationship with the natural world.

To help *G. peckii*, we need to continue learning. The collective body of work with *G. peckii* by the ASB and KCIESC represents a major achievement toward the development and implementation of ex situ conservation work for a vascular plant species. This work involved years of collaborative efforts and dozens of individuals dedicated to biodiversity conservation (see Acknowledgments). This is undoubtedly the most comprehensive ex situ research effort yet contributed to an endangered plant species in Nova Scotia. This effort supports provincial, federal, and global species recovery policies (Hird and Kramer 2013). Moreover, these approaches developed for *G. peckii* can act as a model for ex situ conservation of other species at risk in Wapna'ki and beyond.

Coinciding with this ex situ research was the collection of several hundred other species and a significant expansion of the ASB, which now includes more than five million seeds. The relatively thorough investigations of *G. peckii* described in this chapter have been influential toward the ASB's growth by providing many lessons inherently offered with such detailed trials and tribulations. This includes valuable insights on germination, propagation, outplanting, and on tissue culture techniques in general. These studies have generated an extensive body of information from which to move forward with ex situ conservation and recovery strategies for native species in Wapna'ki.

In the future, if the local climate does not support *G. peckii* in its current range, our research results can provide a basis for any proposed assisted migration efforts. Thousands of *G. peckii* seeds are stored long term in the ASB, with thousands more to be collected routinely and sustainably in the future. Reliable germination, propagation, and outplanting techniques have been identified and are noted in this chapter. This work with *G. peckii* is a prime example of the utility of seed banks and of tissue culture techniques to the conservation of biodiversity.

Acknowledgements Progress for ex situ conservation of *G. peckii* was made possible by numerous funding sources from 2012 to 2024. Thank you to the Nova Scotia Museum Research Grant Program, Sage Environmental Program, Nova Scotia Habitat Conservation Fund, TD Friends of the Environment Foundation Grant, Environment and Climate Change Canada's Science Horizons Internship Program, Morton Centre Legacy Fund, Blomidon Naturalist Society Student Scholarships,

Nova Scotia Co-op Education Incentive Program, The Raddall Research Fund and Acadia University's Honours Summer Research Award. Special thanks to the Arthur Irving Academy Foundation Research Grant Program, which supported the student projects in Phase 2.

Most of this *G. peckii* research would not have been possible without the many Acadia students and staff members involved. We owe thanks to Lauren Bower, Jacqueline Bradbury, Heather Elliott, Emily Evans, Kayoung Heo, Samuel Jean, Gina MacDonald, Hannah Gillam, Lita O'Halloran, Keelan O'Malley, April Muirhead, Rachel Murray, Melanie Preisnitz, Rebecca Ryan, and Nicole White.

Heartfelt thanks to the Friends of the Acadian Forest Society volunteers who spent many hours helping to process seed samples.

Special thanks for the contributions from Ruth Newell (retired E.C. Smith Herbarium Curator) and Diane LaRue (MTRI Researcher) who were instrumental in getting the project started.

Also, a special thanks to local Brier Island residents for their friendliness, and their support and appreciation of *G. peckii* and *G. peckii* researchers.

All seed collections, rhizome, root, and soil sampling were completed with the appropriate permits from the Nova Scotia Department of Natural Resources and Renewables.

References

Adams S (2016) Foliar endophytic diversity of Eastern Mountain Avens, *Geum peckii* Pursh (Rosaceae), from degraded and pristine habitats in Digby County, Nova Scotia, Canada. Acadia University, Wolfville, NS (Honours Thesis)

Bradbury JG (2018) Contributions of biochar and arbuscular mycorrhizal fungi to the growth of Geum peckii (Eastern Mountain avens). Acadia University (Honours Thesis)

Bunn E, Turner SR, Panaia M, Dixon K (2007) The contribution of in vitro technology and cryogenic storage to conservation of indigenous plants. Aust J Bot 55:345–355

Chandra S, Bandopadhyay R, Kumar V, Chandra R (2010) Acclimatization of tissue cultured plantlets: from laboratory to land. Biotech Lett 32:1199–1205

Convention on Biological Diversity (2012) Global strategy for plant conservation: 2011–2020. Botanic gardens conservation international. Richmond, UK

COSEWIC (2010) COSEWIC assessments and update status report on the Eastern Mountain Avens *Geum peckii in* Canada. Committee on the status of endangered wildlife in Canada, Ottawa

Dhiman M, Sharma L, Singh A, Sharma MM (2020) Ex situ conservation using in vitro methods of an endangered plant Sterculia urens Roxb.: a high volume trade plant for Gum Karaya. Ind Crops Prod 158:113015

Dodge RS (2019) Arbuscular mycorrhizal fungi in ex situ populations of Eastern Mountain Avens, Geum peckii Pursh (*Rosaceae*) from Nova Scotia, Canada. Acadia University (Honours Thesis)

Fancy S, López-Gutiérrez JC, Walker AK, LaRue D, Browne R (2020) Evaluating out planting success and mycorrhizal status of endangered *Geum peckii* Pursh (Rosaceae), the Eastern Mountain Avens, in Nova Scotia. Proceed Nova Scotian Inst Sci NSIS 50(2):269

Fancy S (2017) Investigations of *ex situ* strategies for conservation of *Geum peckii* (Eastern Mountain Avens). Acadia University (Honours Thesis)

Ferrol N, Azcón-Aguilar C, Pérez-Tienda J (2019) Arbuscular mycorrhizas as key players in sustainable plant phosphorus acquisition: an overview on the mechanisms involved. Plant Sci 280:441–447

Galbraith D (2010) Updated plant conservation targets for 2020 in the Revised GSPC. Canada's National Focal Point for the Global Strategy for Plant Conservation. pp 1–4

Hird A, Kramer A (2013) Achieving target 8 of the global strategy for plant conservation: lessons learned from the north American collections assessment. Ann Mo Bot Gard 99(2):161–166

Kapoor R, Sharma D, Bhatnagar AK (2008) Arbuscular mycorrhizae in micropropagation systems and their potential applications. Sci Hortic 116(3):227–239

Kulak V, Longboat S, Brunet ND, Shukla M, Saxena P (2022) In vitro technology in plant conservation: relevance to biocultural diversity. Plants 11:503. https://doi.org/10.3390/plants11040503

Kumar K, Rao IU (2012) Morphophysiological problems in acclimatization of micropropagated plants in *Ex vitro* conditions: a review. J Ornamental Hortic Plants 2(4):271–283

LaRue D (2016a) Natural regeneration of the globally rare *Geum peckii* (Rosaceae) on Brier Island. Nova Scotia Rhodora 118(975):310–314

LaRue D (2016b) *In situ* and *ex situ* propagation of the globally rare *Geum peckii* (Rosaceae). Rhodora 118(975):315–319

LaRue D (2014) Summary report on rhizome mapping and collections of *Geum peckii* (Eastern mountain avens) under Scientific Permit Dated May 22, 2013. 1–4

Maunder M, Havens K, Guerrant E, Falk D (2004) *Ex situ* methods: A vital but underused set of conservation resources. *Ex situ* Plant Conservation. Washington, DC:Island. pp 3–20

Mortier E, Jacquiod S, Jouve L, Martin-Laurent F, Recorbet G, Lamotte O (2023) Micropropagated walnut dependency on phosphate fertilization and arbuscular mycorrhiza for growth, nutrition and quality differ between rootstocks both after acclimatization and post-acclimatization. Sci Hortic 318:112081

Murashige T, Skoog F (1962) A revised medium for rapid growth and bioassays with tobacco tissue cultures. Physiol Plant 15:473–497

New York Declaration (2011) Ecological restoration on a global scale: harnessing the power of the world's botanic gardens. BGCI communication. p 1

Paunescu A (2009) Biotechnology for endangered plant conservation: a critical overview. Rom Biotechnol Lett 14(1):4095–4103

Pence V (2011) Evaluating costs for the in vitro propagation and preservation of endangered plants. In Vitro Cell Dev Biol Plant 47:176–187

Poirier J (2016) The relationship between water table depth and health of Eastern Mountain Avens (*Geum peckii)* on Brier Island, Nova Scotia. St. Francis Xavier University (Honours Thesis)

Radomir A-M, Stan R, Florea A, Ciobotea C-M, Banuta FM, Negru M, Neblea MA, Sumedrea DI (2023) Overview of the success of in vitro culture for *ex situ* conservation and sustainable utilization of endemic and subendemic native plants of Romania. Sustainability 15:2581. https://doi.org/10.3390/su15032581

Schenck NC (1982) Methods and principles of mycorrhizal research. The American Phytopathological Society, St Paul, p 210

Sharma GK, Jagetiya S, Dashora R (2015) General techniques of plant tissue culture. Lulu Press, Raleigh, North Carolina

Shurgurkar MV, Thengane SR, Poonawala IS, Jana MM, Nadgauda RS, Mascarenhas AF (1996) A simple method of propagation and rhizome formation in *Dendrocalamus strictus* Nees. Curr Sci 70(10):940–943

Smith SE, Read DJ (2008) Mycorrhizal Symbiosis (3rd ed). Elsevier

Uibopuu A, Moora M, Saks Ü, Daniell T, Zobel M, Öpik M (2009) Differential effect of arbuscular mycorrhizal fungal communities from ecosystems along management gradient on the growth of forest understorey plant species. Soil Biol Biochem 41(10):2141–2146

Chapter 16
Managing Gull Numbers: History and Challenges

Mark L. Mallory, Jennifer F. Provencher, Alain Belliveau, Claire E. Wilson, J. Sherman Boates, Dave Shutler, and Sarah E. Gutowsky

Abstract American herring gulls (*Larus smithsoniansus*; hereafter herring gulls) are widely distributed, common seabirds with generalist diets. As is the case for many large gull species, herring gulls exploit human-altered landscapes for feeding and breeding, and these synanthropic birds can create a suite of human–bird conflict situations, ranging from mild annoyances (e.g., faeces on vehicles) to major risks (e.g., striking aircraft). At Brier Island, Nova Scotia, a large colony of nesting gulls uses a diversity of human and natural foraging areas and foods during the breeding season. Some gulls nest in proximity to the globally rare eastern mountain avens (*Geum peckii*), and there are concerns that altered soil chemistry through nutrient additions from avian faeces, or physical alterations to soil and plants from nesting gulls, may deleteriously affect this plant. In this chapter we summarize some of the issues around managing gulls, and describe some of the initial, largely unsuccessful efforts to manage them at Brier Island.

Keywords Colony · Seabird · Generalist · *Larus* · Herring gull

M. L. Mallory (✉) · C. E. Wilson · J. S. Boates · D. Shutler · S. E. Gutowsky
Department of Biology, Acadia University, 15 University Avenue, Wolfville, NS B4P 2R6, Canada
e-mail: mark.mallory@acadiau.ca

C. E. Wilson
e-mail: claire.wilsonodriscoll@novascotia.ca

J. S. Boates
e-mail: sherman.boates@acadiau.ca

J. F. Provencher
Science and Technology Branch, Environment and Climate Change Canada, Carleton University, Raven Road, Ottawa, ON K1A 0H3, Canada

A. Belliveau
E.C. Smith Herbarium, K.C. Irving Environmental Science Centre, Acadia University, 32 University Avenue, Wolfville, NS B4P 2R6, Canada

C. E. Wilson
Nova Scotia Department of Natural Resources and Renewables, 136 Exhibition Street, Kentville, NS B4N 4E5, Canada

N. M. Hill et al. (eds.), *Wetland Restoration for Endangered Species Recovery*, Environmental Contamination Remediation and Management,
https://doi.org/10.1007/978-3-031-71344-6_16

16.1 Introduction

As their name implies, seabirds are common organisms found in or near marine environments (Gaston 2004), but among the broad group of species known as "seabirds", many gull species (Laridae) are well-adapted to living in terrestrial or freshwater habitats (Hobson et al. 2015). Consequently, while a suite of changes in the marine environment have posed population-level threats to many seabird species over the past centuries (Dias et al. 2019), some gull species have been able to adjust much of their diet and nesting habitat selection to rely more on coastal, human-altered habitats, notably in industrialized areas influenced by agriculture, waste management, or food processing (Oro et al. 2013; Newsome et al. 2015; Meléndez-Arteaga et al. 2022). In this sense, many gull species have become undomesticated animals that benefit by living with humans (synanthropes), and in some cases are as common in urban or agricultural environments as are well-established synanthropic species such as rock doves (*Columba livia*; Hernández-Alonso et al. 2023), house sparrows (*Passer domesticus*; Leu et al. 2008), and raccoons (*Procyon lotor*; Prange et al. 2003). Managing species that can adapt to, and in many cases exploit, human-altered landscapes has become an important and challenging discipline in applied biology (e.g., Adams 2005; Magle et al. 2012).

The ability of gulls to exploit human-altered habitats has led to considerable conflict globally between gulls and people, at levels that vary from nuisance (e.g., gulls harassing people for food) to serious safety threats (e.g., gulls at airports striking planes). Although peoples' tolerance of, and affection for, gulls vary, there are several key, common interactions that have led gulls to often be placed in the category of nuisance or "undesirable" species (Earn et al. 1998; see Sect. 16.3).

In this chapter, we review some of the reasons gulls may become nuisance species, and some measures taken to try and control synanthropic colonies or other local aggregations. We focus principally on American herring gulls (*Larus smithsoniansus*; hereafter herring gull), the main gull nesting at Brier Island (44.2° N, 66.3° W), and the focus of management attempts, although many of the environmental or political effects of large numbers of gulls are similar among species.

16.2 Brief History of Gulls in Eastern North America

16.2.1 General Numbers and Trends

The number of gulls in eastern North America has fluctuated dramatically through time (e.g., Cotter et al. 2012), at least to the extent that we have been able to monitor their populations. In fact, if one looks back farther than the past few decades, one constant is that gull numbers seem to change substantially and regularly, as has their relationship with people (Anderson et al. 2016). The drivers behind these processes have been diverse, often including competing effects of augmenting food supplies (i.e., allowing populations to grow) versus persecution (i.e., controlling numbers), and have clearly had positive or negative effects on other nesting bird species. For example, gulls have been persecuted as threats to agricultural crops, air travel, or have been exploited for feathers in fashion, all of which may have led to local or regional declines in numbers at times. In contrast, gulls have benefited from fisheries (and attendant discard products), open landfills, and various agricultural crops (e.g., berries; reviewed in Anderson et al. 2016), which have provided food subsidies and allowed numbers to increase. A consistent element through most of this story is that human activities have been primarily responsible for major shifts in numbers of gulls in eastern North America. There have been several detailed reviews on trends in large gull population size in eastern Canada and the north-eastern United States (e.g., Drury 1973; Tufts 1973; Nisbet et al. 2013; Anderson et al. 2016). In Maritime Canada, the general pattern appears to have been one of low gull numbers near the end of the nineteenth century, which then increased through much of the twentieth century, followed by declines through the early twenty-first century and continuing today (Cotter et al. 2012, MacKinnon and Kennedy 2014, Anderson et al. 2016).

16.2.2 Gull Species Breeding in This Region

Eastern North America has a rich history supporting large numbers of coastal marine birds both in the breeding season (e.g., Brown et al. 1975; Gaston et al. 2009; Ronconi et al. 2022) as well as in winter (e.g., Anderson et al. 2019; Gutowsky et al. 2023a). Some gulls, such as the lesser black-backed gull (*Larus fuscus*), may be found in much of this region in winter, but as of 2023, breeding colonies are rare (but increasing; Burger et al. 2020). Moreover, some species, such as glaucous gulls (*L. hyperboreus*), Iceland gulls (*L. glaucoides*), and ivory gulls (*Pagophila eburnea*) arrive from their Arctic breeding regions and winter in the north-east, principally around Newfoundland and Labrador (e.g., Spencer et al. 2016; Baak et al. 2021). Seven relatively common gull species occur year-round and nest along marine coastal regions of eastern North America: great black-backed gull (*L. marinus*), herring gull, ring-billed gull (*L. delawarensis*), laughing gull (*Leucophaeus atricilla*), Bonaparte's gull (*Chroicocephalus philadelphia*), little gull (*Hydrocoloeus minutus*), and black-legged kittiwake (*Rissa tridactyla*) (Howell and Dunn 2007). The kittiwake is a wholly marine species, only spending time on land at its nest site and typically never involved in conflict with humans (except those that get caught as fisheries bycatch; Hatch et al. 2020). Other than kittiwakes, most of the gull species (including the smaller ones such as laughing, Bonaparte's, and little gulls) can be pests in some situations (e.g., feeding at landfill sites, contributing faeces to municipal water supplies, collisions with aircraft; Burger 1981, 1983; Belant 1997), but in general, the gulls most commonly associated with human–bird conflict are the larger *Larus* species (Blokpoel and Tessier 1992; Belant 1997).

16.3 Human–Wildlife Conflict Associated with Gulls

16.3.1 Many Large Gull Species Can Be "Nuisances"

Large *Larus* gull species are often nuisances and challenging to manage because they quickly exploit human alterations to landscapes. To take advantage of these alterations, large gulls have some characteristics that enable rapid individual or group adjustments to anthropogenic changes. First, gulls have relatively good problem-solving skills and are considered "intelligent" among bird species (e.g., Lamarre and Wilson 2021), which not only allows them to respond to new situations, but presumably also to habituate or acclimate quickly. Second, they fill a generalist niche in their environment (e.g., O'Hanlon et al. 2017), albeit with considerable individual specialization in behaviour (e.g., Westerberg et al. 2019), but collectively having plastic and adaptable diets (e.g., Moreno et al. 2010). This means that the variety of foraging options in anthropogenically altered landscapes (e.g., fish plants, landfills, agricultural fields, waste containers, fast food parking lots) are diverse and overlapping, and may change in importance through a year. Third, colonial nesting provides the safety advantages of many eyes watching for predators, as well as potential information transfer on locations of ephemeral food resources (Coulson 2002). Thus, many gull colonies can be located very close to urban areas, e.g. Morris et al. (2011), even on roof tops (Perlut et al. 2016), and can thrive in those situations.

So why are gulls considered a nuisance, or undesirable? First, when gull colonies form, especially near urban or industrial centres, numbers of other (often considered more desirable) bird species decline because their eggs and young can become prey for gulls (Kress 1983; Scopel and Diamond 2017). Second, the presence of gulls at trash facilities, parking lots, food outlets, airports, golf courses, or urban parks, and the associated issues with noise (disturbed gull colonies can emit sounds levels sufficiently high to cause damage to human hearing; Blokpoel and Neuman 1997), faeces, or safety in those locations, has meant gulls are often viewed with disdain, and may be subject to population control measures (Thomas 1972; Belant 1997; Wu et al. 2017; Campbell 2019; Trotter 2019). Third, gulls are biovectors and can transport and concentrate nutrients and contaminants near their aggregation sites (Blais et al. 2007; Kickbush et al. 2018). They do this when individuals from a colony ingest food at dispersed feeding areas, digesting it as they return, and then concentrate and defecate those nutrient-laden materials near their colony or loafing area (Geizer et al. 2021; see also Chap. 12). Consequently, nutrient subsidies from gulls can dramatically alter chemical properties of soils or freshwater near their colonies, and indeed influence nutrient cycles (e.g., Otero et al. 2018) potentially leading to algal blooms (Martín-Vélez et al. 2019), or losses of vegetation (Klimaszyk and Rzymski 2016; Simpson et al. 2021), as well as having other repercussions, such as accumulation of toxic chemicals (Blais et al. 2007). Fourth, during the breeding season, gulls may attack people who come too close to their nests. Similarly, gulls may "attack" people to steal food from them (Thomas 1972), an observation that has become more common in an era of social media. Furthermore, at the more extreme range of "nuisance", gulls may aggregate near airports and are a clear hazard to safe air travel as they collide with aircraft (e.g., Burger 1983, 1985); consequently, in some areas there are effective lethal control measures (i.e., shooting of adults) in place (e.g., Dolbeer et al. 1993). Lastly, gulls can vector pathogens (e.g., Alm et al. 2018; Navarro et al. 2019; Benskin et al. 2009), as in the recent case of SARS CoV-2 (Boklund et al. 2021), or in the case of highly pathogenic avian influenza, where they were identified as patient zero for North America (Caliendo et al. 2022). These issues are not just North American; there are many analogous issues and solutions to gull-human conflict that are found elsewhere in the world, notably in Europe (Brooks and Lebreton 2001; Oro and Martínez-Abraín 2007; Rock et al. 2016).

16.3.2 Options to Control Nuisance Gulls

In situations where aggregated gulls become a problem, wildlife managers may undertake efforts to deter the birds from attending an area or having numbers build through time (non-lethal control) or killing gulls to reduce numbers or conflict (lethal control). In deterrence efforts, which date back decades in eastern North America (Anderson et al. 2016), a common approach has been to target eggs in nests, which usually involves removing, crushing, poking holes in, or spraying oil on eggs to keep them from hatching (Blokpoel and Tessier 1987). While this has worked in some cases, gulls are persistent and will often renest, and thus egg collection or treatment needs to be repeated within a breeding season. Another approach is to erect exclosures (nets, fences) over nesting areas to keep gulls from a site, but these birds are highly adaptable and can move in response. There are some excellent, published examples (unfortunately in response to a multitude of problems) of gull management around Lake Ontario (see Blokpoel and Tessier 1987): in addition to egg management, some nesting areas received multiple deterrence efforts that included playback of avian distress cries, tethered raptors that could fly at and scare gulls, and pyrotechnics to scare away birds. Harassment of gulls to deter them from gathering at preferred nesting or feeding sites has been widely used, including the development of automated scaring systems, but gulls can habituate to the timing of these tactics, and that randomized or variable deterrence approaches are more effective (Ronconi and Cassady St. Clair 2006; Thiériot et al. 2012). In other cases, particularly at airports where safety is the primary priority, gulls have been shot (Burger 1983; Donehower et al. 2007; Jardine et al. 2021). Efforts have been made to shoot with non-lethal rubber shot, but this has generally proven to be less effective than selective culling of birds (Thiériot et al. 2012). Because lethal methods to manage wildlife raise ethical concerns and often meet with public disapproval (e.g., Lute and Attari 2017), generally lethal control is only used where gulls pose the most risk to human safety (Burger 1983), or where removal of gulls is necessary to facilitate recovery of other wildlife species (Kress 1983).

16.4 Brier Island Gulls

16.4.1 Herring Gulls

The American herring gull is a well-studied, large, common, generalist gull that is widely distributed across North America from the Caribbean to the subarctic and is part of a closely-related complex of species across the North Atlantic (Weseloh et al. 2020). There are many local, often synanthropic populations in Atlantic Canada (Cotter et al. 2012). In many respects the herring gull is a quintessential model of an adaptable species with traits that allow it to co-exist well with people. For example, herring gulls can nest on city rooftops (Perlut et al. 2016), and they eat an incredible diversity of items, including insects, earthworms, zooplankton, fish, birds (eggs, chicks, adults) and small mammals (i.e. direct predation), as well as fruit, carrion, and garbage (i.e. scavenging; Weseloh et al. 2020). They notoriously forage in landfills (Jardine et al. 2021; Lato et al. 2021, as well as around marine industrial sites and fishing vessels (Skov and Durinck 2001; Gutowsky et al. 2021, 2023b). In a coastal, urban area, one may be just as likely to see herring gulls at the beach, in a city parking lot, at a landfill, or following a tractor in fields outside the city (e.g., Schwemmer et al. 2008). Consequently, in many regions, herring gulls have been the direct target of control programmes (i.e., management actions; Drury and Kadlec 1974; Burger 1983; Seif et al. 2018). In some situations, these programmes have had very clear, negative consequences for local gull numbers, largely achieving management efforts of reducing conflicts with the birds (Anderson et al. 2016).

16.4.2 The Brier Island Herring Gull Colony

The herring gull colony in the Big Meadow Bog at Brier Island is the largest in Nova Scotia, with recent estimates suggesting the colony supports approximately 3000 breeding pairs (Chap. 12). Herring gulls have bred at Big Meadow Bog since the 1950s, colonizing the site after land management ditched and drained the bog to support agriculture (Hill et al. 2019). Big Meadow Bog is also one of the few locations in the world where an endangered plant grows, the eastern mountain avens (*Geum peckii*; LaRue 2016), and concern has been raised that addition of nutrients from bird faeces may alter soil chemistry and degrade the plant's habitat (Ellis et al. 2006), potentially leading to its disappearance (NSDLF 2021). For this reason, research has been underway to determine effects of gulls on local water (Kickbush et al. 2018) and soils (McIntyre et al. 2022), along with some initial efforts to deter gulls from nesting (see Sect. 16.4.3).

Because gulls are generalists at the species level but specialists at the individual level, a gull colony can be made up of birds that have individual preferences for locations, habitats, and food types they target. This variation among individuals can be apparent both in the breeding season and in winter. Taken together, gulls breeding at a colony can exploit a variety of food resources, and have exceptional flexibility should one resource become unavailable or a new one presents itself. This has been demonstrated in research on herring gulls at Brier Island. Recent tracking studies on some of these gulls showed that they clearly exhibit high behavioural flexibility and adaptability in their movements and foraging locations, and consequently in the main prey taken (Shlepr et al. 2021; Gutowsky et al. 2021). Gulls in this region can even adjust their daily activity schedules to match the work schedule of local fish plants and fishing vessels (Gutowsky et al. 2023b), increasing their exposure to food sources and decreasing time and energy spent foraging. For example, Fig. 16.1 shows movements of five gulls tracked from Brier Island in 2014 using solar-recharged GPS devices. While breeding, these five birds made foraging trips from the colony to local fish plants and mink farms (Gutowsky et al. 2021), yet clearly each bird had individual tendencies, with some spending the majority of their time due east of the colony, and others exploiting resources at sites to the north-east or south-east (Fig. 16.1). These same five gulls showed large variation in how they spent their non-breeding months as well, with some spending much of the winter around Brier Island and others migrating farther south to winter in various regions spread out from Rhode Island to Chesapeake Bay (Fig. 16.1). Some gulls tracked from Brier Island selected marine habitats close to shore, whereas some specialized on urban habitats and others avoided them (Anderson et al. 2019). How this translates to interactions with industry or dietary preferences during the non-breeding period are not yet known. Regardless, this individual variation and behavioural flexibility clearly poses a challenge to management, because adjustments to reduce anthropogenic food subsidies (e.g., reduced access to mink farm waste or aquaculture) may simply lead to gulls switching to other nearby food sources (e.g., fish plants, landfills). Note that, although improved human practices of reducing available waste around landfills or fisheries is thought to have led to a reduction in herring gull numbers (e.g., Cotter et al. 2012; Anderson et al. 2016; Wilhelm et al. 2016), clearly some local colonies such as Brier Island have flourished during this period of region-wide anthropogenic food reductions.

16.4.3 Recent Gull Management Efforts at Brier Island

Given concerns over gulls and nutrient inputs at the Big Meadow Bog colony, and the potential for deleterious effects on endangered plant eastern mountain avens, efforts were undertaken in 2021, 2022 and 2023 to deter gulls from nesting, with very little success. Two approaches were used: a form of deterrence through habitat manipulation, and egg removal.

First, in 2021 and 2023, at the highest density nesting area within the entire colony that had been used for multiple years by the gulls, researchers set up an experimental grid of bamboo stakes ("bamboo plot") and an adjacent area of equivalent size without stakes ("control plot") to test whether bamboo canes would deter gulls from nesting, or otherwise affect gull nesting behaviour. This method deterred gulls

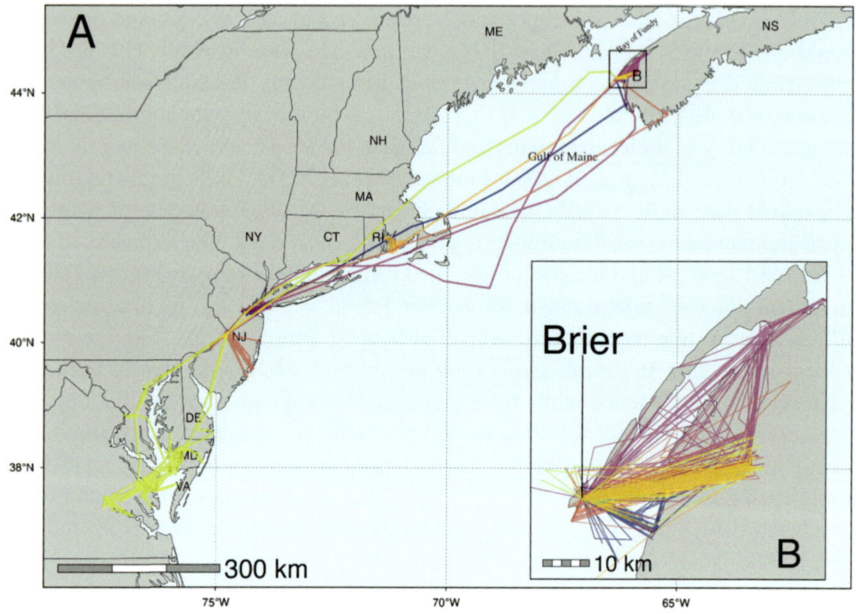

Fig. 16.1 Movements of five American herring gulls (*Larus smithsoniansus*) breeding at Brier Island and tracked by GPS devices: **a** during winter non-breeding months (November to April), and **b** during the spring and summer breeding months (May to August). This map is generated from a subset of previously published data from Shlepr et al. (2021) and Gutowsky et al. (2021)

from the breeding sites of Arctic tern (*Sterna paradisaea*) and reduce predation by gulls with promising results (Boothby et al. 2019). In April 2021, a time when gulls were prospecting or choosing potential nesting sites, the bamboo plot and control plot were set up in Big Meadow Bog. The bamboo plot was built by driving 3 m tall bamboo stakes on a slight angle ~ 30 cm into the ground and 1 m apart. In total, approximately 385 bamboo canes formed a grid that was about 15 × 25 m. The control plot consisted of a similar adjacent part of the gull colony, with no bamboo canes.

In 2023, the bamboo plot and the control plot were each enlarged to approximately 20 × 50 m, and the bamboo plot was modified so that half of the grid remained the same with stakes at 1 m apart, and the other half was "tightened" by placing an additional stake in the middle of each grid square, so that stakes were erected more densely. This was undertaken following results from year one indicating that gulls were not deterred by the 1 × 1 m grid and in fact were walking, flying and nesting in the grid with apparent ease; therefore half the grid was tightened so that the distance between stakes was reduced (and less than the wingspan of the gulls), with the intent of making it more difficult for gulls to fly in and out (Fig. 16.2). This principle follows the well-used tactic of placing plastic spikes on ledges and statues in cities to deter birds (e.g., rock pigeons) from aggregating and defecating on them (e.g., Alderson and Greene 1995). Herring gulls exhibit breeding site fidelity, meaning they will nest

Fig. 16.2 Bamboo stakes were driven into the ground 1 m apart in an experimental grid (no stakes were used in a control grid) to test if this deterred nesting in a preferred part of the bog. In fact, gulls nested readily in the experimental area, walking or sometimes flying directly to their nests. Photo by Alain Belliveau

very near to the same spot in consecutive years, although they are not tied to reusing an old nest (Weseloh et al. 2020). Nonetheless, for this experiment, we assumed that gulls returned to the same basic "patch".

If the bamboo stakes significantly deterred gulls from nesting, we expected to see fewer nests within the experimental plot than in the control. However, in 2022, 38 herring gull nests were found in the control plots and 31 inside the bamboo plots; in 2023, 36 nests were found in the control plot and 36 found in the bamboo plot. Clearly in both years, gulls nested readily in the "deterrence" plot, where they also bred successfully (Fig. 16.3). In 2021, when the first bamboo grids were set up during nest prospecting or nest territory establishment by the gulls, we observed that while gulls vacated the area while the canes were being installed, gulls either flew into or walked into the grid area within minutes of when the researchers left the grid area. In addition, we watched gulls fly directly down to nest sites among the stakes, by gliding over the grid and then tucking their wings and dropping vertically through the openings. Gulls were also observed landing outside the plot and walking into nest sites, dive-bombing researchers among the stakes, and flying up from their nests through the stakes and out of the grid, all strongly suggesting that the gulls did not view these stakes as a substantial deterrent. Indeed, it seems possible that gulls may have perceived the stakes as offering protection for their nests, such as from avian predation (Weseloh et al. 2020). The fact that gulls might be attracted to the bamboo grid raised concern for us and we did not further experiment with bamboo canes in areas where eastern mountain avens was present, as this could further exacerbate the risk to the plant. Consequently, repeating the experiment but further reducing the distance between stakes may still be insufficient to preclude nesting in this site, and an exclosure with netting that reaches down to ground level may be required to be effective. Depending on the size of the mesh in the netting, that technique, when used in a relatively remote and biodiverse environment, might come with deleterious collateral effects, because key functional aspects of regional ecology (e.g., pollination) could be deleteriously affected, other non-target wildlife could get caught in netting, or other species could be prevented from nesting as well (e.g., Angkaew et al. 2022). Other techniques, such as the use of monofilament lines (e.g., Blokpoel et al. 1997), might have similar negative effects.

In 2023, at another plot immediately adjacent to eastern mountain avens plants and where they likely occurred historically in greater numbers, gull eggs were removed from all nests at roughly weekly intervals during the breeding season. The primary purpose of this was to protect existing eastern mountain avens plants from direct physical harm (e.g., trampling, birds pulling up plants for nest material), with a secondary goal of testing whether gulls were deterred by the sustained disturbance

Fig. 16.3 A herring gull chick in its nest; a second egg in the background can be seen pipping (small hole through the shell) but the chick has not hatched out yet. Photo by Claire Wilson O'Driscoll

from nesting in the area. Throughout the season, gulls persisted in their attempts nest (total of 29 clutches; some could have been relaying) despite weekly human intrusions and egg removal. Visits to the site in spring 2024 would be required to provide information on whether total breeding failure in this plot led to reduced breeding effort there in subsequent years. More likely, we expect that egg removal activities will need to be continued annually, as done elsewhere (e.g., Blokpoel and Tessier 1987; Maxson et al. 1996; Blokpoel et al. 1997). On that point, we note that from an ethical perspective, members of the team involved in egg removals had very mixed, negative feelings about destroying nests, rather than using a method that forced birds to nest elsewhere.

Currently it is unclear how gull management at the site to protect eastern mountain avens will move forward. Destroying nests in specific areas appeared to be somewhat effective in the short term, but we did not have regular (e.g., daily) counts to determine what happened to gulls that had lost their nests; they may still defend their nest site for days, weeks, or even the entire breeding season following nest failure (Weseloh et al. 2020; Gutowsky unpubl. data). Certainly in 2023, the research team saw fresh faeces in the egg removal area and were dive-bombed by gulls for weeks after egg removals. Moreover, weekly trips to remove eggs incur financial and temporal costs (travel expenses, wages) that may be unsustainable (unless a local person could be contracted to undertake this work); and even if successful, removal of eggs and nests from specific targeted areas does not address the larger issue of nutrient inputs to the bog and associated habitat impacts. Other techniques, such as use of raptors or intense human harassment, may similarly prove unsuccessful if not sustained. It is hoped that, with the gradual rewetting of the bog following restoration efforts (Hill et al. 2019), some of the gull nesting habitat will be inundated with water and become unsuitable, but it seems likely that they may simply shift to nest elsewhere in the bog. Depending on conservation goals, local public opinion and federal regulatory considerations (given that overall herring gull numbers may be in decline), it might be worth considering selective culling of nesting adults in future.

16.4.4 A Silver Lining from Management Efforts

Colonial herring gulls at Brier Island are central place foragers during the breeding season (Lato et al. 2021), foraging generally within 50 km of their colonies (Fig. 16.1). Because of their generalist, opportunistic foraging strategy integrating food resources available from within this region, gulls accumulate chemical markers of environmental conditions within this area (e.g., Shlepr et al. 2021). In turn, these markers can be tracked through time to monitor environmental conditions of the region (e.g., Bond 2016). For example, environmental exposure to mercury is monitored through herring gull eggs (Burgess et al. 2013) or adults (Mallory et al. 2018), as are a suite of legacy organochlorines and emerging chemicals (e.g., Gebbink et al. 2011; Chen et al. 2012). More recently, plastic debris has become an international concern, and collections of adult birds provide one means of tracking wildlife exposure to this particulate contaminant (e.g., Caldwell et al. 2020; Jardine et al. 2021). Indeed, researchers have found that it is not just plastic that is being ingested by gulls from poorly managed human trash, but other items such as glass, metal, and drywall may also be consumed (Seif et al. 2018). Seabird egg and adult collections are a cornerstone of contaminant monitoring in the Arctic (Mallory and Braune 2012; Bianchini et al. 2022) and the Great Lakes (Hebert et al. 2000) and have been undertaken extensively in Atlantic Canada (Goodale et al. 2008; Burgess et al. 2013). These programmes provide an ongoing, robust index on levels and effects of continuing anthropogenic contamination to the natural environment. Therefore, we suggest that if an egg collection and gull culling programme at Brier Island is implemented, aimed to protect the endangered eastern mountain avens, it could feed into helpful national environmental monitoring.

16.5 Conclusion

Herring gulls are intelligent, adaptable, and resilient seabirds; the group of colonies in eastern North America has experienced centuries of anthropogenic changes and numbers of birds have increased or decreased in response. Although overall numbers of herring gulls appeared to be in decline early in the twenty-first century, the colony at Brier Island has been stable for decades, if not growing, relying on mink farms, fish plants, and fisheries in the region as principle foraging sites. However, the colony is breeding close to the globally rare eastern mountain avens, and to protect that plant population, gull management actions have been initiated. To date, those have largely been unsuccessful. Erecting stakes to deter nesting in a prime location had little effect on breeding birds, and continuous egg removal kept birds from rearing chicks close to the rare plant for one year, but gulls will likely return and try to nest in that area in subsequent years.

Keeping large gulls away from key human landscapes has proven challenging around the world. At Brier Island, it seems likely that more aggressive management may be needed to protect the plants, which may include consideration of lethal control in future. If this occurs, we strongly recommend that egg and adult gull collections be considered for inclusion in regional or national, long-term monitoring programmes for particulate and chemical contamination.

Acknowledgements Thanks to the students and research assistants who greatly contributed to the work described in this paper. Financial support was provided by Acadia University and Nova Scotia Department of Natural Resources and Renewables. Thanks to Randy Milton and John Smol for helpful comments on this manuscript.

References

Adams LW (2005) Urban wildlife ecology and conservation: a brief history of the discipline. Urban Ecosyst 8:139–156

Alm EW, Daniels-Witt QR, Learman DR, Ryu H, Jordan DW, Gehring TM, Santo Domingo J (2018) Potential for gulls to transport bacteria from human waste sites to beaches. Sci Total Environ 615:123–130

Alderson C, Greene A (1995) Bird-deterrence technology for historic buildings. J Preserv Tech 26(2/3):18–30

Anderson JGT, Shlepr KR, Bond AL, Ronconi RA (2016) Introduction: a historical perspective on trends in some gulls in eastern North America, with reference to other regions. Waterbirds 39(1):1–9. https://doi.org/10.1675/063.039.sp106

Anderson CM, Gilchrist HG, Ronconi RA, Shlepr KR, Clark DE, Weseloh DC, Robertson GJ, Mallory ML (2019) Winter home range and habitat selection differs among breeding populations of herring gulls in eastern North America. Move Ecol 7:8. https://doi.org/10.1186/s40462-019-0152-x

Angkaew R, Round PD, Ngoprasert D, Powell LA, Limparungpatthanakij WY, Gale GA (2022) Collateral damage from agricultural netting to open-country bird populations in Thailand. Conserv Sci Pract 4:e12810

Baak JE, Patterson A, Gilchrist HG, Elliott KH (2021) First evidence of diverging migration and overwintering strategies in glaucous gulls (*Larus hyperboreus*) from the Canadian Arctic. Anim Migrat 8(1):98–109

Belant JL (1997) Gulls in urban environments: landscape-level management to reduce conflict. Landscape Urban Plan 38:245–258. https://doi.org/10.1016/s0169-2046(97)00037-6

Benskin CWH, Wilson K, Jones K, Hartley IR (2009) Bacterial pathogens in wild birds: a review of the frequency and effects of infection. Biol Rev 84(3):349–373. https://doi.org/10.1111/j.1469-185X.2008.00076.x

Bianchini K, Mallory ML, Braune BM, Muir DC, Provencher JF (2022) Why do we monitor? Using seabird eggs to track trends in Arctic environmental contamination. Environ Rev 30(2):245–267

Blais JM, Macdonald RW, Mackay D, Webster E, Harvey C, Smol JP (2007) Biologically mediated transport of contaminants to aquatic systems. Environ Sci Technol 41:1075–1084. https://doi.org/10.1021/es061314a

Blokpoel H, Tessier GD (1987) Control of ring-billed gull colonies at urban and industrial sites in southern Ontario, Canada. Proc East Wildl Damage Control Conf 3:8–17

Blokpoel H, Tessier GD (1992) Control of ring-billed gulls and herring gulls nesting at urban and industrial sites in Ontario, 1987–1990. Proc East Wildl Damage Control Conf 5:51–57

Blokpoel H, Neuman J (1997) Sound levels in 3 Ring-billed Gull colonies of different size. Colon Waterb 20:221–226

Blokpoel H, Tessier GD, Andress RA (1997) Successful restoration of the Ice Island Common Tern colony requires on-going control of Ring-billed Gulls. Colon Waterb 20:98–101

Boklund A, Hammer AS, Quaade ML, Rasmussen TB, Lohse L, Strandbygaard B, Jørgensen CS, Olesen AS, Hjerpe FB, Petersen HH, Jensen TK, Mortensen S, Calvo-Artavia FF, Lefevre SK, Nielsen SS, Halasa T, Belsham GJ, Bøtner A (2021) SARS-CoV-2 in Danish mink farms: course of the epidemic and a descriptive analysis of the outbreaks in 2020. Animals 11:164. https://doi.org/10.3390/ani11010164

Bond AL (2016) Diet changes in breeding herring gulls (*Larus argentatus*) in Witless Bay, Newfoundland and Labrador, Canada, over 40 years. Waterbirds 39(sp1):152–158

Boothby C, Redfern C, Schroeder J (2019) An evaluation of canes as a management technique to reduce predation by gulls of ground-nesting seabirds. Ibis 161:453–458. https://doi.org/10.1111/ibi.12702

Brooks EN, Lebreton JD (2001) Optimizing removals to control a metapopulation: application to the yellow legged herring gull (*Larus cachinnans*). Ecol Model 136(2–3):269–284

Brown RGB, Nettleship DN, Germain P, Tull CE, Davis T (1975) Atlas of eastern Canadian seabirds. Cdn Wildl Serv, Ottawa, Canada, 220 pp

Burger J (1981) Feeding competition between laughing gulls and herring gulls at a sanitary landfill. Condor 83(4):328–335

Burger J (1983) Bird control at airports. Environ Conserv 10:115–124. https://doi.org/10.1017/s0376892900012200

Burger J (1985) Factors affecting bird strikes on aircraft at a coastal airport. Biol Conserv 33:1–28. https://doi.org/10.1016/0006-3207(85)90002-3

Burger J, Gochfeld M, Kirwan GM, Christie DA, de Juana E (2020). Lesser black-backed gull (*Larus fuscus*), version 1.0. In: del Hoyo J, Elliott A, Sargatal J, Christie DA, de Juana J (eds) Birds of the world. Cornell Lab of Ornithology, Ithaca, NY, USA. https://doi.org/10.2173/bow.lbbgul.01

Burgess NM, Bond AL, Hebert CE, Neugebauer E, Champoux L (2013) Mercury trends in herring gull (*Larus argentatus*) eggs from Atlantic Canada, 1972–2008: temporal change or dietary shift? Environ Pollut 172:216–222

Caliendo V, Lewis NS, Pohlmann A, Baillie SR, Banyard AC, Beer M, Brown IH, Fouchier RAM, Hansen RDE, Lameris TK, Lang AS, Laurendeau S, Lung O, Robertson G, van der Jeugd H, Alkie TN, Thorup K, van Toor ML, Waldenström J, Yason C, Kuiken T, Berhane Y (2022) Transatlantic spread of highly pathogenic avian influenza H5N1 by wild birds from Europe to North America in 2021. Sci Rep 12:11729. https://doi.org/10.1101/2022.01.13.476155

Caldwell A, Seavey J, Craig E (2020) Foraging strategy impacts plastic ingestion risk in seabirds. Limnol Oceanogr Lett 5:163–168. https://doi.org/10.1002/lol2.10126

Campbell MO (2019) Links between human quality of life and gulls in urban areas. Multidim Approach Quality Life Issues: Spatial Anal 359–371

Chen D, Letcher RJ, Burgess NM, Champoux L, Elliott JE, Hebert CE, Martin P, Wayland M, Weseloh DC, Wilson L (2012) Flame retardants in eggs of four gull species (Laridae) from breeding sites spanning Atlantic to Pacific Canada. Environ Pollut 168:1–9

Cotter RC, Rail J-F, Boyne AW, Robertson GJ, Weseloh DVC, Chaulk KG (2012) Population status, distribution, and trends of gulls and kittiwakes breeding in eastern Canada, 1998–2007. Can Wildl Serv Occas Paper 120, 93 pp

Coulson J (2002) Colonial breeding in seabirds. In: Biology of marine birds, CRC Press, New York, pp 87–113. https://doi.org/10.1201/9781420036305-7

Dias MP, Martin R, Pearmain EJ, Burfield IJ, Small C, Phillips RA, Yates O, Lascelles B, Borboroglu PG, Croxall JP (2019) Threats to seabirds: a global assessment. Biol Conserv 237:525–537. https://doi.org/10.1016/j.biocon.2019.06.033

Dolbeer RA, Belant JL, Sillings JL (1993) Shooting gulls reduces strikes with aircraft at John F. Kennedy International Airport. Wildl Soc Bull 21(4):442–450

Donehower CE, Bird DM, Hall CS, Kress SW (2007) Effects of gull predation and predator control on tern nesting success at Eastern Egg Rock, Maine. Waterbirds 30:29–39. https://doi.org/10.1675/1524-4695(2007)030[0029:eogpap]2.0.co;2

Drury WH (1973) Population changes in New England seabirds. Bird-Band 44:267–313

Drury WH, Kadlec JA (1974) The current status of the Herring Gull population in the northeastern United States. Bird-Band 45(4):297–306

Earn DJ, Rohani P, Grenfell BT (1998) Persistence, chaos and synchrony in ecology and epidemiology. Proc Roy Soc Lond Ser B: Biol Sci 265(1390):7–10

Ellis JC, Fariña JM, Witman JD (2006) Nutrient transfer from sea to land: the case of gulls and cormorants in the Gulf of Maine. J Anim Ecol 75:565–574. https://doi.org/10.1111/j.1365-2656.2006.01077.x

Gaston AJ (2004) Seabirds: a natural history. T & AD Poyser, 222 pp. https://doi.org/10.5040/9781472596840

Gaston AJ, Bertram DF, Boyne AW, Chardine JW, Davoren G, Diamond AW, Hedd A, Montevecchi WA, Hipfner JM, Lemon MJ, Mallory ML (2009) Changes in Canadian seabird populations and ecology since 1970 in relation to changes in oceanography and food webs. Environ Rev 17:267–286

Gebbink WA, Letcher RJ, Burgess NM, Champoux L, Elliott JE, Hebert CE, Martin P, Wayland M, Weseloh DC, Wilson L (2011) Perfluoroalkyl carboxylates and sulfonates and precursors in relation to dietary source tracers in the eggs of four species of gulls (Larids) from breeding sites spanning Atlantic to Pacific Canada. Environ Int 37(7):1175–1182

Geizer HD, Klapstein SJ, Mallory ML, O'Driscoll NJ (2021) Total mercury, methylmercury, phosphate, and sulfate inputs to a bog ecosystem from herring gull (*Larus smithsoniansus*) guano. Ecotoxicol Environ Saf 226:112845. https://doi.org/10.1016/j.ecoenv.2021.112845

Goodale MW, Evers DC, Mierzykowski SE, Bond AL, Burgess NM, Otorowski CI, Welch LJ, Hall CS, Ellis JC, Allen RB, Diamond AW (2008) Marine foraging birds as bioindicators of mercury in the Gulf of Maine. EcoHealth 5:409–425

Gutowsky SE, Studholme KR, Ronconi RA, Allard KA, Shlepr K, Diamond AW, McIntyre J, Craik SR, Mallory ML (2021) The influence of multiple industries on the behaviour of breeding gulls from four colonies across the eastern Gulf of Maine, Canada. Wildl Biol 2021:1–13. https://doi.org/10.2981/wlb.00804

Gutowsky SE, Robertson GJ, Mallory ML, McLellan NR, Gilliland SG (2023a) Redistribution of wintering American Common Eiders (*Somateria mollisima dresseri*). Avian Conserv Ecol 18(2)

Gutowsky SE, Baak JE, Craik SR, Mallory ML, Knutson N, d'Entremont AA, Allard KA (2023b) Seasonal and circadian patterns of herring gull (*Larus smithsoniansus*) movements reveal temporal shifts in industry and coastal island interaction. Ecol Solut Evid 4:e12274. https://doi.org/10.1002/2688-8319.12274

Hatch SA, Robertson GJ, Baird PH (2020) Black-legged Kittiwake (*Rissa tridactyla*), version 1.0. In: Billerman SM (ed) Birds of the world. Cornell Lab of Ornithology, Ithaca, NY, USA. https://doi.org/10.2173/bow.bklkit.01

Hebert CE, Norstrom RJ, Weseloh DC (2000) A quarter century of environmental surveillance: the Canadian Wildlife Service's Great Lakes herring gull monitoring program. Environ Rev 7(4):147–166

Hernández-Alonso G, Ramos-Madrigal J, van Grouw H, Ciucani MM, Cavill EL, Sinding MHS, Gopalakrishnan S, Pacheco G, Gilbert MTP (2023) Redefining the evolutionary history of the rock dove, *Columba livia*, using whole genome sequences. Mol Biol Evol 40(11):msad243

Hill NM, Chinner K, Toms B, Garbary DJ (2019) Recovery succession of native plant communities at Big Meadow Bog after disturbance and eutrophication from herring gulls. Wetlands 39:533–548. https://doi.org/10.1007/s13157-018-1099-1

Hobson KA, Blight LK, Arcese P (2015) Human-induced long-term shifts in gull diet from marine to terrestrial sources in North America's coastal Pacific: more evidence from more isotopes (δ2H, δ34S). Environ Sci Technol 49(18):10834–10840

Howell SNG, Dunn J (2007) A reference guide to gulls of the Americas. Houghton Mifflin Company, New York, 516 pp

Jardine AM, Provencher JF, Pratte I, Holland ER, Baak JE, Robertson GJ, Mallory ML (2021) Annual plastic ingestion and isotopic niche patterns of two sympatric gull species at Newfoundland, Canada. Mar Pollut Bull 173:112991. https://doi.org/10.1016/j.marpolbul.2021.112991

Kickbush JC, Mallory ML, Murimboh JD, Rand J, Klapstein SJ, Loder AL, Hill NM, O'Driscoll NJ (2018) The influence of avian biovectors on mercury speciation in a bog ecosystem. Sci Total Environ 637:264–273. https://doi.org/10.1016/j.scitotenv.2018.04.224

Klimaszyk P, Rzymski P (2016) The complexity of ecological impacts induced by great cormorants. Hydrobiology 771:13–30. https://doi.org/10.1007/s10750-015-2618-1

Kress SW (1983) The use of decoys, sound recordings, and gull control for re-establishing a tern colony in Maine. Colon Waterb 185–196

Lamarre J, Wilson DR (2021) Waterbird solves the string-pull test. Roy Soc Open Sci 8(12):211343

LaRue D (2016) Natural regeneration of the globally rare *Geum peckii* (Rosaceae) on Brier Island, Nova Scotia. Rhodora 118:310–314. https://doi.org/10.3119/15-39

Lato KA, Madigan DJ, Veit RR, Thorne LH (2021) Closely related gull species show contrasting foraging strategies in an urban environment. Sci Rep 11:23619. https://doi.org/10.1038/s41598-021-02821-y
Leu M, Hanser SE, Knick ST (2008) The human footprint in the west: a large-scale analysis of anthropogenic impacts. Ecol Applic 18(5):1119–1139
Lute ML, Attari SZ (2017) Public preferences for species conservation: choosing between lethal control, habitat protection and no action. Environ Conserv 44(2):139–147
MacKinnon CM, Kennedy AC (2014) Decline in breeding of the Great Black-backed Gull, *Larus marinus*, and the Herring Gull, *L. argentatus*, on Boot Island, Nova Scotia, 1986 to 2010. Can Field-Nat 128(2):165–172
Magle SB, Hunt VM, Vernon M, Crooks KR (2012) Urban wildlife research: past, present, and future. Biol Conserv 155:23–32
Malcata Martins B, O'Driscoll NJ, Mallory ML, Canário J (2021) A review of freshwater invertebrates as biomonitors of methylmercury: the importance of more complete physical and chemical reporting. Bull Environ Contam Toxicol 107:801–808. https://doi.org/10.1007/s00128-021-03274-9
Mallory ML, Braune BM (2012) Tracking contaminants in seabirds of Arctic Canada: temporal and spatial insights. Mar Pollut Bull 64(7):1475–1484
Mallory ML, Provencher JF, Robertson GJ, Braune BM, Holland ER, Klapstein S, Stevens K, O'Driscoll NJ (2018) Mercury concentrations in blood, brain and muscle tissues of coastal and pelagic birds from northeastern Canada. Ecotoxicol Environ Saf 157:424–430
Martín-Vélez V, Sánchez MI, Shamoun-Baranes J, Thaxter CB, Stienen EW, Camphuysen KC, Green AJ (2019) Quantifying nutrient inputs by gulls to a fluctuating lake, aided by movement ecology methods. Freshwater Biol 64(10):1821–1832
Maxson SJ, Mortensen SA, Goodermote DL, Lapp CS (1996) Success and failure of Ring-billed Gull deterrents at Common Tern and Piping Plover colonies in Minnesota. Colon Waterb 19:242–247
McIntyre JA, O'Driscoll NJ, Spooner I, Robertson GJ, Smol JP, Mallory ML (2022) Scavenging gulls are biovectors of mercury from industrial wastes in Nova Scotia, Canada. Chemosphere 304:135279. https://doi.org/10.1016/j.chemosphere.2022.135279
Meléndez-Arteaga J, Bregnballe T, Frederiksen M (2022) Identifying spatial drivers of long-term population growth in three large gull species: the importance of mink farms and urban areas. Avian Conserv Ecol 17(2):10
Moreno R, Jover L, Munilla I, Velando A, Sanpera C (2010) A three-isotope approach to disentangling the diet of a generalist consumer: the yellow-legged gull in northwest Spain. Mar Biol 157:545–553
Morris RD, Weseloh DV, Wires LR, Pekarik C, Cuthbert FJ, Moore DJ (2011) Population trends of Ring-billed Gulls breeding on the North American Great Lakes, 1976 to 2009. Waterbird 34(2):202–212
Navarro J, Grémillet D, Afán I, Miranda F, Bouten W, Forero MG, Figuerola J (2019) Pathogen transmission risk by opportunistic gulls moving across human landscapes. Sci Rep 9(1):10659–10665. https://doi.org/10.1038/s41598-019-46326-1
Newsome TM, Dellinger JA, Pavey CR, Ripple WJ, Shores CR, Wirsing AJ, Dickman CR (2015) The ecological effects of providing resource subsidies to predators. Glob Ecol Biogeogr 24:1–11. https://doi.org/10.1111/geb.12236
Nisbet ICT, Veit RR, Auer SA, White TP (2013) Marine birds of the eastern United States and the Bay of Fundy. Distribution, numbers, trends, threats, and management. Nuttall Ornithol Monogr 29
NSDLF (2021) Recovery plan for the eastern mountain avens (*Geum peckii*) in Nova Scotia [Final]. Nova Scotia Dept Lands Forestry, Kentville, Nova Scotia. Nova Scotia Endangered Species Act Recovery Plan Series
O'Hanlon NJ, McGill RA, Nager RG (2017) Increased use of intertidal resources benefits breeding success in a generalist gull species. Mar Ecol Progr Ser 574:193–210

Oro D, Martínez-Abraín A (2007) Deconstructing myths on large gulls and their impact on threatened sympatric waterbirds. Anim Conserv 10(1):117–126

Oro D, Genovart M, Tavecchia G, Fowler MS, Martínez-Abraín A (2013) Ecological and evolutionary implications of food subsidies from humans. Ecol Lett 16(12):1501–1514. https://doi.org/10.1111/ele.12187

Otero XL, De La Peña-Lastra S, Pérez-Alberti A, Ferreira TO, Huerta-Diaz MA (2018) Seabird colonies as important global drivers in the nitrogen and phosphorus cycles. Nat Commun 9:246–247. https://doi.org/10.1038/s41467-017-02446-8

Perlut NG, Bonter DN, Ellis JC, Friar MS (2016) Roof-top nesting in a declining population of Herring Gulls (*Larus argentatus*) in Portland, Maine, USA. Waterbirds 39:68–73. https://doi.org/10.1675/063.039.sp113

Prange S, Gehrt SD, Wiggers EP (2003) Demographic factors contributing to high raccoon densities in urban landscapes. J Wildl Manag 67:324–333

Rock P, Camphuysen CJ, Shamoun-Baranes J, Ross-Smith VH, Vaughan IP (2016) Results from the first GPS tracking of roof-nesting Herring Gulls *Larus argentatus* in the UK. Ringing Migrat 31(1):47–62

Ronconi RA, Cassady St. Clair C (2006) Efficacy of a radar-activated on demand system for deterring waterfowl from oil sands tailings ponds. J Appl Ecol 43:111–119

Ronconi RA, Lieske DJ, McFarlane Tranquilla LA, Abbott S, Allard KA, Allen B, Black AL, Bolduc F, Davoren GK, Diamond AW, Fifield DA, Garthe S, Gjerdrum C, Hedd A, Mallory ML, Mauck RA, McKnight J, Montevecchi WA, Pollet IL, Pratte I, Rail J-F, Regular PM, Robertson GJ, Rock JC, Savoy L, Shlepr KR, Shutler D, Symons SC, Taylor PD, Wilhelm SI (2022) Predicting seabird foraging habitat for conservation planning in Atlantic Canada: integrating telemetry and survey data across thousands of colonies. Front Marine Sci 9:816794

Schwemmer P, Garthe S, Mundry R (2008) Area utilization of gulls in a coastal farmland landscape: habitat mosaic supports niche segregation of opportunistic species. Landscape Ecol 23:355–367. https://doi.org/10.1007/s10980-008-9194-y

Scopel LC, Diamond AW (2017) The case for lethal control of gulls on seabird colonies. J Wildl Manag 81:572–580. https://doi.org/10.1002/jwmg.21233

Seif S, Provencher JF, Avery-Gomm S, Daoust PY, Mallory ML, Smith PA (2018) Plastic and non-plastic debris ingestion in three gull species feeding in an urban landfill environment. Arch Environ Contam Toxicol 74:349–360. https://doi.org/10.1007/s00244-017-0492-8

Shlepr KR, Ronconi RA, Hayden B, Allard KA, Diamond AW (2021) Estimating the relative use of anthropogenic resources by Herring Gull (*Larus argentatus*) in the Bay of Fundy, Canada. Avian Conserv Ecol 16(1):2. https://doi.org/10.5751/ace-01739-160102

Simpson LT, Canty SWJ, Cissell JR, Steinberg MK, Cherry JA, Feller IC (2021) Bird rookery nutrient over-enrichment as a potential accelerant of mangrove cay decline in Belize. Oecologia 197:771–784. https://doi.org/10.1007/s00442-021-05056-w

Skov H, Durinck J (2001) Seabird attraction to fishing vessels is a local process. Mar Ecol Prog Ser 214:289–298

Spencer NC, Gilchrist HG, Strøm H, Allard KA, Mallory ML (2016) Key winter habitat of the ivory gull *Pagophila eburnea* in the Canadian Arctic. Endang Sp Res 31:33–45

Thiériot E, Molina P, Giroux J-F (2012) Rubber shots not as effective as selective culling in deterring gulls from landfill sites. Appl Anim Behav Sci 142(1–2):109–115

Thomas GJ (1972) A review of gull damage and management methods at nature reserves. Biol Conserv 4:117–127. https://doi.org/10.1016/0006-3207(72)90012-2

Trotter S (2019) Birds behaving badly: the regulation of seagulls and the construction of public space. J Law Soc 46(1):1–28

Tufts RW (1973) The birds of Nova Scotia, 2nd edn. Nova Scotia Museum, Halifax, Nova Scotia

Weseloh DV, Hebert CE, Mallory ML, Poole AF, Ellis JC, Pyle P, Patten MA (2020) Herring Gull (*Larus argentatus*), version 1.0. In: Billerman SM (ed) Birds of the world. Cornell Lab of Ornithol, Ithaca, NY, USA. https://doi.org/10.2173/bow.hergul.01

Westerberg K, Brown R, Eagle G, Votier SC (2019) Intra-population variation in the diet of an avian top predator: generalist and specialist foraging in great black-backed gulls *Larus marinus*. Bird Study 66(3):390–397

Wilhelm SI, Rail J-F, Regular PM, Gjerdrum C, Robertson GJ (2016) Large-scale changes in abundance of breeding herring gulls (*Larus argentatus*) and great black-backed gulls (*Larus marinus*) relative to reduced fishing activities in southeastern Canada. Waterbirds 39:136–142

Wu B, Wang XC, Dzakpasu M (2017) Genetic characterization of fecal impacts of seagull migration on an urban scenery lake. Water Res 117:27–36

Chapter 17
Summary and Conclusions

Nicholas M. Hill, Sarah Hines, and Nelson J. O'Driscoll

17.1 Summary of Chapter Conclusions

Work in this book was instigated due to an endangered plant that was losing its ideal habitat because of historical ditching which changed the hydrology of Big Meadow Bog. A multidisciplinary approach was taken to the restoration of Big Meadow Bog and the recovery of the Eastern Mountain Avens as is evident below in the summaries for each chapter.

Chapter 2, "The Lost Backyard" describes the historical context of the bog and its relationship to indigenous people. The chapter starts with "Lnu", the indigenous people at the end of the glaciation, fishing beside a sea lagoon that would in millennia to come, become Big Meadow Bog. The Meadow was a commons to the Mi'kmaw and functionally so, to the settlers. It provided berries and waterfowl and much more. When this commons was lost and the Meadow was ditched as a private agricultural venture, the community lost its backyard and way of life. Wetland alterations the world over result in a loss of the traditional uses and familiarity with wetland areas whether they are river shores, lakes, marshes or peatlands.

Chapter 3 explores the genetic world of this endangered hexaploid species and reinforces the taxonomic decision that together with the Avens in New Hampshire, *Geum peckii* is a good species and genetically distinct from the *Geum radiatum*, the Spreading Avens, of the southern Appalachians. The Nova Scotian gene pool is further separate from that in New Hampshire but importantly, the authors find a lack of distinction between the clusters spread throughout the wetlands and old

N. M. Hill (✉)
UNESCO Southwest Biosphere Association, 399 Wyman Road, Sand Beach, NS B5A 5G4, Canada
e-mail: nick@swnovabiosphere.ca

S. Hines · N. J. O'Driscoll
Acadia University, Wolfville, NS B5A 5G4, Canada

N. M. Hill et al. (eds.), *Wetland Restoration for Endangered Species Recovery*, Environmental Contamination Remediation and Management,
https://doi.org/10.1007/978-3-031-71344-6_17

sheep pastures of Brier Island. They hypothesize that the genetic differentiation between populations in the White Mountains has resulted from the isolation brought about by the large distance and elevation differences between these populations. The lack of differentiation in Nova Scotia does not mean a lack of genetic diversity but their results simplify both the contemplated ex situ cultivation as well as facilitated migration. The Kozela chapter is a slow, careful read, and it provides guidelines for applied conservation:

- Treat Brier Island as one population
- Maintain gene flow processes within and between sites to guard against the strong inbreeding depression demonstrated at the seed germination stage
- Determine how the considerable clonal diversity (plants are usually different genetic individuals not just ramets) that is greatest in the highest integrity habitats comes about: under what conditions and times does seedling establishment succeed?

Chapter 4 explores the ecology and conservation of the Eastern Mountain Avens. The ecology of Avens rests upon its evolutionary and genetic history. This chapter fills in some of the gaps highlighted as priorities in the Recovery and Action Plans. We find that while Avens tolerates waterlogged areas, a three-year transplantation experiment shows least growth in these areas, reinforcing that the wet fens are the post competition, realized niches for the species. Avens, the authors maintain, has given up long distance dispersal since it has no hooks or barbs on its seeds (achenes) that enable other species in its genus to attach to the fur of mammals and constantly find new disturbed areas. Avens has a stress-tolerant life history strategy (Grime 1979) and rather than focus on regeneration in disturbed areas or on competitive interactions with other plants in high biomass areas, it survives through slow growth and investment in its rhizome tissues in its local habitat. At a global level, through misjudged land-use decisions, we have set up innumerable disturbance experiments and everywhere weedy generalists succeed and these slower-growing specialist organisms are being eliminated. At Big Meadow Bog and with the Avens, we are fighting for the survival of a globally imperilled specialist through the restoration of the ditched bog.

The ditching of the Meadow gave the advantage to more competitive trees and shrubs that moved into the bog from the edges. It also led to an invasion of the bog by thousands of nesting Herring Gulls and this enrichment created ideal conditions for exotic and native weedy generalists. The ditching (including consequent woody plant encroachment) and the gulls were identified as the greatest threats in the Recovery Strategy but climate change was not mentioned in the threat section and while it was considered in Appendix C, the Occurrence, Causal Certainty and Severity of this threat for Avens were all judged Uncertain.

Chapter 5 examines the evidence for regional climate change and concludes that; "Brier Island, on the outer edge of the Bay of Fundy, sits in an ocean hotspot that has warmed at least 2 °C since the 1990s". In the light of a rapidly warming regional climate, ex situ conservation should be regarded as an insurance policy for endangered plant species (see Chap. 15). Warming, apparently driven by changes in ocean currents and temperatures, has been particularly pronounced on Brier Island so threatening the survival of the boreal ecosystems that developed in isolation where Fundy meets the Gulf of Maine.

Chapter 6 looks back at the geological and climatological past. The Avens is confined to fens on Jurassic basalt in Canada, an observation that is strengthened by poor performance of transplants in ombrotrophic (raised bog) stations removed from groundwater influence. Climate changes over the past million years rearranged multiple times both temperate and boreal North America and determining the distribution of present day *Geum peckii.* The loss of seed dispersal capability of Avens referred to above is consistent with the extremely disjunct distribution of Avens. This chapter takes an ontological look at Big Meadow Bog from a core extracted from the centre of what was a raised bog. The development of this wetland was initially abiotic. It was a sea lagoon that slowly filled with the same grey sands that cover the beach at Pond Cove then at one point, perhaps 2500 years ago, the development of the bog began to be directed by the mosses that form peat. As the core neared the surface, a layer of black spruce trees was found, trees that died standing up. The authors contend that the climate had then changed, and we were witnessing the impact of a "regional wetting event" about 1700 years ago when a period of high rainfall flooded the treed swamp and created open fen conditions favourable to the Eastern Mountain Avens. The reconstruction of events over the last 2000 years of Big Meadow Bog was made possible by microscopic examination of each layer of peat that contained the resistant, millimetre long leaves of *Sphagnum* mosses that were identified to species or section: *Sphagnum magellanicum, S. affine* and the Section *Acutifolia.* The look back shows the dynamism of wetlands, their hydrological regimes and biological communities.

Chapter 7 shows how the ditch blocking (see Chap. 10) has rewetted the bog from 2017 when the on-the-ground restoration began to two years later in 2019. In the centre of the Meadow, the impact of rewetting was greatest near the former central ditch and not all bog areas showed water retention raising a concern over a loss of peat structure over time since ditching. In the lagg areas—the margins of the bog—most areas were successfully brought to the target water table state outlined in Chap. 10, within 20 cm of surface in the growing season where we typically observe Avens. At the outset, hydrogeologist Jonathan Price was asked how precise could we in restoring be about water levels and their we could have from the restoration, said it was like were nudging the ecosystem with a sledgehammer. The hydrological results shown in Chap. 7 are considerably better than feared but now ecological restoration will need time and adaptive management.

Chapter 8 examines the shallow water level dynamics among four peatland classes. Mean water levels, monthly fluctuations and water table depth durations ranged along a spectrum from highly consistent among the three undisturbed fens to highly variable among the three disturbed bogs. Mean water levels at undisturbed fens were at or very near the surface (< 5 cm) most of the time and monthly fluctuations were minimal (< 15 cm), while in disturbed bogs, mean water levels were considerably deeper (18–40 cm), monthly fluctuations were nearly three times higher, and conditions were dry or very dry most of the time. Geum peckii was absent or poorly represented in both of these hydrological setting. The authors suggests that mean water levels – 10 to – 20 cm, levels in the saturated root zone for about 75% of the time and in the dry root about 25% of the time are likely most conducive to the success of Geum peckii.

Chapter 9 contrasts the greenhouse gas emissions from the ditched Big Meadow with that of a pristine "Reference Bog". The total emissions (as CO_2 equivalents) related to ditching where the Meadow had recovered from eutrophication from herring gull guano were double that of the pristine system but emissions from the current gull nesting areas were double that of the recovered areas. Phosphate accounted for most of the variation of either carbon dioxide (CO_2) or nitrous oxide (N_2O). These stark differences in greenhouse gas emissions relating both to ditching and to eutrophication underline the climate importance of the hydrological restoration and to efforts to reduce the population of nesting gulls. They also point to the need to maintain a monitoring programme for herring gulls, phosphate levels and greenhouse gases.

Chapter 10 details the techniques used for the water table restoration in Bog Meadow Bog. The restoration was based on comparable restoration sites in the United Kingdom and the lessons learned from a ditch block trial initially concluded at the site. Between 2016 and 2018 a total of 123 ditch blocks were constructed, 3700 m of ditch being re-profiled as well as shrub and trees being cut across 34 ha of the bog. A qualitative follow up review of the effectiveness of hydrology restoration was completed in 2023.

Chapter 11 finds that the structure of bird communities at Big Meadow Bog was shaped by herring gulls (*Larus smithsoniansus*), due mainly to their high number ($n = 3162$) in the north-western portion of the bog. Common yellowthroat, American redstart, song sparrow, and black-throated green warbler were the next most common species at BMB after herring gulls. Differences in the bird communities associated with BMB and the reference system, Lost Bog, were most pronounced at NWBMB, where there were the fewest species, the lowest diversity and evenness, the lowest numbers of individuals apart from herring gulls, and the fewest species of conservation concern. Recovery of bird communities in the northern part of BMB was suggested to take several more decades and will depend upon a reduction in the herring gull population.

Chapter 12 suggests that the contribution of nutrients from gulls at Big Meadow Bog has not shown "excessive" deleterious effects on local habitat such as tree die-offs, although in nesting areas there is some altered vegetation (e.g. ruderals, invasives). The authors also suggest that concentrations of "ornithogenic" subsidies of other contaminants, such as organochlorines that are in the mink and their feed may be elevated in the bog along with plastics and microplastics. They suggest that these contaminants and plastics may alter the health of biota that consume them and should be a priority in future research.

Chapter 13 showed that MeHg, DOC, and PO_4^{3-} concentrations showed seasonal trends with highest concentrations in summer in the water samples taken from the northern outflow of Big Meadow Bog. Highest annual concentrations were observed in the first year following the water table restoration and declined in the subsequent

5 years. The highly elevated levels of phosphate may have exacerbated the DOC and MeHg concentrations in water through increased eutrophication.

Chapter 14 describes a preliminary survey of foliar endophytic fungi from Eastern Mountain Avens. Endophytic fungal species richness changed with observed changes to habitat condition, suggesting habitat may influence the fungal assemblage present within leaves. A core fungal assemblage was documented for the first time in this plant host, from multiple sites throughout Brier Island. By approaching plant conservation from a mycological perspective, we provide the first mycobiome assessment of *G. peckii*, an understudied component of the habitat.

Chapter 15 explores the role of ex situ (off site) conservation for *G. peckii*. Conservation work often focuses on site restoration, but this work summarizes a decade of efforts towards preserving seeds and tissue at a secondary location in the Acadia Seed Bank. Recognizing *G. peckii* as a globally rare, endangered species it is critical to ensure the genetic material is preserved for the future. This work also included a project to take material from seed bank, propagate with tissue culture methods and outplant seedlings back into a habitat adjacent to the natural habitat. This was the first time in Nova Scotia a seed bank was used to propagate material and introduce it to a natural habitat. To enhance plant success the relationship between *G. peckii* and Arbuscular Mycorrhizal Fungi (a group of plant symbionts) was both confirmed and investigated. Overall, this chapter shows how ex situ conservation is possible as a "disaster plan" for the endangered *G. peckii*.

Chapter 16 examines the techniques and complications of managing gull populations, and unsuccessful efforts to manage them at Brier Island. Despite nationally declining numbers of herring gulls, the colony at Brier Island has been stable or increasing for decades possibly due to the availability of anthropogenic foraging sites (mink farms, fish plants, and fisheries). Most attempts to deter gulls from specific sites in Big Meadow Bow have been unsuccessful, including erecting stakes to deter nesting and continuous egg removal. More aggressive management may be needed to protect the rare plants at this site.

17.2 Conclusion

This book summarizes the ongoing attempts to restore Big Meadow and our struggle to save the Endangered *Geum peckii*. The Meadow was a commons to the Mi'kmaw and continued as one during the first two centuries after settlement. Property law, however, allowed the owner of the Meadow in 1958 to dig two long ditches the length of the bog (~ 2 km) leading to ecological ramifications that continue six decades later. At a global level, land-use disturbances have led to an increase in both light and nutrient availability so that weedy generalists succeed at the expense of slower-growing specialists. In most cases, we move on and look past the disturbances to other areas for conservation and let regulatory agencies determine actions. In this case, the impact of the wetland alteration on a major population of an endangered species triggered the Species at Risk Act (https://laws-lois.justice.gc.ca/eng/acts/s-15.3/page-1.html). The Action Plan for Avens (*Geum peckii*) listed the following as High Priority for Big Meadow Bog:

> Finalize restoration plan, including engineering design, for increasing water levels in order to recover ombrotrophy, restore the bog margin (lagg) hydrology, restore open fenland for Eastern Mountain Avens and increase Eastern Mountain Avens populations and spatial distribution. (ECCC 2018)

There are two philosophical approaches to the restoration of natural ecosystems. The more traditional one is a pragmatic mixture of reinstating key processes with the addition of key species, or the removal of those that are undesirable. In contrast, a rewilding approach restores degraded systems only by returning the processes that shape the community and ecosystem functions (Mutillod et al. 2024). It does not dictate a successional path or the involvement of particular species although in some cases, species such as top-level carnivores or ecological engineers (e.g. beavers) may be introduced as processes in their own right. We have seen rewilding through succession at Big Meadow Bog in Herring Gull nesting areas which were abandoned from one through three decades: left to themselves, these nesting areas went from being hyper-eutrophic and populated above and below-ground by exotic weeds to becoming dominated by woody plants that include the Ericaceae—indicative of low nutrient conditions (Schwery et al. 2014)—without a trace of exotic weeds (Hill et al. 2018). Disturbance was likely what invited Herring Gulls to colonize the bog since they had not previously been noted in the bog by lifetime resident of Brier Island Raymond Robicheau (personal communication in 2019 to N Hill), although a small nesting colony was observed in the 1980s in a bog near the Reference Bog where gravel was being extracted from a ridge (Joyce De Vries, personal communication in 2018 to N Hill). The rewilding approach accepts that restoration is under the control of dominant processes and it rests on time and succession for species to come and go. In the latest example of rewilding, a lone beaver added its dam during the winter of 2023–2024 to the northern outflow of Big Meadow Bog.

The rewilding approach leaves the restoration pathway and time table to nature, once the processes have been reinstated. There is little evidence, however, that the density of gulls has been reduced by the restoration of the original hydrology and there is direct evidence that the gulls are fertilizing this ombrotrophic wetland at agricultural rates of phosphorous application. Furthermore, Chap. 13 presents a model where phosphorus increases microbial activity and creates the redox conditions that lead to the formation and mobilization of toxic methyl mercury. Two goals were identified in the Recovery Strategy (2010) that guided the restoration: (1) to recover the hydrology of the lagg that supported Avens populations, and (2) to recover ombrotrophy in the raised bog. These are goals consistent with the restoration of ecological services of the bog and they are intimately related to the survival of Avens. Avens is outcompeted in high shrub biomass communities which are associated with low water tables. Rare plants as a rule are excluded when habitat become enriched in a limiting nutrient (e.g. P in freshwater wetlands). In addition, in Big Meadow, the ecological engineer, Sphagnum, is inhibited at high fertilization levels. If gulls continue to pollute what should be the ombrotrophic centre of the bog, a bog plant community cannot form and nutrients from the central bog can flow towards the Avens habitats at the lagg of the bog. The current warming exacerbated by anthropogenic greenhouse gas production jeopardizes the ability of a group of plants and animals to survive as habitat changes force these species to relocate or be extirpated by organisms better able to compete (Vitt et al. 2010).

17.3 Uncertainties and Future Research Directions

If "land restoration is the acid test of our ecological understanding" (Bradshaw 1987), there are critical gaps and uncertainties. Some of the questions for which we need further research are assembled below by category.

The influence and management of Herring Gulls at this site still has many unanswered questions. It is clear that gull populations are difficult to manage, and most attempts have so far been unsuccessful. We do not know if active aggressive management is needed to remove or affect the population of Herring Gulls? We also do not know how the historical deposition of gull guano will affect the newly flooded system over the next several decades. The initial post restoration data shows a decline in nutrients and mercury species in the outflow water, however, the nutrient concentrations are still enormously greater than expected for a bog site. The influx of nutrients also brings with it a myriad of research questions. Such as: what is the pathway and residence time in the bog for phosphorous, and is phosphorous mobilizing methyl mercury into food webs? Is the dominance of the nitrogen-fixing shrub, Sweet Gale (*Myrica gale*) in lagg communities that were flooded following the hydrological restoration (ditch blocking), a result of increased light following the loss of Ericaceae at high water tables, is it fuelled by an accumulation of mineralized nutrient no longer taken up by Ericaceae, or is an ingress of phosphorous from the adjacent high density gull areas in the central bog, a part of the process?

In terms of the recovery of Avens in the lagg of Big Meadow Bog, we are reintroducing Avens using a scarification and sowing approach but we do not know how seedling establishment occurs in nature and what processes provide the requisite scarifications. We do know, however, that there is a diversity of genets at the patch level and that selection against selfed progeny is severe (Chap. 3). The Recovery Strategy and Action Plan (ECCC 2010, 2018) highlighted the importance of the hydrological restoration at a time before there was evidence for substantial climate warming. We now know that the Gulf of Maine has warmed more than most coastal marine areas and that the average temperature of Brier Island has increased by almost 2 °C from one climate normal to the next (1961–1990 to 1991–2020; Chap. 5). Will this heating result in shrubification and a loss of Avens through competitive exclusion? Climate change is also resulting in rising sea levels; at what point, however, will sea-level rise threaten low-lying laggs of Big Meadow that support Avens? At what point should we engage plan B: ex situ conservation and facilitated migration (Vitt et al. 2010, Chap. 15)?

The boardwalk and associated educational signage were initiated to help reconnect the community to the Meadow. Last year, almost 5000 people walked the boardwalk into the bog. This has become for many a part of their daily routine, taking them past the northern outflow saltmarsh and into the bog. The boardwalk raises both the community's and visitors' appreciation of wetlands and of the rare plants that are indicators of wetland health.

17.4 Significance of Work for Adaptive Management Practices

This multidisciplinary work highlights the need for monitoring and adaptive management in wetland restoration. The multiple ecological and physical factors at Big Meadow Bog mean that any site interventions may take years or decades to see the full effect. The research and infrastructure built at Big Meadow Bog have resulted in a site of national importance for the ongoing study of bog ecosystems and restoration management. Hydrological weirs have been installed in the summer of 2024 and these combined with accessible boardwalk mean that future sampling, monitoring, and fundamental research at this site can be easily facilitated.

This book shows what a multidisciplinary team learned as it put the benchmark research together, implemented the on-the-ground restoration of blocking the ditches, followed the consequences of the restoration, and assessed the scientific gaps and unanswered ecological questions. This book largely focuses on science but there is another story that runs parallel to this of how the village of Westport has engaged and taken "ownership" of the Meadow and its recovery. Over the past 50 years, the Brier Island community has come to value and take pride in the whales—Rights, Humpbacks and Finbacks—through ecotourism and this has developed into an ethic towards the sea. Now this pride of place and nature has extended to Big Meadow Bog. The community has built a boardwalk that reconnects village with the Meadow and through the boardwalk, people can follow the restoration story and the Endangered Eastern Mountain Avens.

This restoration is a valuable case study of how many groups can work together (government, researchers, community members) on a comprehensive exploration of restoration to save an endangered plant species. Many things were learned during this process, and we continue to learn from the ongoing management of the Big Meadow Bog ecosystem and interactions with community.

References

Bradshaw AD (1987) Restoration: an acid test for ecology. In: Jordan WR, Gilpin ME, Aber JD (eds) Restoration ecology: a synthetic approach for ecological research. Cambridge University Press, Cambridge, UK, pp 23–30

Environment Canada (2010) Recovery strategy for the Eastern Mountain Avens (*Geum peckii*) in Canada. Species at Risk Act Recovery Strategy Series. Environment Canada, Ottawa, vii + 17 pp. + Appendices

Environment Climate Change Canada (2018) Action plan for the Eastern Mountain Avens (Geum peckii) in Canada. Species at Risk Act Action Plan Series. Environment and Climate Change Canada, Ottawa, v + 23 pp.

Grime JP (1979) Plant strategies and vegetation processes. Wiley, Chichester

Hill NM, Chinner K, Toms B, Garbary DJ (2018) Recovery succession of native plant communities at Big Meadow Bog after disturbance and eutrophication from herring gulls. Wetlands 39:533–548. https://doi.org/10.1007/s13157-018-1099-1

Mutillod C, Buisson E, Mahy G, Jaunatre R, Bullock JM, Tatin L, Dutoit T (2024) Ecological restoration and rewilding: two approaches with complementary goals? Biol Rev 2024:000–000. https://doi.org/10.1111/brv.13046

Schwery O, Onstein RE, Bouchenak-Khelladi Y, Xing Y, Carter RJ, Linder HP (2014) As old as the mountains: the radiations of the Ericaceae. New Phytol. https://doi.org/10.1111/nph.13234

Vitt P, Havens K, Kramer A, Sollenberger D, Yates E (2010) Assisted migration in plants: changes in latitudes, changes in attitudes. Biol Conserv 143(1):18–27

GPSR Compliance

The European Union's (EU) General Product Safety Regulation (GPSR) is a set of rules that requires consumer products to be safe and our obligations to ensure this.

If you have any concerns about our products, you can contact us on ProductSafety@springernature.com

In case Publisher is established outside the EU, the EU authorized representative is:

Springer Nature Customer Service Center GmbH
Europaplatz 3
69115 Heidelberg, Germany

Batch number: 10756376

Printed by Printforce, the Netherlands